The Conclusive Argument from God

The Conclusive Argument from God

Shāh Walī Allāh of Delhi's Ḥujjat Allāh al-Bāligha

Translated by

Marcia K. Hermansen

BRILL

LEIDEN | BOSTON

This paperback was originally published in hardback as Volume 25 in the series *Islamic Philosophy, Theology and Science. Texts and Studies*.

Cover illustration: Drawing of the Jāmiʿ Masjid in Delhi in the 19th century, Germany 2019, unknown artist.

The Library of Congress has cataloged the hardcover edition as follows:

Walī Allāh al-Dihlawī, 1702 or 3-1762 or 3.
[Ḥujjat Allāh al-bālighah. English]
The conclusive argument from God = Shāh Walī Allāh of Delhi's Ḥujjat Allāh al-bāligha / translated by Marcia K. Hermansen.
p. cm. — (Islamic philosophy, theology, and science, ISSN 0169-8729 ; v. 25)
Includes bibliographical references (p.) and indexes.
ISBN 900410298 1 (cloth : alk. paper)
I. Islam—Early works to 1800. I. Hermansen, Marcia K. II. Title. III. Series.
BP160.W313 1995
297'.2—dc20 95-22093
CIP

Typeface for the Latin, Greek, and Cyrillic scripts: "Brill". See and download: brill.com/brill-typeface.

ISBN 978-90-04-44477-5 (paperback, 2021)
ISBN 978-90-04-10298-9 (hardback, 1996)
ISBN 978-90-04-44476-8 (e-book, 2021)

This book is printed on acid-free paper and produced in a sustainable manner.

This translation is dedicated to Fazlur Rahman, an inspiring teacher who wanted Shāh Walī Allāh's thought to be more widely known.

CONTENTS

Acknowledgments xiii
Translator's Introduction xv

The Conclusive Argument from God. *Ḥujjat Allāh al-Bāligha*
VOLUME I
The General Principles from Which Are Derived the Beneficial Purposes Considered in the Rulings of the Divine Law

Shāh Walī Allāh's Preface 3
Shāh Walī Allāh's Introduction 11

BOOK I The First Investigation: The Causes of Religious Obligations and Requital 31

1. Absolute Origination, Creation of the Material World and Divine Management 33
2. The World of Images (*'Ālam al-Mithāl*) 37
3. The Highest Council 43
4. God's Customary Way of Acting (Sunna) 49
5. The True Nature of the Spirit (*Rūḥ*) 53
6. The Inner Dimension of the Imposition of Religious Obligations 57
7. How Religious Obligation is Derived from What is Divinely Decreed 60
8. Why Religious Obligation Necessitates Requital 69
9. The Variations in the Natural Dispositions of People .. 73
10. The Causes of the Thoughts Which Motivate Actions 78
11. The Adhering of Acts to the Soul and Its Being Held Accountable for Them 80
12. The Connection of Actions to Psychological Attitudes 84
13. The Causes for Requital 87

BOOK II The Second Investigation: The Manner of Requital during Life and after Death 91

14. Punishment for Acts in This Life 93
15. The True Nature of Death 97
16. The Variation of the States of People in the Intermediary World 101
17. Concerning Some of the Secrets of the Events of the Day of Assembly 106

BOOK III The Third Investigation: The Supports of .... Civilization (The *Irtifāqāt*) 113

18. The Manner of Discovering the *Irtifāqāt* 115
19. The First *Irtifāq* 119
20. The Art of the Proper Manner of Livelihood (The Second *Irtifāq*) 121
21. The Management of the Household 123
22. The Art of Economic Transactions (*Muʿāmalāt*) 127
23. The Governing of the City-State (the Third *Irtifāq*)..... 129
24. The Conduct of Kings 132
25. The Policy towards Assistants 134
26. The Fourth *Irtifāq* (the Caliphate) 137
27. The Agreement of People on the Principles of the *Irtifāqāt* 140
28. The Conventions Current Among People 142

BOOK IV The Fourth Investigation: Human Felicity 145

29. Concerning the True Nature of Felicity 147
30. The Differences among People in Felicity 151
31. The Divisions among People in their Manner of Obtaining This Felicity 153
32. The Principles to Which the Attainment of the Second Way Goes Back 156
33. The Way of Acquiring These Qualities 161
34. The Veils Preventing the Manifestation of the Sound Original Nature (*Fiṭra*) 165
35. The Way of Removing These Veils 168

BOOK V The Fifth Investigation:
Piety and Sin ... 171

36. Preface: Concerning the True Nature of Piety and Sin ... 173
37. The Unity of God (*Tauḥīd*) ... 175
38. Concerning the True Nature of Associationism *(Shirk)* 179
39. The Categories of Associationism (*Shirk*) ... 184
40. Belief in the Divine Attributes ... 190
41. Belief in Predestination ... 197
42. Belief that Worship is the Right of God, May He be Exalted ... 203
43. Honoring the Emblems of Allāh, May He be Exalted ... 210
44. The Inner Meanings of the Ablution and Full Bath ... 214
45. The Inner Dimensions of Prayer ... 218
46. The Inner Meanings of the *Zakāt* ... 221
47. The Inner Dimensions of the Fast ... 224
48. The Inner Dimensions of the Pilgrimage ... 227
49. The Inner Meanings of Various Types of Pious Actions ... 229
50. The Ranks of Sin ... 232
51. The Harmful Influences of Sins ... 236
52. The Acts of Disobedience of a Person which Have an Impact on Himself ... 238
53. The Sins which are Between a Person and Other People ... 241

BOOK VI The Sixth Investigation:
The Regulation of Religion ... 245

54. The Need for the Guides to the Paths and the Founders of Religions ... 247
55. The True Nature of Prophecy and Its Characteristics ... 250
56. The Explanation of the Fact that the Basis of Religion (*Dīn*) is One and the Religious Laws (*Shara'ī'*) and Codes (*Manāhij*) Vary ... 257
57. The Causes of the Revelation of the Divine Laws Relating to One Age Rather than Another, and to One People Rather than Another ... 262

58. The Causes for Being Held Accountable for the Codes (*Manāhij*) 271
59. The Secrets of the Ruling (*Ḥukm*) and the Reason for Legislation (*ʿIlla*) 275
60. The Beneficial Interests (*Maṣāliḥ*) which Require the Obligatory Religious Acts, the Pillars, the Proper Behaviors, and Their Like 280
61. The Inner Meanings of the (Appointed) Times 287
62. The Inner Meanings of the Numbers and the Fixed Quantities 292
63. The Inner Meanings of Making Up (*Qaḍā'*) (for Missed Religious Obligations) or Receiving Dispensations (*Rukhṣa*) (to Diminish Them) 299
64. The Establishment of the Supports of Civilization (*Irtifāqāt*) and the Reformation of Conventions 303
65. Rules which Bring into Effect Other Rules 310
66. Rendering the Ambiguous Precise, Solving the Problematic, and Deriving Rulings from All-Embracing Principles and So On 316
67. The Facilitation (of Religious Duties – *Taisīr*) 322
68. The Inner Meanings of Encouragement (to Righteousness) and Deterrence (from Sin) 327
69. The Ranks of the Community with Regard to Arriving at the Desired Perfection or Its Opposite 334
70. The Need for a Religion which Abrogates the Other Religions 340
71. Fortifying the Religion Against Distortion 346
72. The Causes for the Differing of the Religion of Our Prophet, from the Religion of the Jews and Christians 353
73. The Causes of Abrogation 357
74. The Explanation of What Had Been the Condition of the People of the Jāhiliyya which the Prophet Reformed 361

BOOK VII The Seventh Investigation: The Derivation of the Sharīʿa Laws from the Reports of the Prophet 371
75. The Explanation of the Categories of the Prophet's Knowledge 373

76. The Difference Between the Beneficial Purposes (*Maṣāliḥ*) and the Religious Laws 376
77. The Way the Community Received the Divine Law from the Prophet 383
78. The Ranks of the Books on Hadith 387
79. The Manner of Understanding What is Meant by the Utterance (*Kalām*) 396
80. The Manner of Understanding the Legal (*Sharʿī*) Meanings From the Qur'ān and the Sunna 400
81. Judging Among Divergent Hadiths 405
82. The Causes of the Disagreement of the Companions and the Successors Concerning Applied Jurisprudence 414
83. The Causes for Disagreement Among the Schools of the Jurists 427
84. The Difference Between the People of the Hadith and Those Who Exercise Personal Opinion 437
85. An Account of the Condition of People Before the Fourth Century and After 451

Bibliography 479

Indices 483

Qur'ānic Citations 483

Hadith Citations-English 488

Subjects and Terms 496

Proper Names 503

ACKNOWLEDGMENTS

In a work which has seen this long to come to fruition, there is naturally a long list of persons and institutions who have contributed to the effort materially, intellectually, or morally over the years.

Among those whom I will mention here are Dr. Fazlur Rahman, who first suggested the need for this translation and to whom it is dedicated. The University of Chicago and the Social Sciences and Humanities Research Council of Canada both of which supported my graduate study. Among friends and colleagues Alan and Sylvia Godlas, Barbara von Schlegell, Hoda Boyer, J.M.S. Baljon, Bernard Weiss, Clemens and Judy Roothaan, and Michael Sells; in Montreal, Farrukh Amin, Aftab Ahmad, Wael Hallaq and Salwa Farahian and many other friends and colleagues in Chicago, Canada and San Diego who contributed in various ways. Any faults and errors which may remain, however, are my responsibility.

It is my sincere hope that this translation lives on as a tribute to Dr. Rahman and as a contribution to the study of Islamic thought and its challenges in the contemporary period.

TRANSLATOR'S INTRODUCTION

The importance of the Ḥujjat Allāh al-Bāligha *and its situation in the Islamic intellectual tradition.*

Ḥujjat Allāh al-Bāligha (The Conclusive Argument from God) is considered the master work of the eighteenth century Indian scholar, Shāh Walī Allāh of Delhi (1703–1762). Drawing on his wide erudition, the author brings together in this work the intellectual and spiritual disciplines of the Islamic heritage in his age in order to elucidate the wisdom and inner meanings behind the interpretation of the hadith reports of the Prophet Muhammad.

In his preface Shāh Walī cites the Qur'ānic verse, "Indeed, the conclusive argument is from God," (Qur'ān 6: 149), from which the title of this work is derived. The word "*ḥujjat*" conveys the sense that one party to a debate or argument has presented the peremptory or convincing proof, argument, or formulation.[1] This argument, according to Shāh Walī Allāh, "refers to the inner meaning of religious obligation and requital and the inner dimensions of the divine laws revealed for mercy and guidance."[2] He explains that he has bestowed this title, *Ḥujjat Allāh al-Bāligha* (The Conclusive Argument from God), since his book is like a branch spreading out from this conclusive argument.

In *Ḥujjat Allāh al-Bāligha,* Shāh Walī Allāh attempts to elucidate the deeper levels of meaning of traditional symbols and practices while integrating mystical, intellectual, and traditional textual approaches to their interpretation.

He understood his mission to be the integration of the increasingly fragmented and disparate articulations of the Islamic intellectual tradition in his own lifetime and thus he sought for a mode of

---

[1] In the Qur'ānic reference the phrase is set in the context of God's proof or argument in response to the unbelievers who declare that if God had willed He could have guided everyone. A number of commentators on the Qur'ān understand this verse as referring to the assertion of the omniscience and omnipotence of God, which at the same time entails moral responsibility on the part of humans.

[2] *Ḥujjat Allāh al Bāligha* translation, 10.

expression which would simultaneously draw on the tools of demonstrative proof of the philosophers and theologians, Sufi mystical insight into the higher realities of things, and the authority of the words of the transmitted textual sources of the Qur'ān and the hadith.[3]

The first volume of the *Ḥujjat Allāh al-Bāligha* lays out the theoretical foundations for the interpretation and application of the corpus of prophetic sayings, the hadith, against the background of human purposefulness. In the earlier chapters of the work a metaphysical structure is elaborated which is based on the internal dynamics within systems of experience. This internal dynamic is presented by Shāh Walī Allāh in the form of levels or systems which are initially understood to be composed of parts in conflict. This conflict must be resolved or harmonized toward a higher purpose through achieving or restoring harmony and balance within the system. Once balance is achieved the inherent perfection of the ideal form implicit in the person, species, society, and so on, is fulfilled, and this leads to the entire form or system expanding or moving up to a new, higher, order.

Shāh Walī Allāh's theory of these systems is thus simultaneously one of a conflict of forces, whether two or many; and of one great force that is a drive towards the greatest perfection according to the highest salutary purpose (*maṣlaḥa kulliyya*). This one great purpose that guides the entire universe toward progress through attaining harmony makes conflict intelligible and gives overcoming it a religious significance.

The concept of *maṣlaḥa* (beneficial purpose) is central to Shāh Walī Allāh's thought as the point of integration of the benefits arising due to the natural order of things, and the salutary outcomes which religious (sharīʿa) legislation intends to achieve. The term *maṣlaḥa* is used in Islamic jurisprudence (*fiqh*), particularly in the school of Mālik ibn Anas in the sense of "public interest" when a case may be judged on this basis if no primary Islamic sources ruled on it. al-Ghazzālī recognized both this narrower sense of the term and a broader concept of *maṣlaḥa* as furthering the ultimate purpose of the sharīʿa in the maintenance of religion, life, offspring, reason, and property.[4] For Shāh Walī Allāh the term

[3] Shāh Walī Allāh, *al-Tafhīmāt al-Ilāhiyya* I:110–111.

[4] Majid Khadduri, "maṣlaḥa" in *The Encyclopaedia of Islam* New Edition, VI (Leiden: EJ Brill, 1954–), 738–740. He is citing al-Ghazzālī's *al-Mustaṣfā* I (Cairo, 1356 A. H.), 139–140.

maṣlaḥa conveys yet a broader sense of the intended benefit or salutary purpose of the divine injunctions. At the highest level, this beneficial interest consists of fulfilling the one great universal purpose (*al-maṣlaḥa al-kulliyya*) of the cosmic order.

In his exposition of the underlying wisdom of the Islamic legal system, in particular the corpus of hadith reports, Shāh Walī Allāh develops an understanding of layers of signification which draws on the theories of the Islamic legal and linguistic theoreticians but goes beyond purely linguistic elements of meaning to incorporate more subtle and even mystical aspects of the reference and efficacy of symbols. His theory is implicitly dynamic in that these religious symbols have not only been modified with the passage of historical time but will continue to respond to developments in the world.

The structure of the first volume of the *Ḥujjat Allāh al-Bāligha*, which is subtitled, "The General Principles from Which Are Derived the Beneficial Purposes Considered in the Rulings of the Divine Law" is developed through seven sections comprised of Eighty-One Chapters.[5] This first volume begins by laying out the metaphysical, psychological, sociological, and moral basis of his system. The latter portion of the volume addresses the Islamic religion more specifically through a consideration of basic religious practices and their inner meanings, the role of prophecy and the divine injunctions, and the means for their derivation. It concludes with a discussion of the prophetic reports and the hadith sciences. Since the second volume, which is not translated here, applies this theory to selected hadiths following, for the most part,[6] the topical arrangement of classical works of hadith and Islamic law, it is likely that this constituted Shāh Walī Allāh's work as originally planned.

The additional four chapters translated here which discuss the issue of juristic disagreement likely were originally prepared as part of a separate treatise entitled "Fairness in Explaining the Cause of Juristic Disagreement" and were later appended to Volume One.

[5] According to al-Sayyid Ṣābiq the original plan for Volume One was a work of seven sections and seventy chapters. He concludes, therefore, that the final section consisting of four chapters on Juristic Disagreement was appended later and that some subsections of the original chapters came to be treated as chapters on their own. *Ḥujjat Allāh al-Bāligha* I (Cairo: Multazim al-Ṭab' wa'l-Nashr, 1952), 296.

[6] The major exception is a section discussion Sufi moral perfection, *iḥsān*.

Since they are normally printed as part of the text and since they contain some material not included in the separate treatise, they have been translated here and included as Chapters Eighty-Two through Eighty-Five.

In terms of the plan for the first volume, the first section, "The Causes of Religious Obligations and Requital" concerns the metaphysical aspects of causation beginning with creation. It begins with the process of creation and continues through the various ways in which God's will is implemented in the physical and supraphysical world. In chapter five and subsequent chapters the human soul and the psychological dimensions of the person are described together with the consequences of this for requital. A central dimension of Shāh Walī Allāh's psychological and moral framework is the idea that human beings are composed of both a higher and a lower side, which he terms the angelic and the animalistic components. Each individual also exhibits an intrinsic nature depending on the relative strength of each component and the way in which the two combine. The course of human moral and spiritual development is put into practice through following religious injunctions at the most necessary and basic level, and then may be taken further, in ways only tangentially indicated in this work, through higher development based on Sufi spiritual practices.[7]

The second section of the present work treats requital for thoughts and actions in this life and subsequently in the next life.

The third section of the work deals with Shāh Walī Allāh's theory of civilizational development in which he presents four basic stages of human advancement, for which he coined the term *irtifāqāt*. Shāh Walī Allāh's concept of the *irtifāqāt* has provoked much interest among contemporary scholars of his thought. The term in his usage is idiosyncratic and its precise meaning varies contextually. The term "*irtifāq*" is derived from the Arabic root "*r-f-q*"—to be gentle, soft, gracious, courteous, or civil—and in the VIIIth form of the Arabic verb this root conveys the idea of "resting on (as a support) or deriving a use from." ʿAbdel ʿAal, in his thesis,[8] reviews some of the translations or explications offered for this

[7] These more esoteric dimensions of the author's psychological and spiritual theories may be found in the works, *Hamaʿāt* amd *Alṭāf al-Quds*.

[8] Khalīl ʿAbdel Ḥamīd Abdel ʿAal, "God. the Universe. and Man in Islamic Thought: The Contribution of Shāh Walīullah of Delhi (1702–1762)." Ph. D. Dissertation. University of London, 1970.

term as follows: "*al-tadbīrāt al-nāfi'ah*" or "the useful management of human affairs" which was offered by the editors of the Arabic text of *Hujjat Allāh al-Bāligha;* 'Abdul Hamid Haleopota's definition of *irtifāqāt* as "every trait; characteristic, and institution that comes under the subject of sociology,"[9] Aziz Ahmad who regards *irtifāqāt* as the stages of the history of the growth of human societies,[10] and Sabiḥ Aḥmad Kamālī who sees the *irtifāqāt* as "civilization and its devices" and a theory of natural law.[11] Jacques Berque concludes that it seems possible that "the plural of the verbal noun, *irtifāqāt*, may be translated by 'uses, commodities, services.'" Berque further reads a socio-economic slant into the *irtifāqāt* by equating them with "the services constituting the institutional section of the "collective good" (*al-maṣlaḥa al-ʿamma*).[12] It appears, therefore, that this term may encompass all of these connotations depending on context and does not have a simple English equivalent. The core of Walī Allāh's explanation of the *irtifāqāt* presents the development of human societies through four *irtifāqāt* or stages of increasingly refined order and elaboration of arts of civilized life. These four stages are roughly equivalent to: following natural or instinctive laws, integrating family life and social transactions, developing a local political order, and finally the extension of this to the international level, which Walī Allāh identifies with the Islamic Caliphate.[13]

Interpreters of his thought have been primarily interested in assessing the originality of this idea and whether it presages a sociological interest and method–either through an influence, perhaps of Ibn Khaldūn, or as an intuition of stages of societal evolution which anticipates Comptian thought. Shāh Walī Allāh's understanding of these developments is, in fact, consistent with the pattern underlying all of his explanations of historical phenomena, i.e., the concept of growth towards an ideal, already-posited form, modeled

---

[9] From Halepota's thesis, "Practical Theology and Ethics of Shāh Walī Allāh."

[10] 'Azīz Ahmad, *Studies in Islamic Culture in the Indian Environment*, (Oxford: Oxford University Press, 1964), 206.

[11] Sabih Ahmad Kamali, *Types of Islamic Thought*, ('Alīgarh: 'Alīgarh University Press, 1966), 123, 124. The references above are cited in 'Abdel 'Aal's thesis, "God. the Universe and Man," 392–397.

[12] Jacques Berque, "Un contemporain islamo-indien de Jean-Jacques Rousseau," in *L'Islam au temps du monde* (Paris: Sindbad, 1984), 113-146, 135.

[13] Shāh Walī Allāh, *Ḥujjat Allāh al-Bāligha*, Arabic text (Cairo: Multazim al-Ṭabʿ wa'l-Nashr, 1952), 80–104.

on Hellenized Sufi concepts of human development and prophetic cycles. Within this framework Walī Allāh at times comments on specific persons, movements or phenomena but his overall understanding of the general principles precedes the inductive method and the thrust is toward confirming one synthetic vision rather than deriving social theory based on the observation of discrete phenomena.

The fourth section of the work discusses the ways to achieve ultimate human felicity through the cultivation of four main virtues; purity, humbling oneself before God, magnanimity, and justice. Subsequently three barriers or veils to this felicity, the veils of custom, conventions, and misunderstanding of the nature of God, and the ways to overcome them are treated.

The fifth discussion of piety and sin treats piety with respect to conceiving and honoring God and performing the various religious practices of Islam with the correct understanding and attitude. This is reminiscent of works on the inner meanings of religious practices in the tradition of al-Ghazzālī's *Revivification of the Religious Sciences.*[14]

The sixth section investigating the policies of religion treats a number of important topics: prophecy, the development of religious tradition in a historical context, the situation of specific religious rulings within such a context in terms of their role and interpretation, and finally the relationship of Islam to other religious traditions.

The development of Shāh Walī Allāh's ideas focuses on affirming the eternal principles of the sharīʿa which are derivable from the particular historical pronouncements of the Prophet enshrined in the corpus of hadith. This formulation of the metaphysical status of legal derivation is his original contribution in the *Ḥujjat Allāh al-Bāligha.* In order to effect this integration of disparate types of reasoning, Shāh Walī Allāh has to deal with theoretical issues of language and referent.

Shāh Walī Allāh represents each particular religious revelation throughout human history as having its origin in a primordial, archetypal religion which he terms the "*dīn*". This form of religion exists as an ideal form at a higher level in the World of Images. This ideal form of the *dīn* is the one most in harmony with the ideal form of the human species, which Shāh Walī Allāh desig-

[14] al-Ghazzālī, *Iḥyā' ʿUlūm al-Dīn* (Beirut: Dār al-Maʿrifa, 1982).

nates by the term "*fiṭra*", or inherent nature. Actualized manifestations of the ideal form known as "*milal*" (sing. *milla*—religious community, formal religion) descend in successive revelations depending on the particular material and historical circumstances of the peoples to whom they are sent down.

Each succeeding revelation "reforms" the elements previously existing into a new pattern or gestalt which embodies the primordial form of the ideal religion (*dīn*) in an altered form suitable for those prepared to receive it. As these human systems have developed throughout history, religion has adapted in its form, beliefs, and spiritual practices to the customs, previously held beliefs, and temperaments of the nations to whom it has been revealed. His theory thus envisions an interaction between the human material and the actualized form (*milla*) of the ideal religion which responds to and shapes it; in that the more developed and complex the human system, the higher the religious form which it will require to achieve its ideal balance.

There are therefore, according to his system, two sources of religious legislation. The first consists of a branch based on the universal beneficial purposes for the human species as embodied in its ideal natural constitution (*fiṭra*). This substratum of beliefs and practices suited to the basic constitution of all peoples he terms the "*madhhab ṭabīʿī*."[15]

The second is composed of those systems of laws (*sharīʿa*s) revealed in response to particular historical circumstances not initially encompassed in the requirements of the human constitution. Walī Allāh explains how the specific rulings of a religion instituted by a prophet are based on several factors. In addition to remaining as close to the "natural way" as possible, the prophet preserves among people the element of previous revelations still remaining which are beneficial, adding at the same time, new legislation that is necessary for their development and that recognizes changes which have occurred since previous revelations. Finally, there exists a category of contingent, accidental factors in legislation which arise due to demands on the prophet to provide answers on particular occasions and in the context of specific circumstances.

According to Shāh Walī Allāh's explanation, one who wishes to be well-versed in the religious sciences has to understand the

[15] Translation pp. 281, 285, 290.

requirement to carry out religious obligation on the basis of either or both of these branches. A key principle here is his treatment of the concept of *maẓinna*. pl. *maẓānn* or *maẓinnāt* which can be understood as derived from the Arabic root "*ẓ-n-n*" meaning to think or believe, thus a *maẓinna* is "a place where something is expected or thought to be, a supposed location or instantiation" or "a mark or indication of something". Certain "*maẓann*" are symbols or acts which were set from pre-eternity as most suitable for the natures of all humans. The relationship of the *maẓinna* to the principle for which it stands is compared by Shāh Walī Allāh to that of a word to its referent or the mental image to the reality which is being thought about.

> Then when God inspired this knowledge to the Highest Council and revealed to them that the anticipated sources (*maẓinnāt*) would stand in the position of the principles, and that they were their embodiments and representations, and that religious obligations could not be imposed on a people except through them, a certain consensus (*ijmāʿ*) obtained in the Holy Enclave that they (the laws) were like the word in relation to the reality which is its referent, and like the mental image in relation to the external reality from which it is derived, and like the painted portrait in relation to the one of whom it was drawn in its portraying him, and like the written form in relation to the words which it puts down. For, in all of these things, when the relation between the signifier and the signified is strengthened there arises between them a permanent bond and a close association with each other, and in a certain scope, one is the other.[16]

It is thus in Shāh Walī Allāh's concept of *maẓinna* that we find the basis of his theory of change, variation, and abrogation of religious rulings. While the relationship of the symbol to its referent is fixed within any context, it is possible that just as there are different languages, there may be different sets of *maẓānn*, even corresponding to successive revelations of the one true *dīn*. According to this theory of change, the symbols for the best interests (*maṣāliḥ*) of the human race will vary with its ages and customs. "The anticipated sources (*maẓānn*) of things conducive to benefit, (*maṣāliḥ*) differ in accordance with the differences in the eras and customs."[17]

---

[16] Ch. 58.
[17] Ch. 57.

Some of these *mazānn* may be considered natural rulings or symbols, as they are derived through the branch of rulings based on the beneficial purposes (*maṣāliḥ*)—i. e., those things which are in accordance with the nature (*fiṭra*) and purpose of the human race. Violating rules connected with such *mazānn* leads to a natural punishment or harm coming to a person even if no religious law had been revealed explicitly requiring them. In the case of the religious laws connected with historical contexts; the religious symbols embodied in their rulings are also derived from particular historical contexts and situations. These, then, come to have a more general validity at the higher spiritual level and from there come to have an effect of the form of the entire human species which has existed from pre-eternity at the archetypal plane of the World of Images. In this manner a symbol or anticipated source of benefit derived from the sharī'a branch of legislation becomes "naturalized" so that being requited on the basis of these rulings has a natural effect, as well as being based on legislative decrees. Some of these religious laws and symbols may be explained or understood through reason while others are not rationally comprehensible, and thus textual sources are required in order to know of them. This conclusion reinforces Shāh Wali Allāh's argument for the importance of the study of the hadith or saying of the Prophet, an argument central to this entire volume of the *Ḥujjat Allāh al-Bāligha*.

Part seven then proceeds to the central topic of the volume, the hadith reports of the Prophet Muhammad, by treating the traditional elements of the discipline, reviewing the methods of evaluating and interpreting the hadith reports, and surveying the historical development of this field and its major works.

The final section comprises four chapters concerning juristic disagreement within the four Sunni legal schools (*madhāhib*) which were likely appended to the original work. Shāh Walī Allāh's position on *ijtihād* and *taqlīd* as expressed in these chapters and elsewhere will be discussed later in this introduction.

Biography

Shāh Walī Allāh was the great intellectual figure of eighteenth-century Islam in India and a prolific writer in Arabic and Persian. Biographical material and anecdotes concerning his life and family may be found in his brief autobiography *Al-Juz' al-Laṭīf fi Tarjuma al- 'Abd al-Ḍa'īf* and in his work *Anfas al-'Ārifīn* which features accounts of his father, his uncle, and his spiritual teachers in India and the Hijaz. Shāh Walī Allāh was born on the 4th of Shawwāl, 1214 A.H. or February 21, 1703. He was the first child of his father's second marriage entered into when the latter was already sixty years old. His father and spiritual guide, Shāh 'Abd al-Raḥīm (d. 1719) was an extremely learned man and a practicing mystic of the Naqshbandiyya order, as well as the Chistiyya and Qādiriyya orders. In Shāh Walī Allāh's anecdotes about his father in the *Anfas al-'Ārifīn* we get a sense of the deep reverence in which he held him. His father was in charge of his own *madrasa* or teaching academy in Delhi and was a very well-known scholar. For a time Shāh 'Abd al-Raḥīm was engaged to work on the compilation of legal rulings commissioned by the Mughal ruler Aurangzeb (d. 1707) and known as the *Fatāwā 'Ālamgīrī*. Apparently he wished to avoid being involved with the court but had difficulty in getting himself excused from the task. There is a suggestion that conflict with some of the other scholars working on the text finally led to his being excused and offered a land grant, which he refused.

Shah 'Abd al-Raḥīm decided to marry a second time at the late age of sixty years, as he had had a mystical intimation that a son would be born to him who would reach a high mystical attainment. One of his disciples, Shaikh Muḥammad of Phulat, offered the hand of his daughter to his teacher. When criticized by some for this late marriage, Shāh 'Abd al-Raḥīm disclosed that he knew that he would have more than one child, and, in fact, he lived until the seventeenth year of Shāh Walī Allāh's life and had another son, who was named Ahlullāh. A number of anecdotes recounted in Shāh Walī Allāh's works indicate that his mother also possessed exceptional religious devotion and mystical insight.

The ancestors of Shāh Walī Allāh had settled in the town of Rohtak, near Delhi in the thirteenth century, soon after the conquest of Delhi. On his father's side his ancestry could be traced back to the caliph 'Umar and on his mother's side to 'Alī, fourth caliph and son-in-law of the Prophet.[18]

---

[18] Ghulam Hussain Jalbani, *Life of Shāh Waliyullāh* (Lahore: Ashraf, 1978), 4. J.M.S. Baljon, *Religion and Thought of Shāh Walī Allāh* (Leiden: E.J. Brill, 1986), 1.

Shāh 'Abd al-Raḥīm devoted considerable attention to the education of his precocious son. Shāh Walī Allāh began his schooling at the age of five. When he was seven he began to pray and fast and completed his first reading of the Qur'ān. At this same age he began to read Persian treatises, and by the time he was ten he was able to read and study independently. At some time before or after his second marriage, Shāh 'Abd al-Raḥīm moved from Agra to Delhi. The family lived in a house in the district of Kotla Firūz Shāh, where Shāh 'Abd al-Rahīm's *madrasa* was located.[19] With his father, young Walī Allāh studied hadith works such as the *Mishkāt al-Maṣābiḥ*[20] and the *Ṣaḥīḥ al-Bukhārī*, works on Qur'an interpretation, Islamic jurisprudence, and theology. In addition, he was exposed to works of Sufism by such masters of Ibn 'Arabī's school of the Unity of Existence as 'Abd al-Raḥmān Jāmī (1492), and Fakhruddīn 'Irāqī (1289).[21] Besides religious subjects, his studies included Astronomy, Mathematics, Arabic and Persian Language and Grammar, and medical science (*ṭibb*) from which many concepts and theories influence his works.

Shāh Walī Allāh was married during his fourteenth year to the daughter of his maternal uncle. According to his autobiography, this was at the insistence of his father who had foretold that there was some secret reason for conducting the marriage with all haste. It turned out that shortly after the marriage a number of family members died in succession which would have resulted in a long delay and possibly cancellation of the match. When Shāh Walī Allāh was fifteen years old his father accepted him as a disciple in the Naqshbandiyya order and he began to perform the practices of that order. He also completed his course in Islamic studies in that year and was permitted by his father to teach others. When Shāh 'Abd al-Raḥīm was on his deathbed he gave his son permission to initiate others in Sufism and to give them spiritual guidance. For twelve years after his father's death in 1719 Shāh Walī Allāh taught and studied the religious sciences and continued in meditative

---

[19] Sayyid Athar Abbas Rizvi, *Shāh Walī Allāh and his Times* (Canberra: Ma'rifat Publishing House, 1980), 213.

[20] *Mishkāt al-Maṣābīḥ* by-al-Marghīnānī, trans. James Robson (Lahore: Ashraf, 1963). A large number of the hadith cited by the author in this work may be found in the *Mishkāt*.

[21] "Al-Juz' al-Laṭīf fi-Tarjamat al-'Abd al-Ḍa'īf" (Persian original) in *Journal of the Asiatic Society of Bengal* 14 (1912): 161–175 with English translation by M. Hidayat Ḥusain.

discipline. Then, in about April 1731, he departed India to perform the pilgrimage to Mecca and Medina where he stayed for some fourteen months, returning to India in December 1732.

This stay in the Hijaz was an important formative influence on his thought and subsequent life. While in the Holy Cities he studied hadith, *fiqh*, and Sufism with various eminent teachers whom he mentions in the *Anfas al-ʿĀrifīn*, the most important influences being Shaikh Abū Ṭāhir al-Kurdī al-Madanī (d. 1733)[22], Shaikh Wafd Allāh al-Makkī, and Shaikh Tāj al-Dīn al-Qalaʿī' al-Ḥanafī (d. 1734). These teachers in Mecca exposed Shāh Walī Allāh to the trend of increased cosmopolitanism in hadith scholarship which began to emerge there in the eighteenth century due to a blending of the North African, Hijazi, and Indian traditions of study and evaluation.[23] While in the Holy Cities, Shāh Walī Allāh developed a particular respect for Mālik's work the *Muwaṭṭa'*, on which he later was to write two commentaries, *Musawwā* and *Muṣaffā*. During his stay in Mecca and Medina, Shāh Walī Allāh had many mystical experiences, dreams, and visions of Prophet Muhammad in which his questions were answered and he was instructed to carry out a mission of teaching. He felt that these visions confirmed his religious purpose and his exalted state, and later included them in a work entitled *The Emanations of the Two Holy Cities*.[24]

Shāh Walī Allāh's writing career began in earnest on his return from pilgrimage. While the *Ḥujjat Allāh al-Bāligha* was composed sometime during the decade after his return, it seems that the inspiration to compose such a work came to him while on the pilgrimage. At that time, he saw a vision of the grandsons of the Prophet, Ḥasan and Ḥusain, holding a broken pen out to him, then repairing it, and later bestowing upon him a robe of the Prophet. From this he understood that he had a mission to restore the Islamic

[22] Son of the famous hadith scholar and Sufi of the Hijāz, Shaikh Ibrahīm al-Kurānī (d. 1690). See A.H. Johns, "al-Kurānī, Ibrahīm" in *The Encyclopaedia of Islam* New Edition. V. (Leiden: EJ Brill, 1954–), 242–243.

[23] Voll, John O., "Hadith Scholars and Tariqahs: An 'Ulema' Group." in *Journal of Asian and African Studies* 15 (July–October 1980): 264–273.

[24] *Fuyūḍ al-Ḥaramain*. Arabic with Urdu translation (Karachi: Muḥammad Saʿīd, n. d.). The practice of writing accounts of both inner and outer event of the pilgrimage seems to have been a tradition of Indian Naqshbandīs of the Mujaddidī line, see for example, *Ḥasanāt al-Ḥaramain* which features the visions of Khwāja Muḥammad Maʿṣūm (1662) and was compiled by his son Muḥammad 'Ubaid Allāh. Persian version edited and annotated by Muḥammad Iqbāl Mujaddidī, (Dera Ismail Khan, Pakistan: Maktaba Sirājiyya, 1981).

sciences through the study of the reports of the Prophet.[25] The revitalization of hadith studies as a discipline supporting attempts at social and moral reconstruction has been cited as a feature of eighteenth-century Islam.[26] In the preface to the *Ḥujjat Allāh al-Bāligha*, Shāh Walī Allāh writes that he delayed for some time in undertaking such an ambitious project, but the urgings of one of his closest disciples, Muḥammad 'Āshiq of Phulat (1773), and his recognition of the desperate need for such a work led him to embark on the work.[27]

It was sometime after his return to India that he concluded a second marriage from which four sons and one daughter were born. From his first marriage he had one son and a daughter.

After Shāh Walī Allāh's death in 1762, his teachings were carried on by his descendants, in particular his sons, Shāh 'Abd al-'Azīz (d. 1823) and Shāh Rafī' al-Dīn (d. 1818), and his grandson Shah Ismā'īl Shahīd (d. 1831). The influence of this notable family has been termed a "Walī Allāhī movement."[28] While the works of his descendants show some influence of Shāh Walī Allāh's thinking, they do not appear to have the same grasp of universal principles as the original. The extent of Shāh Walī Allāh's influence and the shift in interests and intellectual positions among his descendants still remains to be studied.

---

[25] *Ḥujjat Allāh* translation. p. 6 This vision is recounted in the author's book, *Fuyūḍ al-Ḥaramain*. "The Emanations of the Two Sacred Mosques," pp. 65–66. an *al-Tafhīmāt al-Ilāhiyya*, II, p. 300.

[26] The connection of this movement with reformist tendencies in Sufism in this period was first noted by Fazlur Rahman in *Islam* (Chicago,: University of Chicago, 1979), 206, who coined the term Neo-Sufism for such developments. For a discussion of scholarship on "Neo-Sufism" see R.S. O'Fahey. *Enigmatic Saint*. (Evanston: Northwestern U. P., 1990), 1–9 and the subsequent article by R.S. O'Fahey and Bernd Radtke, "Neo-Sufism Reconsidered" in *Der Islam: Zeitschrift fur Geschichte und Kultur des Orients*. LXX (1, 1993): 52–87 which attempts to refute the "neo-Sufi" hypothesis on the basis of historical and textual evidence from Africa and the Middle East. Perhaps the South Asian context evidences stronger trends in this direction.

[27] *Ḥujjat Allāh* translation p. 8.

[28] This understanding seems to have been developed by 'Ubaydullāh Sindhī in *Shāh Walī Allāh aur unkī Siyāsī Taīrīk* (Lahore, 1970) and remains influential in South Asian studies on the legacy of Shāh Walī Allah, for example Aziz Ahmad "The Waliullahi Movement," In *Studies in Islamic Culture in the Indian Environment*. (Oxford: Oxford University Press. 1964).

The Historical Context

Shāh Walī Allāh's life and work must be set against the background of the troubled age in which he lived. The stable and powerful leadership of the Mughal ruler Aurangzeb ended in 1707 when Shāh Walī Allāh was still a child. During the remainder of his life of slightly less than sixty years, ten monarchs sat upon the throne of Delhi. The power of the Mughal rulers declined steadily after Aurangzeb and the power and control of the central government gradually weakened, as both internal and external forces pressed on the empire, pushing it toward instability and decay. From 1708 until 1716 the Sikhs plundered the northwest until they were driven back by the Mughals. The Hindu Marathas from the south invaded suburbs of Delhi in 1738, and the Mughals were forced to give up the province of Malwa to them, thus dividing their empire. By 1750 they were involved in the Punjab and Uttar Pradesh, and they managed to attack Delhi in 1757.

In 1739 the disastrous invasion of Nadir Shah, King of Persia, occurred in which the imperial treasury was looted and the citizens of Delhi were put to the sword.

The power of another outside force, the Afghan Rohillas under Aḥmad Shāh Abdālī, was also on the rise and an invasion of India from the northwest began so that by the 1740s the Punjab was occupied. At the battle of Panipat in January 1761, the Maratha forces were defeated by the Afghans, but Abdālī's army mutinied and he was forced to retire, ending anticipations of a renewed Muslim power in Delhi.

In his work, *Shāh Walī Allāh ke Siyāsī Maktūbāt*,[29] Professor Khaliq Ahmad Nizami published a collection of Shāh Walī Allāh's letters to various Muslim nobles and leaders, urging them to either strengthen the current administration, or in the case of outside Muslim leaders such as Abdālī, to defeat Hindu threats to Muslim rule such as the threat of the Marathas. This involvement of Shāh Walī Allāh in the politics of his time seems to have been prompted by the urgency of the situation, and was more in the nature of a respected religious leader and scholar speaking out of conscience, than that of an active political force. It was not uncommon for Sufis of the Naqshbandiyya order to take stands on political issues

[29] Delhi: Nadwat al-Muṣannifīn, 1969.

in this way, and to have contacts with the ruling class and nobles. While Shāh Walī Allāh occasionally comments on injustices and the need for a better government, his political thought consists principally of presenting idealized models of the state and the qualities of its rulers and officials.

Another feature of the Muslim society of his time was the conflict between Sunni and Shiʿi factions at the court and in the society at large. The Sunni Muslims condemned Shiʿa practices such as the Muḥarram processions as excessive, and the issue of who should have been the successor of the Prophet was actively debated by representatives of each belief—Shāh Walī Allāh was very staunch in his defense of the Sunni position and his repudiation of Shiʿa views of the caliphal succession. His enterprise of the reconciliation of divergent positions among Muslims, therefore, did not include these Shiʿī elements.

Shāh Walī Allāh's Views on Ikhtilāf, Ijtihād, and Taqlīd

The subject of juristic disagreement or *ikhtilāf* was often treated by the Muslim jurists once legal schools began to form.[30] Among the most influential of these works on Shāh Walī Allāh's formulations are al-Shāfiʿī's *Risāla* and al-Suyūṭī's (1505) works on *ijtihād*.

Shāh Walī Allāh's position on following the legal schools seems to have modified over the course of his lifetime. Maẓhar Baqāʾ categorizes these developmental stages chronologically as 1) inherited tendencies, 2) the youthful outcome of his own reflections, 3) influences of his stay in the Hijāz, and 4) the effects of the practical environment in which he taught in India.[31]

As previously indicated, his family background would have stressed the Ḥanafī school which was by far the dominant school in Central Asia and India. While his father was a prominent Ḥanafī jurist, he was known to have disagreed with the school on a number of issues.[32]

---

[30] For a list of some early *ikhtilāf* works see, al-Ṭaḥawī, *Ikhtilāf al-Fuqahāʾ* ed. M. Ḥasan Maʿṣūmī (Islamabad: Islamic Research Institute, 1971), 24–30. A French translation of another early ikhtilāf work is Gérard Lecomte, *Le traité des Divergences du Ḥadīth d'Ibn Qutayba d. 889* (Damascus: Institut Français de Damàs, 1962).

[31] *Uṣūl-i-Fiqh aur Shāh Walī Allāh* (Islamabad: Idāra Taḥqīqāt Islāmī, 1979), 24–31.

[32] Baqā', 24–25.

Before making the pilgrimage in 1731 Shāh Walī Allāh seems to have independently arrived at a rejection of *taqlīd*, the acceptance of the rulings of the founders of any one of the main legal schools.[33] He writes in *Fuyūḍ al-Ḥaramain* that personally he had been inclined to reject *taqlīd* but that he was commanded to support it in a vision of the Prophet which he had during his pilgrimage.[34]

This vision may also have reflected the influence of his experiences with his various teachers in the Holy Cities who belonged to the other schools. His own research into the early works, specifically the *Muwaṭṭā'* of Mālik led him to examine the earliest sources and original texts in search of a solution, in other words, towards undertaking his own process of *ijtihād*, or independent investigation, rather than absolute following, *taqlīd*, of one of the schools.

These factors led him to modify his anti-*taqlīd* position to allow following a school for common people,[35] while at the same time trying to mitigate inter-*madhhab* antagonism.[36]

In his time the Hanbalī school was of limited scope but he was acquainted with the literalist trend of Ibn Ḥazm (1036) and the works of Ibn Taimiyya (1263–1328). His most influential teacher, al-Kurdī, followed the Shāfiʿī school and this seems to be the dominant influence on Walī Allāh's theoretical jurisprudence.

Later in his life he saw his mission in the Indian context as one of doing *taṭbīq* or accommodation of the Ḥanafī and Shāfiʿī schools.[37] In a personal note in his own writing dated 1159 A.H./ 1746 found on a manuscript of al-Bukhārī which he used in teaching hadith he says that he used to follow the Ḥanafī school in practice and teach both Ḥanafī and Shāfiʿī fiqh.[38]

---

[33] J.M.S. Baljon, *Religion and Thought of Shāh Walī Allāh* (Leiden: E.J. Brill, 1986), 165 quotes "al-Juz' al-Laṭīf" relative to his position at this time.

[34] *Fuyūd al-Ḥaramain*, Vision 33, 187–8 Baqā', 24–25.

[35] In the treatise *ʿIqd al-Jīd* he includes a section on why people should stay within the four legal schools. *ʿIqd al-Jīd fī-Aḥkām al-Ijtihād wa-l-Taqlīd*. Arabic. (Cairo: Maktaba al-Salafiyya, 1965). See Daud Rahbar article for partial English translation.

13–14. One reason is that these schools have preserved the historical continuity of the tradition going back to the Prophet, the second is that they represent the accurate consensus of the community, and thirdly it is more reliable to follow the sayings of these schools than to follow any contemporary scholar who may have been co-opted or corrupted.

[36] In *Fuyūḍ* Vision 10, 89–91 he states that he was mystically informed that the Prophet dislikes conflict among the schools and considers them equal.

[37] *Musawwā* 1, 12–13. *al-Tafhīmāt* 1, 212.

[38] Baqā', 40.

In *Fuyūḍ al-Ḥaramain* which was composed soon after his return from the pilgrimage, several notes extol the virtues of Hanafism,[39] although elsewhere he says that the Shāfiʿī school is most in conformity with the Sunna.[40] He seems to have envisioned that some sort of accommodated version of both schools was the most suitable *fiqh* for the Indian context, although in theory all four schools were acceptable to him.[41]

Views on Ijtihād and Taqlīd

At one extreme on this issue stand those who disallow any following of legal schools, such as Ibn Ḥazm. By Shāh Walī Allāh's time the issue of whether the gate of *ijtihād* was closed, so that, in fact, only *taqlīd* was possible, had been raised and accepted by some jurists.

His position resembles that of the Shāfiʿī school in that all but the most radical forms of *ijtihād* remain possible.[42] He follows the theory worked out by Shāfiʿī masters such as al-Nawawī (1277) that absolute *ijtihād* remains possible as the level of the affiliated absolute *mujtahid*, while that of the original founders of the schools is no longer in existence. "This theory recognizes the possibility that there were still absolute mujtahids without, however, compromising the superiority of the founders of the madhhabs."[43]

Shāh Walī Allāh was a Ḥanafī in practice, but intellectually and preponderantly in his analysis of cases he gave preference to the Shāfiʿī school.[44] In the treatise *al-Inṣāf* he speaks highly of this school, for example,

> As for the school of al-Shāfiʿī, it has most absolute *mujtahid*s and *mujtahid*s within the school among the legal schools, and it is the

---

[39] *Fuyūḍ* Vision 19, 136–137. In which he is informed of the greatness of the Hanafī school; Vision 31, 179–183. On the necessity of Hanafī *madhhab* in India, and 311–320 In which he is inspired about the superiority of the Hanafī school.

[40] *al-Khair al-Kathīr* Khizāna 10, cited in Baljon, 166, "The madhhab of al-Shāfiʿī, which gets to the root of things, is among the four the most in agreement with the sunna."

[41] *Fuyūḍ*, Vision 10, 89–91.

[42] *Inṣāf*, present translation, chapter four.

[43] Rudolph Peters, "Idjtihād and Taqlīd in 18th and 19th Century Islam," *Die Welt des Islams* XX, (3–4, 1980): 137.

[44] Baqā', 27 and also 502 where he show that Shāh Walī Allāh prefers Shāfiʿī rulings in 70% of cases.

> school with the most developed legal theory and theology and the one which has provided the most Qur'ān interpretation and commentary on the hadith. It is the strongest in chains of hadith reporters and transmission, the best at verifying the statements of the founder, the strongest in distinguishing between the opinions of the founder and the points of view of his associates, and the most scrupulous when giving certain opinions and points of view preference over others.[45]

In terms of his own categories he could be considered a *mujtahid* affiliated (*mujtahid muntasib*) to this school[46] based on his criterion of preferring the analytical principles of Imam Shāfi'ī. According to his own statements and prevailing legal theory, this level of *ijtihād* had died out in the Ḥanafī school after the third century due to the fact that hadith analysis was not a primary activity of this school. Hanbalīs and some Shāfi'īs permit a higher level of *ijtihād* to continue.[47]

Rudolph Peters finds Shāh Walī Allāh's views on *ijtihād* to be more conservative than the comparable but later works of al-Shawkānī (1760–1832) and al-Sanūsī (1787–1859).[48] His views on *ijtihād* also participate in a tradition which Voll and Peters suggest has pan-Islamic roots going back to a circle of pious and learned scholars in the Hijaz, in particular Ibrahīm al-Kūrānī (1690).[49] This, of course, is the group frequented by Shāh Walī Allāh during his two-year stay in the Holy Cities.

It has thus been shown that Shāh Walī Allāh went from a position of rejecting *taqlīd*[50] to an acceptance of generally following the four *madhhabs*, although being able to go outside them on specific cases. These final chapters demonstrate the author's mastery

---

[45] *al-Inṣāf*, 77f.

[46] Baqā', 48. "This, however, cannot be regarded as *taqlīd* since the affiliated *mujtahid* accepts his Imam's ruling with complete understanding of its bases and arguments." Peters, 137.

[47] Wael B. Hallaq, "Was the Gate of Ijtihād Closed?" *International Journal of Middle East Studies* 16 (1984): 30.

[48] Peters studied Muḥammad ibn 'Alī al-Shawkānī's (1760–1832), *al-Qaul al-Mufīd fī adillat al-ijtihād wa-l-taqlīd* and *Irshād al-Fuḥūl* and Muḥammad ibn 'Alī al-Sanūsī's (1787–1859) *Īqāẓ al-Wasnān fīl-'amal bi-l-ḥadīth wa-l-Qur'ān* and *Kitāb al-Masā'il al-'ashar al-musammā Bughyat al-maqāṣid fī khulāṣat al-marāṣid*, 142–144.

[49] Peters, 144–145.

[50] *Fuyūḍ*, 184–189. "Although it went against my temperament the Prophet ordered me to follow in practice one of the four legal schools, and not to go beyond them, although by nature I disliked and rejected *taqlīd*." 187–8.

of both theoretical and practical jurisprudence, his position within a long tradition of debate over juristic disagreement, and his attitude to issues of *ijtihād* and *taqlīd.*

Posterity

Assessing the importance of the work is a complex endeavor. One approach has been to assess the amount of controversial literature which it generated in the period after which it appeared. By this standard, the work did not attain a highly controversial status in which it was either extensively commented on or contested, in its own era or subsequently, even in the author's home of the Indian sub-continent.[51] This lack of commentary, however, could have resulted from a general acceptance of the work and also from a deterioration of the level of related scholarship.

On the other hand, it is interesting that today, a number of the religious movements in South Asia construe Shāh Walī Allāh as an intellectual progenitor. In the Indian subcontinent the Deobandis have perhaps the most direct link to his heritage in combining the spiritual linkage of master disciple with intellectual learning. Shāh Walī Allāh's son, Shāh ʿAbd al- ʿAzīz (1823) was a noted scholar and teacher with a wide-circle of pupils, some of whom are linked directly with the establishment of the Deoband *madrasa*, an institution which came to symbolize a mode of thought and practice in South Asian Islam. Briefly this could be characterized as an acceptance of the mystical elements of the Islamic intellectual and practical tradition combined with a rejection of those practices which were more associated with purely local customs and with the practice of the less educated masses.[52]

Those who have a more reformist and puritan outlook such as the *Ahl al-ḥadīth* and even the followers of Maulānā Maudūdī (d. 1978) find in Shah Walī Allāh's elucidation of the sharīʿa and call for reform a precursor to their own beliefs.[53] On the basis of Shāh

[51] Maḥmūd Aḥmad Barakatī, *Shāh Walī Allāh aur unkā Khāndān* (Lahore; Majlis Ishāʿat-e-Islam. 1976).

[52] On the Deoband *madrasa* see Barbara Daly Metcalf, *Islamic Revival in British India: Deoband, 1860–1900* (Princeton: Princeton University Press, 1982).

[53] For example in his *Tajdīd wa Iḥyā' al-Dīn* (Rāmpūr, 1954). At the same time it must be acknowledged that Maudūdī tries to disengage the Sufism of Shāh Walī

Walī Allāh's legal works it is clear that he did not go so far as these reformers who are prepared to reject following the four legal schools entirely.[54] In fact the movement of his grandson Shāh Ismā'īl Shahīd (1831), known as the "Wahhābī Movement" in India, is one direction in which his reformist ideas were taken.

The third and most prevalent inclination in contemporary Muslim South Asia is that of the *Ahl al-Sunna* or Bareilvis, who are more oriented to popular religious practices and Sufism. They support the Ḥanafī *madhhab* to such an extent that Abū Ḥanīfa achieves almost a saintly status. The *Ahl al-Sunna* would find some of the author's statements in support of Hanafism in India to be in agreement with their position. Closest to the contemporary *Ahl al-Sunna* position would be that of the group of his successors best exemplified by his closest disciple and cousin, Muḥammad 'Āshiq (1773).[55] 'Āshiq most fully embodies Shāh Walī Allāh's Sufi inclinations in which there is an indication that he established his own eclectic Sufi order and repertoire of practices.[56]

The bifurcation of the Sufi and the reformist tendencies among his successors is most striking in this figure of his grandson, Shāh Ismā'īl Shahīd, who wrote a work on highly theoretical mysticism, *'Abaqāt*,[57] and then followed with *Taqwiyat al-Imān*,[58] a reformist programmatic for ending heretical innovation in the practice of popular Islam in India.

Outside of the Indian sub-continent, *Ḥujjat Allāh al-Bāligha* and *al-Inṣāf fī Bayān Sabab al-Ikhtilāf*, and *'Iqd al-Jīd* are the most

---

Allāh from other aspects of his thought and activities. *A Short History of the Revivalist Movement in Islam* English trans. by al-Ash'arī. (Lahore: Islamic Publications, 1972), 105ff and appendices.

[54] Maẓhar Baqā', *Uṣūl-e-Fiqh aur Shāh Walī Allāh* (Islamabad: Idāra Taḥqīqāt Islāmī, 1979), 35–6.

[55] Muḥammad 'Āshiq, to whom the present work is dedicated, compiled a biography of Shāh Walī Allāh in the Sufi tradition which combines many anecdotes known from his other works with accounts of his virtuous and miraculous deeds. *al-Qaul al-Jalī*. Persian (Delhi: Shāh Abū'l-Khair Akādmī, 1989). Urdu translation by Maulānā Ḥāfiẓ Taqī Anwār 'Alavī (Kakorvī. Kakorī, India: Maktaba Anwārī, 1988).

[56] The outline of the practices and metaphysics of this *ṭarīqa* or Sufi path are outlined in *al-Tafhīmāt al-Ilāhiyya* II pp. 5–98.

[57] A little known work of Sufi metaphysics in the tradition of the school of Ibn 'Arabī. Ismā'īl Shahīd, Shāh. *'Abaqāt*. Urdu translation by Munāẓir Aḥsan Gīlānī, Hyderabad. India: Al'-Lajnat al-'Ilmiyya, n.d. English translation by G.H. Jalbani (Hyderabad: Zeb Adābī Markaz, 1982).

[58] *Taqwiyat al-Īmān*. (Deoband: Asad Book Depot, n.d.) English translation published in *Journal of the Royal Asiatic Society* 13 (1852): 310–372.

readily available of Shāh Walī Allāh's works. These works were composed in Arabic and thus able to reach a wider reading public of which the author was in contact as part of a cosmopolitan network of learned scholars concerned with the revival of hadith learning among Muslims in the eighteenth century.[59]

Ḥujjat Allāh al-Bāligha has been edited by a prominent Islamic scholar from Egypt, al-Sayyid Ṣābiq[60] and is currently popular among the present generation of reformers in the Arab Middle East, South and South East Asia. At the same time, Islamic modernists such as Muhammad Iqbal (1938) and Fazlur Rahman (1988) have seen in Shāh Walī Allāh a thinker who responded to the crisis of his time with an attitude of moderation and a search for the spirit behind specific injunctions of the tradition.[61]

In any particular field to which he addresses himself, Shāh Walī Allāh is clearly building on the tradition, so that one can understand his Sufism as extending the ideas of al-Ghazzālī (1111) and Ibn al-ʿArabī (1240), his political and social theories as developed within the context of Islamic philosophical ethics, and so on. A distinctive element in his work is the integration of these mystical or philosophical branches of the Islamic intellectual tradition with the more literally textually based elements of the religious curriculum, hadith studies, *uṣūl al-fiqh*, and so on. It may be instructive to compare our assessment of Walī Allāh's work as it was received by his successors to that of Ibn Khaldūn (1404), to whom he is also likened. In the case of Ibn Khaldūn, who was also a synthetic thinker, his successors failed to respond to the implicit synthetic thrust of his argument and simply expanded upon those fragments which they felt were relative to their concerns.[62] Shāh Walī Allāh made his contribution more than three centuries later on the eve of the modern period. His specific programs for hadith study, Sufism, and so on, were in general not taken up in a concrete way by his successors, but the overall orientations integrating his

---

[59] John O. Voll, "Hadith Scholars and Tarīqahs: An 'Ulema' Group" in *Journal of Asian and African Studies* 15 (July–October 1980): 265.

[60] Cairo, 1952–1953.

[61] Fazlur Rahman, "The Thinker of Crisis Shāh Waliy-Ullah" in *The Pakistan Quarterly* (Summer 1956), 44–48. A. Halepota, "Affinity of Iqbāl with Shāh Walī Allāh." *Iqbal Review* XV (1,1974): 65–72. Halepota, "Shāh Waliyullah and Iqbal. The Philosophers of the Modern Age." *Islamic Studies* 13 (December 1974): 225–234.

[62] Aziz al-Azmeh, *Ibn Khaldūn*. (London: Routledge, 1982) especially Chapter Three pp. 145 ff.

work have allowed subsequent thinkers to construe him as him variously as a reformer and purifier, a mystic, or as a modernizer. All of these interpretations have been evoked without much concern with the concrete application of his ideas.

The more subtle elements of his interpretive methodology may come into their own in the current period in which the intellectual heritage of the Islamic religion is in a period of challenge, politically and philosophically. The challenge of developing a methodology for the interpretation of linguistic and symbolic elements of the traditional sources addressed in this text is one which confronts Muslims today.

His Works

Shāh Walī Allāh wrote a large number of works in various categories[63] including biographical, legal, and mystical, as well as two comprehensive books: the *Ḥujjat Allāh al-Bāligha* and *Al-Budūr al-Bāzigha,* which is the later of the two and expands many of the major themes of the *Ḥujjat Allāh*, although it incorporates more mystical philosophy and does not stress the study of hadith.

In general, his Sufi works belong to the earlier part of his career. The major ones among these are: *Hamaʿāt*, *Saṭaʿāt*, *Fuyūḍ al-Ḥaramain*, *al-Qaul al-Jamīl*, *Lamaḥāt*, *Alṭāf al-Quds*, and *Al-Khair al-Kathīr*. *Hujjat Allāh al-Bāligha* was also composed in the earlier period of his career, after his return from the Hijaz.

Among his later works composed after 1756 are the legal works *Musawwā* and *Muṣaffā*, which are commentaries on Mālik's *Muwatta'*, his treatises on law *ʿIqd al-Jīd*, and *Sharḥ Tarājim Abwāb al-Bukhārī*, and his work on Tafsīr, *Al-Fauz al-Kabir*. His two-volume work, *Al-Tafhīmāt al-Illāhiyya*, is a compilation of brief passages in Arabic and Persian which are largely mystical in content but also address questions of law and theology and comment on the situation of different groups during his time. It appears that

[63] The precise enumeration and dating of this prolific author's works is a lengthy and complex process. For annotated bibliographies attempting to establish a chronology of his works see J.M.S. Baljon, *Religion and Thought*, pp. 8–14, S.A.A. Rizvi, *Shāh Walī Allāh and His Times*, pp. 220–228, and G.M. Qasimi's introduction (in Arabic) to *al-Tafhīmāt al-Ilāhiyya* I: 15–37. Publication details of individual works may be found in the bibliography accompanying this work.

these were letters and discourses which were later collated, and then arranged, to some extent, in thematic order. This work is important for a understanding of the writer's mystical and metaphysical thought.

Two of Shāh Walī Allāh's works, the *Izālat al-Khafā'* and the *Qurrat al-ʿAynayn fi Tafḍīl al-Shaikhayn*, were composed in the latter part of the author's life, in order to address Shiʿa claims concerning the nature of the caliphate and the superiority of ʿAlī.

His other works which were examined in the preparation of this treatise are listed in the bibliography. The renewed interest in his writings can be seen from the fact that a significant number of new editions of his original works, many of which were first published in India in the late 19th century are currently being reissued in reproduction, released in new Urdu translations, or published in new editions in Pakistan.

Indexes

In order to facilitate the further study of the general reader and the specialist, I have included several indexes with this work. One lists the Qur'ānic verses cited by Shāh Walī Allāh, the second lists a recognizable phrase from the hadiths to which the author refers. A subject index has also been prepared to accompany the study of the work.

The footnotes accompanying the text are intended to clarify those points which may be obscure for the modern reader, whether Muslim and non-Muslim. Since the position of the hadith is so central to this work, each hadith cited is referenced to the standard collections based on Wensinck's index to the major Arabic hadith collections.[64] In some cases I was able to update these references to hadith works more readily available today in English translation or in new editions of the Arabic texts. For example, I have referenced M.M. Khan's translation of *Ṣaḥīḥ Bukhārī* in the bilingual edition, Siddiqi's translation of *Ṣaḥīḥ Muslim*, and Robson's translation of the *Mishkāt al-Masābiḥ*, from which the author drew a large number of his references. In cases where a person is using a

[64] Arent Jan Wensinck, *Concordances et indices de la tradition musulmane* (Leiden: EJ Brill, 1933–1938).

different edition, the hadith can usually be located based on the topic which it falls under, and this is cited in the notes in those cases in which the hadith work was topically arranged. Since many hadiths are cited in multiple collections, precise references are given only to Bukhārī and Muslim if these were available, as it was assumed that the reader who wished to check the Arabic in all collections would have no problem going back to Wensinck for the other references.

Ḥujjat Allāh al-Bāligha: The Text

The *Ḥujjat Allāh al-Bāligha* has long been recognized as the author's major work and hence has been copied and edited over the centuries since it was written. In the preparation of this translation two printed editions were used.

(1) The Cairo edition or 1952–53, checked and verified by al-Sayyid Sābiq, based on a text originally printed by Maktaba Amīriyya.* This 1952 edition was published by Multazim al-Ṭabʿ wa-l-Nashr Dār al-Kutub al-Ḥadīth.

(2) The edition of the Kitābkhāna Rashīdiyya, Delhi, in which volumes 1 and 2 are bound together. The edition of 1373 A. H. - 1953/54 was used based on a version of the text which had been edited and verified by a number of Indian scholars.

The notes accompanying these two editions of the text were virtually identical, except that the Cairo edition of 1952 gives the verses of the Qur'ānic citations. Only minor variations or misprints were detected on comparison and these have been indicated in the translation. This edition has been reprinted in Cairo since 1936, and is the most readily available edition in the Arab world today.

References to the second volume of *Ḥujjat Allāh al-Bāligha* are given with the pagination in both Arabic editions, with that edited by al-Sayyid Sābiq first and with reference to the chapter title so that those using Urdu editions will also be able to easily locate the reference.

In addition, a two-volume edition of the Arabic text accompanied by an Urdu translation *Niʿmat Allāh al-Sābigha* by ʿAbd al-Ḥaqq Ḥaqqānī (Karachi: Aṣaḥḥ al-Muṭabiʿ, n.d.) was occasionally consulted.

A copy of Maulānā ʿUbayd Allāh Sindhi's *Urdū Sharḥ Ḥujjat*

* An edition was printed in Cairo by Būlāq in 1877.

Allāh al-Bāligha was helpful in preparing certain of the notes but it only covered the first sixteen chapters of the work. This was published in Lahore (undated) and distributed by Maktūbiyyāt Ḥikmat.

The final four chapters (82–85) of the work are to a large extent duplicated in the author's well-known treatise on juristic disagreement, "*al-Inṣāf fī-Bayān Sabab al-Ikhtilāf*", (*Fairness in Explaining the Cause of Disagreement (Among the Jurists)* This has been published in numerous Arabic editions and a number of annotated translations of this work exist in Urdu.[65]

These final chapters of Volume One deal with the issue of disagreement (*ikhtilāf*) among Muslim jurists and the types of individual reasoning (*ijtihād*) still existing within the Islamic legal system. The editors of this text note that this appended section originally appeared in only one of the manuscript copies available to them[66] although its inclusion has become standard in the subsequent printed versions. At the conclusion of these chapters Shāh Walī Allāh indicated his desire to compose a further text on the subject of juristic disagreement[67] but it seems that ultimately he had to content himself with the material in these chapters plus the addition of some new material. This was then issued as a separate treatise under the title *al-Inṣāf fī-Bayān Sabab al-Ikhtilāf*. The additional material found in *al-Inṣāf* consists primarily of a long quote from al-Khaṭṭābī at the end of chapter three, an extended discussion of the topics of ijtihād and the Shāfiʿi school in chapter four, and a reorganization of material to constitute an additional fifth chapter.[68]

The author composed a further treatise *ʿIqd al-Jīd fī Aḥkām*

---

[65] For a list see the bibliography accompanying this work.

[66] *Ḥujjat Allāh al-Bāligha* I, 296.

[67] At the conclusion of the final one of these chapters included as part of the *Ḥujjat Allāh al-Bāligha*, Shāh Walī Allāh writes, "I have had the intention of composing a treatise entitled 'The Summit of Fairness in Explaining the Cause for Disagreement' . . . but I have not been free to do this until this time, so that when the argument drew near the source of (juristic) disagreement I was led by my inner motivation to explain whatever I easily could about this subject." *Ḥujjat Allāh* I, 340.

[68] ʿAbd al-Faṭṭāh Abū Ghuddah rearranges the text of the final chapter of the *Inṣāf* in his edition (Beirut: Dār al-Nafā'is, 1978). There do seem to be some problems with the consistency of structure and expression of chapter five, perhaps due to its being reworked based on material composed as part of the *Ḥujjat Allāh al-Bāligha*.

al-Ijtihād wa-l-Taqlīd which focused specifically on *ijtihād*.[69] Some pages from this treatise are also found duplicated in the material appended to *Ḥujjat Allāh al-Bāligha*.

---

[69] *'Iqd al-Jīd fī-Aḥkām al-Ijtihād wa-l-Taqlīd*, (Cairo: Maktaba al-Salafiyya, 1965). Partially translated by Daud Rahbar. "Shah Waliullah and Ijtihād." *The Muslim World* 45 (December 1955): 346–358.

THE CONCLUSIVE ARGUMENT FROM GOD
(ḤUJJĀT ALLĀH AL-BĀLIGHA)

SHĀH WALĪ ALLĀH

VOLUME 1

The General Principles from Which Are Derived the Beneficial Purposes Considered in the Rulings of the Divine Law

PREFACE

In the name of God, the Merciful, the Compassionate. Praise be to God who created the human race with a predisposition to the religion of Islam and right-guidance, and formed them with a natural disposition for the clear, tolerant, easy, monotheistic (Ḥanīfī)[1] religion. Then they became beclouded by ignorance and fell to the lowest of the low and suffering overtook them. Then God was merciful to them, gracious to them, and sent the prophets to them so that through them they would be brought out from the darkness to the light, and from the constricted place to the vast expanse, and He made obedience to Him conditional upon obedience to the prophets; how great an honor and exaltation! God made whomever He wished among their followers succeed in carrying on their teachings and in understanding the inner meanings of their laws, so that they became, through the blessing of God, possessors of their (the prophets') secrets and winners of their lights, and there is no honor greater than that. Those counted among them were preferred over a thousand worshippers, and they ascended to Malakūt[2] as great ones, and reached a level where the creatures of God pray for them, even the fish deep in the water. O God, grant blessings and peace upon them and their heirs as long as the earth and heavens remain. Among them may You accord our master, Muḥammad, supported by the noble and clear signs, the best of blessings, the most noble of greetings, and the greatest felicity. May You shower upon his family and Companions the first rain of Your pleasure and reward them with the best of rewards.

This servant, who is in need of the mercy of God, who is the Most Generous One; Aḥmad, called Walī Allāh, the son of ʿAbd al-Raḥīm, may God treat both him and his father with His great

---

[1] This description is based on a hadith of the Prophet. Shāh Walī Allāh later explains in Chapter 74 that by "easy" the Prophet meant not having arduous practices in it, and that by "clear" are meant the reasons behind its rules and goals, and that by "Ḥanīfī" is meant that it is the monotheistic religion of Abraham. Hadiths are cited in Ibn Ḥanbal V:266, VI:116, 233 and variants in Bukhārī, Tirmidhī.

[2] The plane of the angels.

grace and make their descendants blessed and continuing, holds that the main topic in the fields of sciences of certainty (*'ulūm yaqīniyya*)[3] and their chief element, the basis of the religious disciplines and their foundation, is the discipline of the hadith reports. In them is cited what proceeded from the greatest of the messengers, may God send peace and blessings upon him and on his all his family and Companions; whether (this is) a saying, or an action, or an acquiescence,[4] for these are the lamps in the dark night and the signs of guidance and they are like the radiant full moon. The one who follows them and keeps them in mind is rightly-directed and guided and he is given great good.[5] The one who opposes them and turns away is misguided. He has fallen down and his soul only increases in ruin. For the Prophet, may the peace and blessings of God be upon him, forbade, commanded, warned, brought good tidings, coined similitudes and reminded them of God, and these are as extensive as the Qur'ān or more so.[6] This discipline has ranks and its masters are at various degrees. It has outer shells, within them a kernel, and oysters in whose centers are pearls.

The learned scholars, may God have mercy on them, have subsumed under many headings the means for dealing with its elusive and problematic issues.

1) The outermost layer of the shell is the field of recognizing the hadiths as sound, weak, multiply-transmitted, or rare.[7] Such a task has been undertaken by those persons with brilliant understanding among the hadith scholars and the ones who had excellent memories for hadith[8] among the former generations.

2) Following this is the field of (determining) the meanings of

---

[3] The sciences of certainty are those connected with the Islamic religion.

[4] When the Prophet saw persons performing an action and did not stop them.

[5] "*Al-khair al-kathīr*" is a Qur'ānic phrase (2:268) and the title of one of Shāh Walī Allāh's other works concerning mystical philosophy.

[6] This sentence is itself a hadith which affirms the necessity of Prophetic reports for understanding the Qur'ānic teaching since in them the same subjects are discussed and further elaborated. Sindhī, 'Ubayd Allāh, *Urdū Sharḥ Ḥujjat Allāh al-Bāligha.* (Lahore: Maktabiyyat Ḥikmat, n.d.), 7. The hadith as found in Abū Dāwūd III:170 #3050 is worded "I forbade things and these are as extensive as the Qur'ān or more so." Imāra, 33. "*nahaitu 'an ashyā annuhā limithli al-qur'ān au akthar.*"

[7] These are all technical terms for classifying hadith within the discipline of hadith criticism, which are discussed later in Chapter 78.

[8] Having an excellent memory for hadith (*Ḥuffāẓ al-Ḥadith*) is a term used in the science of determining the reliability of transmitters and indicates a high level of reliability.

the obscure ones and rendering precise the problematic ones; the leaders in the literary fields and the experts among the scholars of the Arabic language applied themselves to this.

3) Next is the field of their import for religious law and the derivation of the judgments on applied cases (*al-aḥkām al-far'iyya*), drawing analogies (*qiyās*) on the basis of a ruling (*ḥukm*) reported in a text,[9] and deriving inferences (*istidlāl*) from indirect allusions (*īmā'*) and textual indications which are not explicitly stated (*ishāra*);[10] recognizing the abrogated, the definitive, the preferred, and those definitely established. This is at the level of the inner kernel and the pearl according to the majority of the religious scholars, and the investigators among the legal scholars occupied themselves with this.

4) Moreover, the most subtle of the hadith disciplines in my opinion, the most profoundly rooted and the loftiest beacon of light, the primary among the entirety of the Islamic legal sciences according to me, the highest ranking of them, and the greatest in value; is the knowledge of the inner meanings of religion which investigates the wise principles (*ḥikam*) behind the rulings, their rationale, and the secrets of the properties and fine points of actions. This, by God, is the most worthy of the branches of knowledge in which the one who is qualified may spend his precious moments, and he may consider it a provision for his resurrection after those acts of worship which have been made incumbent upon him, since through this the human being develops insight into what the divine law has commanded. His relationship to these reports will be like that of the one who has mastered metrics to the collections of poetry, that of the one who has mastered logic to the proofs of the philosophers, that of the master of grammar to the speech of the pure Arabs, or that of the master of the principles of law to that of the deductions of the legal scholars. Through this he will be preserved from being like a wood-gatherer[11] in the darkness or being like one who dove into a flood.[12] He won't stumble along

---

[9] "*Al-ḥukm al-manṣūṣ*"—that is, based on a text from the Qur'ān or the hadith.

[10] These more technical aspects of *uṣūl al-fiqh* or the theoretical, especially linguistic, elements of Islamic legal science are discussed later in this work in Chapters 79 and 80.

[11] This phrase (*ḥāṭib lail*) applies in Arabic to those who mix the weighty with the trivial or risky in their discourse just as the wood-gatherer at night gathers bad wood with good timber.

[12] He will be pulled away by the current and drowned.

like the dim-sighted one, or ride on the back of a blind mount, or be like the case of a man who heard the doctor giving an order to eat apples and who equated the bitter wild gourd with them on the basis of their similar shapes. Through it (this discipline) he will become a believer on the basis of clear evidence from his Lord; like someone who is told by a truthful person that poison is lethal, and he believes and is aware of what he is told, then he recognizes through circumstantial evidence that its heat and dryness are both excessive and that they conflict with the sound temperament of man, so that he becomes yet more certain of what he believed.

(In the case of hadith studies) the reports of the Prophet, may the peace and blessings of God be upon him, have affirmed its applications and principles. The sayings of the Companions and the Successors have explained its generalities and details. The scrutiny of those who practice independent reasoning (*mujtahidūn*) has resulted in the elucidation of the beneficial purposes considered under every topic of the divine law. The investigators among their followers have elucidated many important points and the meticulous ones among their adherents have issued eloquent conclusions and averted, may God be praised, having the discussion of this become a breach in the consensus of the community or an inroad for perplexity or confusion. However, few are the ones who have written about this,[13] gone into setting out its foundations, presented in an orderly way its principles and branches, or produced something which puts on weight or which satisfies the hunger (for knowledge),[14] while it is worthy of this. Among the proverbs well-known among people is, "who rides behind you, when you are riding on a lion?"[15]

Why shouldn't it be so, when its inner dimensions are not intelligible except to the person who has mastered the sharīʿa disciplines in their entirety and who is without parallel in all the fields of religious learning. Its method is only made clear to the one whose breast God has opened to divinely inspired knowledge[16] and

---

[13] The inner meanings of the religion, in particular, the wisdom and rationale behind the rulings of the divine law.

[14] Cf. Qur'ān 88:7.

[15] That is, writing about this discipline is as lonely as riding on a lion.

[16] (*ʿilm ladunī*). Knowledge that is granted by God through an act of divine grace. Referred to in Qur'ān 18:65. Annemarie Schimmel. *Mystical Dimensions of Islam* (Chapel Hill: University of North Carolina, 1975), pp. 192–193.

whose heart He has filled with a bestowed secret, and who is, in addition to this, of a brilliant nature and intellectual facility, who is skilled in accurate oral and written exposition, proficient in didactic method and embellishment, who knows how to establish the principles and to build the derivative elements upon them, and how to lay out the rules and produce for them both reasoned and textual evidence.

Indeed, one of the greatest blessings of God upon me is that He bestowed upon me a portion and a share of this knowledge. I will never cease to acknowledge and confess my deficiency, "I don't absolve my soul, for the soul persistently inclines to evil."[17]

While I was sitting one day after the afternoon prayer with my concentration turned to God, suddenly there appeared the spirit of the Prophet, may the peace and blessings of God be upon him, which covered me from above with something which appeared to be a robe thrown over me. It was inspired in my heart at that spiritual event that this was a sign of the manner of expounding religion. At this I found in my breast a light which does not cease to expand every minute.[18] Then my Lord inspired me, after a time, of what He had written for me with the exalted Pen,[19] that some day I would undertake this important matter, and that "the earth would be illuminated with the light of its Lord"[20] and that the rays of light would be reflected at the time of sunset, and that the divine law of Muḥammad would shine forth in this age by being presented in long and loose-fitting robes of demonstrative proof. After that I saw the two Imāms, Ḥasan and Ḥusain, may God be pleased with them, in a dream, while I was in Mecca, in which it was as if they gave me a pen,[21] and said, "This is the pen of our grandfather, the Messenger of God, may God's peace and blessings be upon him."

---

[17] Qur'ān 12:53.

[18] Recounted also in *al-Tafhīmāt al-Ilāhiyya*, II (Hyderabad. Sindh: Shāh Walī Allāh Academy, 1973), 299, where Shāh Walī Allāh specifies that at the time of this vision certain subtle knowledge about the sharīʿa disciplines became clear to him.

[19] That is, what had been decided by God about his destiny.

[20] Qur'ān 39:69, where it refers to the time of the Last Judgment.

[21] This vision is recounted in the author's book, *Fuyūḍ al-Ḥaramain*, Arabic with Urdu translation (Karachi: Muḥammad Saʿīd, n.d.), pp. 65–66. It is also included in *al-Tafhīmāt al-Ilāhiyya*, II:300, where Shāh Walī Allāh adds that Ḥusain repairs the pen of the Prophet and then bestows his grandfather's mantle on him, and "From that time on my breast was expanded for writing about the sciences of the divine law."

For a long time I tried to resolve to set down in writing a treatise about this subject which would be instructive for the beginner and a review for the learned one, which persons from all walks of life could equally comprehend, and which could be discussed in turn by the session and the circle. I was constrained by not finding anywhere around me an objective, reliable religious scholar whom I could consult about ambiguous matters. I was also constrained by my lack of skill in the branches of transmitted knowledge compared to what had previously existed in the fortunate centuries.[22] It discouraged me that I am in an age of ignorance, prejudice, and following the passions, in which every person has a high opinion of his own ruinous opinions; for being contemporary is the basis of disagreement, and whoever writes makes himself a target. Thus, while I would set one foot forward in this, I would draw back the other. I would run one lap, and then I would retreat backwards. Until the most honored of my brothers and the most noble of my friends, Muḥammad known as 'Āshiq,[23] may he continue to be preserved from every mischief or accident, realized the value of this knowledge and its merits. He was inspired with the knowledge that human felicity[24] could not be fulfilled except through tracing its fine points and sublime aspects. He recognized that attaining this would not be possible for him except after struggling with doubts and uncertainties, and by suffering disagreements and contradictions, and that the trail would not be blazed for him except through the effort of a man who would be the first to knock at the door, and to whose summons all the intransigent problems would respond. Thus, he roamed to all the cities he was able, and sought out those people characterized by virtue and he tested their minute differences and wide gaps, but he did not find anyone from whose discourse he could take something useful, or from whom he could bring back a blazing torch. When he saw this, he implored me without letting go, insisted, and held fast to me. It reached that point that whenever I used to make excuses he would remind me of the "hadith of the bridling".[25] Thus he confronted me with the

[22] The Prophet's time and soon after.

[23] Muḥammad 'Āshiq of Phulat (d. 1773), who was his chief disciple.

[24] Human felicity (*sa'āda*) in the sense of the cultivation of the virtues and refinement of the soul.

[25] This is a hadith that whoever is asked about some knowledge which he has and keeps it hidden, will be bridled (*iljām*) on Judgment Day with a bridle of fire. Abū

strongest of arguments until the means of escape failed me and my excuses flowed away in streams. I became convinced that it[26] was one of the great challenges, and that it was due to it that I had been inspired with that vision and that my destiny had already been written, and that it was a matter which faced me from every direction. Thus I turned to God and asked for his guidance,[27] and I requested and petitioned for His help, and I totally left strength and power behind and became like a corpse in the hands of the body-washer in having relinquished power over its own movements. I commenced that (work) to which he[28] had urged and inclined me, and I beseeched God to purify my heart from distractions and to disclose to me the realities of things as they are, to guide my heart and make my tongue eloquent, to preserve me from error in the writing which I was embarking on, to grant me success in holding to the truth in every situation, and to aid me in expressing what was stirring in my breast, and what my intellect was struggling with. Indeed, He is a Near One and One who responds.

I submitted to him that I am mute in the field of rhetoric and lame on the racecourse of demonstrative proof, a picked-over scrap bone; and that it was not easy for me to concentrate on scrutinizing pages due to my preoccupation with endless matters, and that it was not possible for me to attain a high degree in preserving weighty traditions, in order to show this off to everyone. I am only a person on his own, waiting for his death, subject to his times, a student before his destiny, a prisoner of his fate, who takes advantage of what comes to him easily. Thus the one who finds pleasure in being content with this, let him be content, and the one who likes something else, it is up to him to do whatever he likes, so let him do it.

Then, since the indication in God's, may He be Exalted, saying, "Indeed, the conclusive argument is from God,"[29] refers to the motives underlying religious obligation and requital and the inner dimensions of God's laws granted by this mercy and guidance—it is

---

Dāwūd, III:321 #3658, 'Ilm 9. Tirmidhī 'Ilm 3, Ibn Mājah and Ibn Ḥanbal also transmitted it.

[26] Writing this work on the inner meanings of religion and hadith studies.

[27] By performing *istikhāra* which is a special prayer invoking the divine guidance through dreams or intuition.

[28] Muḥammad 'Āshiq.

[29] Qur'ān 6:149. The Qur'ānic verse which inspires the title of the book.

appropriate that this treatise, in its capacity as a branch spreading out from this (conclusive argument) and a new moon rising on its horizon, should be called *Ḥujjat Allāh al-Bāligha* (The Conclusive Argument from God). God suffices me, and what an excellent guardian He is, and there is no power and no strength except in God, the Sublime, the Great.

INTRODUCTION

It may be thought that the rulings of the divine laws do not encompass any aspect of the beneficial purposes (*maṣāliḥ*) and that there is no relationship between human actions and that which Allah makes a requital for them, and that being obligated by the divine laws is like the case of a master who wants to test the obedience of his servant, so he orders him to lift a stone or to touch a tree—something which has no use to it besides being a test, so that when he obeys or disobeys, he is requited for his action. This is a false idea which is refuted by the practice of the Prophet, may the peace and blessings of God be upon him, and the consensus of the generations whose goodness has been attested.[1] The one who is incapable of recognizing that actions are considered in the light of the intentions and the psychological attitudes from which they emerge, only touches knowledge in the sense that a needle touches water when it is submerged in the ocean and then withdrawn. It is better that he should weep for himself rather than boast about his own ideas. The Prophet, may the peace and blessings of God be upon him, said, "Indeed, actions are judged according to the intentions,"[2] and God, may He be Exalted, said, "Their flesh and their blood will not reach Allah, but the devotion from you will reach Him."[3]

Prayer was legislated in order to remember God and converse privately with Him, as God, may He be Exalted, said, "Undertake prayers in order to remember Me,[4] and to be a preparation for the vision of God, may He be Exalted, and seeing Him in the next life, as the Messenger of God, may God's peace and blessings be upon him, said, "You will see the Lord like you see this moon,

---

[1] The first three Islamic generations, according to hadiths. For example, Bukhārī, Riqāq, 7. "The good (*khair*) of my community are my generation, then the second, then the third."

[2] Bukhārī Bad' al-waḥī 1, Īmān 41, etc., Muslim, Abū Dūwūd, Tirmidhī, Nasā'ī, Muwaṭṭa', Ibn Ḥanbal.

[3] Qur'ān 22:37. This verse refers to the fact that it is the intention behind the sacrifice which is important and that therefore it is lawful to eat and distribute to others the meat from animals which have been sacrificed to God.

[4] Qur'ān 20:14.

and do not doubt that you will see Him, thus, if you are able to keep from being prevented from praying before sunrise and before sunset, perform the prayers."[5] The alms tax (*zakāt*)[6] was legislated as a defense against the baseness of miserliness, and to suffice the needs of the poor people, as God, may He be Exalted, said about those who refused to pay *zakāt*, "And let not those who hoard up that which Allah has bestowed on them of His bounty think that it is better for them. Rather, it is an evil for them. That which they hoard will be their collar on the day of Resurrection"[7] and as the Prophet, may the peace and blessings of God be upon him, said, "Inform them that God, may He be Exalted, has made charity a duty for them, taken from the rich among them and then given to their poor ones."[8]

Fasting is ordained in order to subjugate the lower soul, as God the Exalted said, "That you might purify yourselves",[9] and as the Prophet, may the peace and blessings of God be upon him, said, "The fast has the effect of curbing lust."[10]

The pilgrimage was legislated to honor the emblems of God, as God, may He be Exalted, said, "Indeed the first sanctuary placed for people . . ."[11] He said, "As-Ṣafā and al-Marwah are among the holy symbols of God."[12] Retaliation (*qiṣāṣ*)[13] was legislated as a deterrent against killing, as God, may He be Exalted, said, "And there is life for you in retaliation, O men of understanding."[14] The punishments (*ḥudūd*) and the ways of atoning (*kaffārāt*)[15] were legislated as deterrents to disobedience, as God, may He be Exalted, said, "That he may taste the evil consequences of his act."[16] The

[5] *Bukhārī* Mawāqit 16, Tauḥid 24, Muslim, Ibn Mājah, Ibn Ḥanbal.

[6] *Zakāt* means more than alms-giving; it is a kind of tax for purification of wealth.

[7] Qur'ān 3:180. Where according to the hadith, it will manifest itself as a hairless snake.

[8] Literally "is a castration". Bukhārī Zakāt 1, 63; Maghāzī 60, Muslim, Abū Dāwūd, Tirmidhī, Nasā'ī, Ibn Mājah, Dārimī, Ibn Ḥanbal.

[9] Qur'ān 2:183.

[10] *Bukhārī*, Ṣaum,10, Nikāh 2,3. Ibn Mājah, Nasā'ī, Ibn Mājah, Dārimī, Ibn Ḥanbal.

[11] "Was that at Bakka." Qur'ān 3:96.

[12] Qur'ān 2:185. Al-Ṣafā and al-Marwah are two small hills between which pilgrims must run back and forth seven times during the Hajj in commemoration of Hagar's desperate search for water. This part of the larger ritual is known as *sa'y*.

[13] *Qiṣāṣ* refers to appropriate retaliation against the person proven guilty rather than the unlimited blood-feud of pre-Islamic times.

[14] Qur'ān 2:179.

[15] *Kaffārāt* are expiations for certain sins. Examples are fasting, feeding the poor, sacrifice, and freeing a slave.

[16] Qur'ān 5:95.

Jihād was legislated for promoting the word of God and making sedition cease, as God, may He be Exalted, said, "Then fight them until there is no sedition and religion is all for Allah."[17] The rules of economic transactions (*mu'āmalāt*) and marriage relationships were legislated to establish justice among them, and so forth with other issues indicated in the Qur'ānic verses and the hadiths, and more than one of the scholars applied himself to this in every generation.

Therefore the Prophet, may the peace and blessings of God be upon him, on certain occasions explained the inner meanings of designating particular times, as he said about four (*rak'ats*) before the noon prayer,[18] "This is an hour when the gates of heaven open, and I like that a righteous act from me ascends to heaven at that time."[19] It is reported from him, may God's peace and blessings be upon him, about fasting on the day of 'Āshūra',[20] that the reason for its being legislated (originally) was the salvation of Moses and his people from Pharaoh on that day, and the reason for its being legislated among us is our following the practice of Moses, may peace be upon him.[21]

He explained the reasons for many rules and thus he said about the one who wakes up, "He doesn't know where his hand has spent the night,"[22] and about rinsing out the nose with water during the ablutions, "That the devil had spent the night inside his nose."[23] He said concerning sleep, "When he lays down to sleep his joints relax,"[24] and he said concerning throwing the stones,[25] "It is to establish the remembrance of God."[26] He also said, "Asking permission (to enter a home) came because of seeing (something which you weren't supposed to),"[27] and he said about the cat, "It is

---

[17] Qur'ān 8:39.

[18] That is, praying four *rak'ats* of the prayer. A *rak'a* is a division of the prayer consisting of a bow followed by two prostrations.

[19] Tirmidhī 1:297 #476, Witr 16 (Chapter on Prayer just after Noon), Nasā'ī.

[20] The tenth day of the month of Muḥarram.

[21] As in the hadith found in Bukhārī Ṣaum 69, Tafsīr 10:1, 20:79, etc., Muslim, Abū Dāwūd, Dārimī, Ibn Ḥanbal.

[22] Therefore he should wash his hands. Bukhārī Wuḍū' 260, Muslim, Abū Dāwūd, Nasā'ī, Tirmidhī, Ibn Mājah, Muwaṭṭa', Ibn Ḥanbal.

[23] *Mishkāt*, p. 82. Bukhārī. Bad' al-khalq, Muslim, Nasā'ī. Meaning that the person must rinse the nose during the ablution to eliminate accumulated dirt.

[24] Therefore ablution is necessary after sleep. *Mishkāt*, p. 71. Tirmidhī 1:51 #77, Ṭahāra 57, Ibn Ḥanbal.

[25] At the pillar representing the devil during the pilgrimage ceremonies.

[26] *Mishkāt*, p. 563. Tirmidhī II:193 #904 Ḥajj 64, Dārimī, Ibn Ḥanbal.

[27] Bukhārī Istidhān 11, Ibn Mājah, Nasā'ī, Ibn Ḥanbal.

not unclean, rather it is an animal which comes and goes in your homes."[28]

He explained in regard to some instances that the rationale behind them was to prevent a harm, such as the forbidding of sexual intercourse while a woman is nursing, "It was for fear of harm coming to the child"[29] or (that they were) to differentiate (Muslims) from a sect of the unbelievers such as his saying, "It (the sun) rises between the horns of the devil," and at this time the unbelievers prostrate before it.[30]

(Something may be legislated) in order to prevent the entry of distortion such as the saying of 'Umar, may God be pleased with him, to the man who wanted to pray the supererogatory prayer together with the obligatory one, "Through this people have been destroyed before you," so the Prophet, may peace and blessings of God be upon him, said, "God has granted you the true opinion, O Ibn al-Khaṭṭāb."[31] Or (it may be legislated) due to the existence of straitened circumstances as in his (the Prophet's) saying, "Does each one of you have two garments?"[32] Another example is His, may He be Exalted, saying, "God knew that you were deceiving yourselves so He forgave you and pardoned you."[33]

The Prophet, may the peace and blessings of God be upon him, explained in certain instances the inner meanings of the deterring (from sin) and the encouragement (to goodness),[34] so that the Com-

---

[28] Abū Dāwūd I:60 #75, Ṭahāra 38, Tirmidhī, Nasā'ī, Ibn Mājah, Dārimī, Muwaṭṭa', Ibn Ḥanbal.

[29] The actual hadith is, "I was about to forbid sexual intercourse while a woman is nursing (*ghīla*) except that I was told that the Persians and Romans practice it without any harmful effects." Bukhārī Nikāh 140,141, Muslim, Abū Dāwūd, Tirmidhī, Nasā'ī, Dārimī, Muwaṭṭa', Ibn Ḥanbal.

[30] The sun. Therefore the Muslims were forbidden to pray at this time so that they would not resemble the unbelievers. *Mishkāt*, p. 213. Bukhārī Bad' al-Khalq 10, Khan trans. IV:319.

[31] This hadith signifies that there should be some pause between the obligatory prayer and the supererogatory (*nafl*) prayers. Abū Dāwūd I:263 #1007, Ṣalāt 188, with the specification of "People of the Book." Ibn Ḥanbal V:368 has a variant of the hadith.

[32] That is, some people told the Prophet, that they had two garments to wear for saying the prayers. He said that having two garments was not easy for everyone, and that if having two garments were made obligatory, prayer would not be possible for all. Bukhārī, Ṣalāt 4, 9.

[33] Qur'ān 2:187. When Islam was first revealed, eating and sexual intercourse were forbidden at night, as well during the daylight, in Ramaḍān. Some men were secretly going to their wives at night, so to prevent this problem God gave them permission to do so during Ramaḍān.

[34] By mentioning the punishments and rewards for actions.

panions turned to him for advice in confusing matters and he disclosed their ambiguities, taking the matter back to its source. He said, "The prayer of a man in the congregation is greater than his prayer in his house and the market by twenty-five degrees, that is to say, if one of you performs the ablution and does it properly, then goes to the mosque only desiring to pray . . ."[35] He also said, "In the sexual act (with the spouse) of one of you there is a reward." They asked, "O Prophet of God, one of us will satisfy his lust and there will be a reward for him in it?" He replied, "Do you think that if he had fulfilled it illegitimately that there would have been for him in this a sin? Therefore, if he fulfills it legitimately, there will be a reward for him."[36] He also said, "When two Muslims raise swords against each other, then both the killer and the slain will go to hell." They said, "Yes, the killer, but what is wrong with the slain one?" He replied, "Indeed, he was bent on killing his companion."[37] And so on with other instances whose enumeration would be difficult.

Ibn ʿAbbās, may God be pleased with him and his father, explained the inner meaning of the legislation of the major ritual ablution for the Friday prayer and Zaid ibn Thābit the cause for the prohibition on selling fruit before its quality became apparent. Ibn ʿUmar explained the inner meaning of touching only two of the corners of the Kaʿba.[38] The Successors and after them the religious scholars who used independent reasoning (*ijtihād*) continued to explain the rationale of the legislation of rulings according to their beneficial purposes, to make their meanings understood, and to derive for the textually based ruling an anchoring point related to its preventing a harm or bringing about a benefit, as is set out in their books and their legal schools.

Al-Ghazzālī,[39] and al-Khaṭṭābī,[40] and Ibn ʿAbd al-Salām[41] and their

---

[35] Then from this, step by step he receives a reward. *Mishkāt*, p. 220. Bukhārī Adhān 30, Khan trans. I:352.

[36] *Mishkāt*, p. 404. Transmitted by Muslim Zakāt 52, Abū Dāwūd, Ibn Ḥanbal.

[37] Bukhārī Īmān 22, Fitan 10.

[38] By saying that those two corners which were touched had been set on the foundation raised by Abraham.

[39] Al-Ghazzālī (d. 1111), one of the greatest Muslim scholars who wrote on the inner meanings of the sharīʿa.

[40] Al-Khaṭṭābī (d. c.998), a Shāfiʿī hadith scholar, author of *Maʿālim al-Sunan*.

[41] ʿIzz al-Dīn Ibn ʿAbd al-Salām. A Mālikī scholar, author of *Qawāʿid al-Aḥkām fī Maṣāliḥ al-Anām*.

like—may God requite their efforts—produced subtle points and elevated investigations, indeed, as the practice (sunna) of the Prophet required, and on which the consensus (*ijmāʿ*) concurred. The practice of the Prophet also required that the revelation of the decree making compulsory or forbidding be a great cause in its own right—regardless of these beneficial purposes—for the rewarding of the obedient and the punishment of the disobedient. Therefore the matter is not as has been thought—i.e., that the goodness of acts or their repugnance in the sense of entitling the doer to reward and punishment can both be completely arrived at through reason. Nor is it the task of the divine law to report on the characteristics of acts as they are in themselves, rather than instituting their being made compulsory or forbidden; in analogy to a doctor who (only) describes the properties of medicines and the types of illness (without giving prescriptions). This is an incorrect assumption which the practice of the Prophet clearly rejects. How could this be so, when the Prophet, may the peace and blessings of God be upon him, said concerning getting up at night to pray during Ramaḍān, "Until I feared that it would be prescribed for you,"[42] and he said, "The Muslim who harms others the most is the one who asked about a thing which had not been forbidden to the people, so that it became forbidden because of his inquiry,"[43] and so on with other Prophetic reports.

How could this be so, when if it were,[44] in like manner the breaking of the fast in the case of the person who was staying at home and suffered hardships similar to the hardships of the traveler would be permissible due to the straitened circumstances on which the dispensations are based, and the breaking of the fast of the traveler in easy circumstances would not be allowed; and this goes for the rest of the limits which the law-giver set. The sunna also required that it not be permitted to depend on recognizing these beneficial

---

[42] *Mishkāt*, p. 270. Transmitted by Bukhārī Tahajjud 4. Khan trans. 2, 128. During Ramaḍān the Prophet had gone to the mosque to pray the *tahajjud* prayer. He ceased going when others started joining him for fear that this would become a general commandment.

[43] *Mishkāt*, p. 42. Bukhārī Iʿtiṣām 3, Muslim, Abū Dāwūd. As Shāh Walī Allāh explains later in the text, this refers to the fact that the statements of the Prophet made certain rulings obligatory, aside from the consideration of any beneficial purpose or reason for legislation. Khan trans. IX:290.

[44] That the goodness or evil of acts was inherent in the beneficial purposes and did not depend on the divine command.

purposes when complying with the rules of the divine law if these rules have been confirmed by hadith reports. This is due to the inability of many people to recognize on their own many of the beneficial purposes and due to the Prophet's being more dependable, for us, than our own reasoning.

Therefore this discipline continues to be restricted to specialists, and the same conditions were laid down for it which are observed in interpreting the Book of God. Investigation into it based on pure personal opinion, rather than on the Prophet's traditions and the reports of the Companions, is forbidden.

It is apparent from what we have mentioned that the truth of being charged with obedience to the divine laws is analogous to the case of a master whose servants fall ill, so he sets over them a particular person to make them take medicine. If they obey him then they are obeying the master and their master is pleased with them, and he rewards them well and they are delivered from the illness; but if they disobey him, they disobey the master and his wrath will smite them and he will give them the worst of punishments, and they will perish from the illness.

This is what the Prophet, may the peace and blessing of God be upon him, meant when he said, reporting from the angels, that he could be likened to "a person who built a house, and made a feast in it, and sent someone (the Prophet) to invite people, and whoever answered the inviter entered the house and ate the feast, and whoever didn't answer the inviter neither entered the house nor did he eat from the feast,"[45] and when he said, "Indeed an allegory for me and the message with which God has sent me is that of a man who came to a people; and said, 'O people, I saw the army (of the enemy) with my own eyes and I am a simple warner so save yourselves, save yourselves!' Thus a group of his people obeyed him and set out at nightfall, and they decamped at their leisure and were saved; and a group did not believe him and they stayed in their place, and the army fell upon them in the morning and destroyed and exterminated them."[46] He also said reporting from his Lord, "Indeed these are but your acts thrown back upon you."[47]

---

[45] *Mishkāt*, p. 43. Bukhārī I'tiṣām 2.

[46] *Mishkāt*, p. 40. Bukhārī Riqāq 26, I'tiṣām 2, Muslim.

[47] Muslim Birr 55 has a slightly different wording "*innamā hiya a'mālukum iḥṣīkum lakum.*" "These are your acts which are counted against you."

In reference to what we have mentioned here there is a middle position which is between the other two, which is that there is both to the actions (in themselves) and to the revelation of the decree making obligatory and forbidding, an effect in entitling reward and punishment. This resolves the opposing textual indications concerning whether the people of the pre-Islamic period are punished for what they did in the Jāhiliyya or not.[48]

One group of people are those who know in a general way that the rulings have their rationale in the beneficial purposes and that punishments are consequential to actions in that they emerge from psychological attitudes through which the soul is improved or corrupted, as the Prophet, may the peace and blessings of God be upon him, indicated when he said, "Indeed, in the body there is a piece of flesh; if it is in good order so is the whole body, and if it is corrupted the whole body is corrupted, and indeed, it is the heart."[49] However, they think that the codification of this field and the setting out of its principles and branches is impossible; either rationally, due to the covert and obscure nature of its issues; or according to the religious law, because the ancestors did not record it due to the proximity of their era to the Prophet, may the peace and blessings of God be upon him, and their plentiful knowledge. Thus, this became tantamount to an agreement to leave this aside, or to say that there was deemed to be no benefit in writing it down, since acting according to the divine law does not depend on recognizing the beneficial purposes; and these opinions are also false.

A person's saying, "Due to the covert and obscure nature of its issues." If he meant by this that writing about them is basically not possible, then the fact that these problems are obscure does not support this. How can this be so, when the issues surrounding the knowledge of God's unity and attributes are more profound and difficult to comprehend, but God makes this possible for whomever He wills. Likewise is every branch of knowledge whose investigation seems impossible and whose comprehension seems unattainable at first glance. Then, if it is approached with the right

---

[48] This is resolved by understanding that the pre-Islamic people are punished due to the naturally negative effects of wrong acts on their own, according to the beneficial purposes (*maṣāliḥ*) for which humans were created; but they are not held accountable according to the rulings of the divine law of which they had no knowledge. The term Jāhiliyya refers to "the Age of Ignorance" before the coming of Islam.

[49] Bukhārī Īmān 39.

apparatus and gradual advances are made in understanding its premises, mastery over it will be achieved so that it becomes easy to establish its foundations and to derive its ramifications and related matters. But, if he means (that in writing about this there is) difficulty in general, then this is accepted. However, it is through this difficulty that the superiority of certain scholars over others becomes apparent, and that hopes of surmounting difficulties and intimidations are fulfilled; and "mastery of knowledge follows upon intellectual struggles and diligent pursuit of understanding."

His saying, "Because the preceding generations did not record this." We say, "There is no harm in the fact that the preceding generations did not write it down once the Prophet, may the peace and blessing of God be upon him, had laid out its basic principles and delineated its ramifications, and those with legal understanding among the Companions followed his example such as the two Commanders of the Faithful, 'Umar and 'Alī, and like Zaid (ibn Thābit), Ibn 'Abbās, 'Ā'isha, and others, may God be pleased with all of them, who investigated and presented various aspects of it. Then the religious scholars and the travelers on the path of certainty continued to expound whatever they needed from the (knowledge) which God had drawn together in their minds. One of them, when put to the test of debating someone who had stirred up a crisis of doubt would unsheathe the sword of investigation and arise, take a firm decision, and with sincerity roll up his sleeves for hard work, and rout and defeat the armies of the innovators. Therefore, we are further of the opinion that the writing of a book containing an accurate summation of the principles of this field is more useful than the disparate scattered parts since "every type of game is in the belly of the wild ass."[50]

The early generations were free from the need to write on this field due to the purity of their beliefs through the blessing of being Companions of the Prophet, may the peace and blessing of God be upon him, the proximity of his time, the rarity of the occurrence of discord among them, their hearts being content to refrain from scrutinizing what was known to be based on his authority, the lack of their concern with accommodating revealed tradition to reason, and their ability to consult reliable authorities about many of the obscure

---

[50] A proverb meaning that it is better to obtain one comprehensive goal than several separately.

branches of knowledge. Likewise they did not need to record the other hadith fields due to the nearness of their era to that of the first generation, their time overlapping with that of the hadith transmitters, their seeing and hearing them, their ability to consult the reliable authorities, and the rare occurrence of disagreements and the forgery of hadiths. Examples of these hadith fields are the explanation of the rare words in the hadiths, the names (and biographical details) of the transmitters and the ranks of their integrity, the ambiguous hadiths, the principles of the hadiths, the hadiths variants, the legal import (*fiqh*) of the hadiths, and distinguishing the weak ones from the sound and the fabricated from the authentic. No one of these fields was not singled out by being written down nor were its principles and applications determined except after many generations and extended periods of time when the need for it became apparent and the welfare of Muslims came to depend on it.

At that point disagreements had increased among the legal scholars based on their differing about the rationales for legislation behind the (sharīʿa) rulings, and this led to their disputing about these rationales from the aspect of whether they led to beneficial purposes (*maṣāliḥ*) taken into account in the divine law. Adherence to reason in many of the religious investigations arose, and doubts emerged concerning doctrinal and practical principles. This situation led to the coming into being of a movement to establish rational proofs on the basis of the revealed texts, and to make what was revealed congruent with the rational, and to make the traditional interpretation consistent with what they understood to be the meaning as an aid and support to the religion, and as a good effort to reunite the Muslims, and this was counted among the greatest means of drawing nearer to God, and as one of the chief pious deeds.

(As to) the claim that there is no benefit in writing them down. We say: The matter is not as they maintain, rather, in this there are clear benefits.

1) Among them is displaying one of the miracles of our Prophet, may the peace and blessings of God be upon him. When he came with the great Qur'ān the most eloquent people of his time were astounded and not one of them could bring a chapter like it. Then when the time of the first generation passed and the aspects of its (the Qur'ān's) inimitable nature became unclear to the people, the scholars of the community arose and clarified this in order that

a person who was not at that level could understand it. Likewise the Prophet brought a divine law from God, may He be Exalted, which is the most complete of divine laws, containing the beneficial purposes, the like of which humans are incapable of taking into consideration. The people of his time recognized the nobility of what he had brought through a (higher) aspect of wisdom (*ma'rifa*) so that they directly articulated it and it came through in their sermons and discussions. Once their era had passed it was necessary that there exist in the Muslim community those who could elucidate the dimensions of this type of inimitability and the traditions proving that the Prophet's divine law was the most perfect one, and that the bringing of something like it by someone like him was a great miracle, so well-known that it need not be mentioned.

2) Through it a confidence which increases faith is obtained, as Abraham the Friend of God, may peace and blessing be upon him, said, "Yes, but (I ask) in order that my heart might be filled with confidence."[51] This occurs when the display of the proofs and the abundance of ways to knowledge set the mind at rest and bring to an end the disquiet of the heart.

3) The seeker of righteousness,[52] while he exerts efforts in acts of obedience, will recognize the reason for their being legislated and commit himself to preserving their spirits and their lights, so that even a little of them will benefit him, and he is less likely to proceed at random. It was in this sense that Imām Ghazzālī was concerned in his books about the Sufi path to make known the inner meanings of the acts of worship.

4) The legal scholars disagreed about many of the legal ramifications based on their disagreements over the corresponding reasons for legislation which had been derived. The determination of what is the truth in this case is not achieved except through an independent discussion of the beneficial purposes.

5) The innovators[53] cast doubts on many of the Islamic issues saying that they were contradictory to reason, and that anything

---

[51] Qur'ān 2:260. In this verse Abraham is referring his doubts about life being given to the dead.

[52] *Ṭālib al-Iḥsān. Iḥsān* refers to making extra efforts in drawing near to God and purifying the soul, and thus refers to Sufism.

[53] "*Mubtadi'ūn*"—those who inject new (i.e., heretical) elements into the Islamic tradition. In this case he seems to be referring to philosophers or philosophically-oriented theologians.

which contradicts reason must be rejected or interpreted, such as their saying about the punishment of the grave, i.e., that it is denied by sense perception and reason. Then they said something of that sort about the Day of Judgment, and the Ṣirāṭ Bridge, and the scales.[54] So they took up interpretation using remote exegeses and one sect (the Ismāʿīlīs) provoked the crisis of doubt, for they said, "Why is fasting on the last day of Ramaḍān compulsory and fasting on the first day of Shawwāl[55] forbidden?" and other such discussions. And one group mocked the incitements to good (rewards) and the deterrents to sin (punishments) believing that these were purely in order to encourage and instigate and were not based upon any real foundation, until the most miserable one among that group arose[56] and invented the hadith that "the eggplant will do whatever it is eaten for," hinting thereby that the most harmful of things was not distinguished from the beneficial one among the Muslims.[57] There is no way to prevent this corruption except that we explain the beneficial purposes and lay the foundations of the basic principles for them, similar to what was done in the disputations with the Jews, Christians, heretics,[58] and their like.

6) A group of legal scholars claimed that it was permissible to reject a hadith which was contradicted by analogy on any grounds, and thus damage was done to many sound hadiths such as the hadith of the animals which are sold without having been milked for some time[59]

---

[54] These are all aspects of Islamic eschatological beliefs.

[55] The first day of Shawwāl is the *ʿĪd al-Fiṭr*, which follows the month of Ramaḍān.

[56] Ibn al-Rāwandī (d. c. mid-10th century), a Muʿtazilite and heretic who criticized prophecy and religious beliefs as being contrary to reason. On his reputation see Josef van Ess, "Ibn ar-Rewandī or the Making of an Image" in *al-Abḥāth* (27, 1978/9):5-26.

[57] This was a false hadith which Ibn al-Rāwandī circulated to the effect that the Prophet had said that eating eggplant was good for whatever purpose you ate it for, although from the aspect of Islamic medical science (*ṭibb*) it is considered damaging. This was constructed so as to parallel a sound hadith concerning the benefits of drinking water from the well of Zamzam. "māʾ zamzam limā shuriba lahu." Ibn Mājah Manāsik 78.

[58] "Dahriyya"—can refer to holders of various types of heretical beliefs, but generally to those who hold materialistic views such as the eternity of the world.

[59] This refers to the practice of leaving camels or cattle unmilked or tying up their udders some days before they are sold to make them appear more productive. In this case the hadith says that the buyer should have a purchase option of three days and then if he gives the animal back he should also give a *ṣāʿ* of dates. The debate concerns the approving of the purchaser's option and remittance of this set amount of food which is allowed according to al-Shāfiʿī and not allowed according to Abū Ḥanīfa because the amount of dates is fixed while the amount and type of milk may

and the hadith of the two jars.[60] The People of Hadith (*ahl al-ḥadīth*)[61] found no way to make them accept the proof except by showing that it agreed with the beneficial purposes taken into account in the divine law; and so on with other benefits (of writing this down) which cannot be fully enumerated in this discussion.

You will find me, when overwhelmed by the clamor of explanation and when I have examined the setting out of the principles as closely as possible; sometimes forced into the position of holding some views which the majority of the debaters among the Theologians did not, such as the theophany (*tajallī*) of Allah, may He be Exalted, in the planes of the hereafter through images and forms, and like the confirmation of a non-elemental world in which ideas and actions are embodied by forms appropriate to them in character and in which new things come into being before they are created on the earth;[62] the connection of the actions to psychological attitudes, and the being of these attitudes, in reality, a cause for requital in this worldly life and after death; the compelling predestination (*al-qadar al-mulzim*),[63] and so on.

Then be informed that I did not venture to do this except after I saw the Qur'ānic verses and the hadiths and the reports of the Companions and the Successors supporting these views, and I saw groups of the elite of the People of the Sunna, who are distinguished

---

vary, thus conflicting with the answer one would arrive at by analogy. *Kashshāf fī Tarjuma al-Inṣāf* Urdu translation by Muḥammad Aḥsan Siddīqī (Delhi: Mujtabā'ī, 1891), 52. Shāh Walī Allāh refers to this hadith in Chapter 85 of the present work. He returns to this issue in *Ḥujjat Allāh al-Bāligha* II "Forbidden Sales" and in *Musawwa Muṣaffā* I (Karachi: Muḥammad 'Alī Karkhāna-i-Islāmī Kutub, 1980) p. 367. Some hadiths on this topic are Bukhārī Buyū' 23, 26, 28, Shurūṭ 11, Muslim, Abū Dāwūd, Tirmidhī, Nasā'ī, Ibn Mājah, Dārimī. It is also discussed in al-Shāfi'ī, *al-Risāla*, trans. Majid Khaddouri (Baltimore: Johns Hopkins, 1962), 330–1.

[60] This hadith concerns the amount of standing water which must be exceeded if it is to be considered pure in order to perform the ablution, i.e., enough to fill two very large jars. Abū Dāwūd 1 :17 #65, Ṭahāra 33. Tirmidhī, Nasā'ī, Ibn Mājah, Dārimī, Ibn Ḥanbal. *Mishkāt*, p. 96. Shāh Walī Allāh again refers to this hadith in Chapter 83 of the present work.

[61] The "People of the Hadith" here refers to the movement in Islam which opposed the 9th century Mu'tazilites by disapproving of the use of reason in Islamic theology and relying solely on the literal textual evidence of the Qur'ān and the hadith.

[62] This is a reference to the World of Images (*'ālam al mithāl*), discussed in detail in Chapter 2.

[63] The "compelling predestination" refers to the divine determination which manages the material and non-material world and from which nothing can escape. This is discussed in Chapter 41 of the present work.

by divinely inspired knowledge, professing them, and basing their principles on them.

The "sunna"[64] is not in actuality a name for any particular school of theology. Rather the issues concerning which the People of the Qibla (Muslims) disagreed and due to which they became separate sects and distinct factions beyond their following the essentials of the religion, are of two categories.

1) One category (of issues) are those which the Qur'ānic verses spoke of, the sunna confirmed, and the former generations among the Companions and the Successors practiced. But when the pride of every person in his own opinion manifested itself and the paths diverged among them, one group chose the outer meaning of the Book (Qur'ān) and the sunna and held fast to the beliefs of the pious ancestors (*al-salaf*)[65] and did not pay any attention to whether these agreed with rational principles or whether they contradicted them, so that when they used reason in theological discussion it was only to impose their opinion on their adversaries and to refute them, and to increase their own confidence, not to derive any articles of belief from this, and these are the People of the Sunna.

Some people resorted to speculative interpretation (*ta'wīl*) and turning away from the literal meaning when it opposed rational principles according to their claim. Thus they used reason in their discussions in order to verify the matter and to investigate it as it is.[66]

Among this (first) type (of issues) are the questioning in the grave,[67] the weighing of acts, the passing over the Ṣirāṭ Bridge, the vision (of God), and the miracles of the saints. All of this was made clear by the Qur'ān and the sunna and the pious ancestors proceeded according to it; but these lay beyond the scope of reason according to the claim of one group, so they denied them or interpreted them. The other group said, "We believe in this even though we do not know its true meaning and reason does not give us evidence about it."[68] But we (the author) say—we believe in all of this based on a proof from our Lord, and reason attests to it according to us.

---

[64] Sunna refers to the practice of the Prophet. The designation "People of the Sunna and the Community" (*ahl al-sunna wa-l-jamā'a*) came to refer to the main body of Muslims.

[65] "*Al-salaf*" is a term referring to the pious ancestors, the early Muslims, whose practice is considered closest to that of the Prophet.

[66] The Mu'tazilites.

[67] Of the person by two angels regarding his beliefs and actions.

[68] The Ash'arites and the literalists (people of the hadith).

2) The second category (of issues about which Muslims disagreed) are those which were not spoken of in the Qur'ān, nor frequently mentioned in the sunna, nor discussed by the Companions, and they had been left unexplored. Then a group of scholars (*ahl al-ʿilm*) came, and they discussed them and disagreed, and their treatment of them was either deduced from textual indicants such as the superiority of the prophets over the angels, and the superiority of ʿĀ'isha over Fāṭima, may God be pleased with both of them; or it was due to the dependence of principles consistent with the sunna upon them and their connection to them, according to their claim. Examples of these issues are: the general commands,[69] and something of the investigations into the substances and accidents, as the belief in the temporal origination of the world depends on the invalidation of prime matter and the affirmation that there is an individual indivisible atom; and the belief in God, may He be Exalted, creating the world without an intermediary depends on the nullification of the proposition which states that only one emerges from the one; and that belief in miracles depends on denying the necessary and rational connection between the causes and their effects; and that the belief in the physical resurrection depends on the possibility of the restoration of the non-existent; and other matters with which they fill their books. As for the detailed exposition and the interpretation of what they learned from the Qur'ān and the sunna, they disagreed about the details and the interpretation after agreeing on the fundament. For example, they agreed on affirming the two divine attributes of seeing and hearing, then they disagreed; one group saying that these two attributes are reducible to God's knowledge of the seen and heard things,[70] while the others held that seeing and hearing were two independent attributes.[71] The groups agreed that God is Living, Knowing, Willing, Powerful, Speaking, then they disagreed; one group holding that what was meant was only the affirmation of the extreme limits of these meanings in terms of effects and acts, and that there is no difference between these seven attributes and between mercy, wrath, and generosity in this respect, and that the hadiths do not affirm any difference (among

---

[69] "*umūr ʿāmma*." That is speculating as to the particular scope of general injunctions.

[70] The majority of the Muʿtazilites of Baghdad and Basra following al-Naẓẓām.

[71] The People of the Sunna. (Ashʿarites), The Necessary Being (*wājib al-wujūd*) is a description for God based on a proof for His existence.

these attributes)[72] and that the sunna did not confirm any difference. Another group held that these (attributes) are existent things, inhering in the essence of the Necessary Being.[73] They agreed on affirming God's sitting on the Throne, His face, and His laughing, in general; then they differed. One group said that what was meant were only meanings related to these (anthropomorphic descriptions) so that "sitting" means "being in control" and "the face" is the "essence,"[74] while the other group did not deal with them saying, "We do not know how to interpret these words."[75] I don't approve of the claim of superiority of one of the groups over their rivals on the basis that they keep to the sunna. How, (can this be) and if pure adherence to the sunna were meant, then this would entail completely abandoning discussion about these questions since the first generations did not go into them. When the circumstances have required further explanation, then not everything which they derived from the Qur'ān and the sunna has been true or more acceptable; nor is all of what they consider to be dependent on a thing to be incontestably accepted as dependent; nor is everything which they require to be rejected to be incontestably rejected; nor is everything which they forbade argument upon, considering it to be difficult, difficult in reality; nor is everything which they produced in elaboration and explanation better than that which others produced. Since what we mentioned concerning a person's being a Sunnī is in consideration of the first division (of issues) and not the second, you see the scholars among the People of the Sunna disagreeing among themselves in many things of the second type, such as the Ash'ārites and the Māturīdites,[76] and you see the proficient among the ulema[77] in every generation not holding themselves back from every fine point as long as it does not oppose the sunna, even if the preceding ones had not spoken of it. You will find me—when the paths broke them up into branches and schools, and inclinations and preferences divided them—persevering in the evident, main, way and affirming the firm middle of the road, and I

[72] The Mu'tazilites.
[73] The People of the Sunna. Ash'arites.
[74] The Mu'tazilites.
[75] The People of the Sunna, Ash'arites, and literalists.
[76] A school of Islamic theology founded by al-Māturīdī (d. 945).
[77] The ulema are the learned religious scholars of Islam.

will not turn off to the outskirts and the fringes, and I shall remain deaf to their derivations and interpretations.

Then be informed that every field of learning has its experts and that every domain has a prerequisite. Thus, just as it is not for the master of the uncommon words in the hadiths to investigate the soundness and weakness of the hadiths, nor for the one who has an excellent memory for hadith to discuss the legal ramifications and the preferences for certain ones over others; likewise it is not for the student of the inner meanings of the hadith to discuss any of these things. Rather, the limit of his effort and the target of his aspiration should be disclosing the inner meaning which the Prophet, may the peace and blessings of God be upon him, intended in what he said, whether that ruling has remained firm or has been abrogated, or whether another indicant (*dalīl*) contradicts it, so that in the opinion of the legal scholars it should not be held to be preferable.

However, everyone practicing a discipline must unavoidably rely on the best of that discipline. The most sound with respect to the discipline of hadith studies is what was refined after the recording of the hadiths of the (Islamic) cities and the reports of their legal scholars, and the recognition of the hadith which is corroborated by others from the one transmitted by only one chain, and the hadith with the most narrators and the best version from those which are less reliable. Nevertheless, if something of this type comes up in passing, then the investigation into issues by independent reasoning and the verification of the most accurate among them is not an innovation on the part of the scholars, nor does this involve any discredit to any of them.

"Indeed, I only desire the improvement of what I am able and my success is only through God, on Him I rely and to Him I repentantly turn."[78]

Here, I disassociate myself from every opinion which emerged in contradiction to a verse from the Book of God, or a practice established on the authority the Prophet of God, or the consensus of the generations attested by good, or which the majority of the *mujtahids*[79] have chosen and the great majority of the Muslims. If anything like that occurs then it is an error; may God, may He be

---

[78] Qur'ān 11:88.

[79] A *mujtahid* is an Islamic legal scholar who practices *ijtihād*, or the effort to derive rulings based on analogical reasoning from textual indications.

Exalted, have mercy on the one who awakens us from our drowsiness or alerts us to our negligence.

As for those who investigate through inference and derivation from the discussion of the first generations, and who embrace the method of debate and argument, we do not need to agree on everything which they voiced, for we are humans and they are humans, and the affair between us and them is a contest for superiority.

Therefore I designed this book in two parts: one of them is the part about the general principles through which are systematized the beneficial purposes considered in the divine laws. Most of them were accepted among the religious communities present in the era of the Prophet, may the peace and blessings of God be upon him, and there was no disagreement among them concerning these principles, and the people at that time did not need to ask about them. Thus the Prophet, may the peace and blessings of God be upon him, pointed them out, as he did the basic rules from which the ramifications branched off, so that the hearers were able to trace back their ramifications to them (the basic rules) due to similar things having been practiced among Arabs affiliated with the religion of Ishmael,[80] and in Judaism, Christianity, and Magianism.

I saw that the particulars of the inner dimensions of the divine laws go back to two fundamentals; the study of piety and sin and the study of the policies regulating religion. Then I saw that the true nature of piety and sin can only be thoroughly understood when before this, one has been made aware of the discussions concerning requital, the development of the supports of civilization (*irtifāqāt*), and the felicity of the human species. Then I saw that these discussions depended mainly on issues already clearly established in this science so that their rationale was no longer discussed, for either they were affirmed as true due to the agreement of the religions upon them until they had become well-attested, or due to the good opinion held of the one who taught them, or due to the proofs cited in a science higher than this science. I (therefore) have avoided giving lengthy proofs for the existence of the soul, its enduring nature, and its being blessed and pained after the separation from the body, because this is a topic already dealt with exhaustively in the available books. I have only mentioned these

[80] Ishmael was the prophet sent to the Arabs and the pre-Islamic Arabs preserved a distorted version of his message, according to Shāh Walī Allāh. See Chapter 74.

topics when I found that they were not treated at all in those books available to me, or that they were lacking in delineation and systemization, which I was able to bring out. I did not mention matters which are generally accepted except those which I felt people had not gone into, or those for which revealed textual indications had not been sufficiently advanced. Inevitably I have mentioned in this (first) section issues which must be accepted in this field without presenting their rationale. For example the manner of requital in the life after death, then the supports of civilization to which human beings are naturally predisposed, and neither the Arabs nor the non-Arabs ever adopted them solely due to the dictates of reason. Next the explanation of human felicity and misery commensurate with the (human) species and in accordance with what will be manifested in the next life, then the principles of piety and sin which happen to be identical in all religions. Next, what is required for governing the Muslim community in terms of imposing the limits of God (*ḥudūd*)[81] and the divine laws, followed by the manner of deriving the divine laws from the words of the Prophet, may the peace and blessings of God be upon him, and learning these from him.

The second section (volume two) will explain the inner meanings of the hadith under the topics of faith, knowledge, purification, prayer, *zakāt*, fasting, pilgrimage, righteousness (iḥsān),[82] economic transactions, management of the household, the governing of cities, manners of lifestyle, and miscellaneous topics. This is the time to set out for the goal. Praise be to God, in the beginning and at the end.

---

[81] The *ḥudūd* are punishments for specific sins such as theft and adultery stipulated in the Qur'ān.

[82] This term refers to inner development, particularly of mystical training.

BOOK I

THE FIRST INVESTIGATION: THE CAUSES OF RELIGIOUS OBLIGATIONS AND REQUITAL

CHAPTER 1

Absolute Origination, Creation of the Material World and Divine Management[1]

Be informed that God has three attributes in relation to the bringing into being of the world, each presupposing the other. One of them is absolute origination (*ibdāʿ*) which means the bringing into being of something from nothing, so that a thing comes out from the concealment of non-being without there being any matter. The Prophet of God, may the peace and blessings of God be upon him, was asked about the beginning of creation. He replied, "There was God and there was nothing before Him."[2]

The second attribute is the creation of the material world (*khalq*), which is the bringing into being of something from something else, as He created Adam from clay and "He created the jinn from smokeless fire."[3] Both reason and revealed tradition indicate that God, may He be exalted, created the world into species and genera and made certain specific properties particular to each species and genus. Thus, for example, the species of man has its specific properties of speech, a bare skin, an upright stature, and the faculty of understanding speech; while the species horse has the specific properties of neighing, having hair on its skin, a curved body, and the inability to understand speech. Similarly the specific property of poison is to kill the man who takes it, the specific property of ginger is warmth and dryness, that of camphor is coolness, and all the species of minerals, plants, and animals consistently follow this pattern.

The habit of God has been that these specific properties are never divorced from the things by which they were particularized and

---

[1] *Ibdāʿ*, *khalq*, and *tadbīr* are terms found in the Qur'ān which describe God's creative activity. In his mystical works Shāh Walī Allāh adds a fourth term, *tadallī*, "descending" to indicate the sending of knowledge and guidance or the perfection of souls as a further facet of God's creative activity. The term "*tadallī*" (descent) is found in Qur'ān 53:8, "He drew nearer and descended."

[2] *Mishkāt*, p. 129. Bukhārī Tauḥīd 22, Bad' al-khalq 1, Ibn Ḥanbal.

[3] Qur'ān 55:15.

the distinctive features of individuals are specified by these characteristics and individuated by certain of their general possibilities. Similarly, the distinguishing characteristics of the genera are specified by the specific properties of their species. The significances of those names which are logically ordered according to generality and specificity such as "inanimate matter," "plant," "animal," "man," and "this particular individual" are intermingled and intertwined at first glance; but subsequently reason perceives the distinctions among them and ascribes every specific property to that to which it belongs.

The Prophet, may the peace and blessings of God be upon him, explained the specific properties of many types of things and ascribed their effects to them, as in his sayings, "talbīna[4] gives rest to the heart of the sick person,"[5] or "in Nigella seed there is a cure for every illness except death,"[6] and in his saying about camels' urine and milk "a cure for diarrhea,"[7] and in his saying that there is "heat" in a purgative grass (*shubrum*).[8]

The third of these creative attributes of God is the divine management (*tadbīr*) of the world of engendered things, and this goes back to rendering the occurrences of this world in conformity with the system which His Divine wisdom approves, so that it will attain the beneficial purpose (*maṣlaḥa*)[9] which His Divine generosity required. For example, He sent down rain from the clouds and with it brought forth the plants of the earth for the people and herds to eat, and this became a cause for their living until the appointed time. Abraham, may peace be upon him, was thrown into the fire, and God made it "cool and wholesome" so that he would remain alive. Similarly, Job, may peace be upon him, had a confluence of diseased matter in his body, so God brought forth a spring containing a cure for his illness.[10] Likewise God, may He be exalted, looked upon the people of the earth and found them abominable, both Arabs

---

[4] A broth made of milk, meal or bran and perhaps some honey.

[5] *Mishkāt*, p. 888. Bukhārī Ṭibb 8, Khan trans. VII:401. Muslim. Salām 90, Ibn Ḥanbal.

[6] *Mishkāt*, p. 945. Bukhārī Ṭibb 7. Khan trans. VII:400. Muslim, Tirmidhī, Ibn Mājah, Ibn Ḥanbal.

[7] *Musnad* Ibn Ḥanbal I:293.

[8] Ibn Mājah, Ṭibb, 12, also Tirmidhī, Ṭibb 30, Ibn Ḥanbal II:369.

[9] This refers to the universal beneficial purpose (*al-maṣlaḥa al-kulliya*), the guiding purpose of the universe. It is a central concept of Shāh Walī Allāh's thought explained in the introductory notes (pp. xvi–xvii).

[10] This episode in Job's story is found in Sūra 38:42, 43 and the incident of Abraham and the fire in Qur'ān 21:69.

and non-Arabs, so He revealed to His Prophet the need to warn them and struggle against them in order to bring out whomever He willed from the darkness into the light.

The further elaboration of this is that when the potentialities placed in created things, wherefrom they are never separated, collide and conflict with each other, the wisdom of God requires the production of various developmental states (*aṭwār*), some of them substances and some of them accidents. The accidents either are acts or volitions which occur on the part of living creatures or else they are something other than these two. These states (which are produced) have no defect in them in the sense of the failure to come into being of something which its cause requires or the coming into being of something contrary to what it requires.

The thing, when considered in relationship to the cause requiring its existence, inevitably is good; as cutting is good in so far as the substance of iron requires it, even though it may be evil in so far as the destruction of the human constitution is concerned. However, there may be evil in these states in the sense of the (possibility of the) coming into being of another thing which, in view of its effects, would be more in conformity with the beneficial purpose (*maṣlaḥa*), or in the failure of something to come into being which would have praiseworthy effects.[11] When the causes for this evil are mobilized, the mercy of God to His servants and His kindness to them and His omnipotence and His omniscience, require that He manipulate these potentialities and the things bearing them through "contraction," "expansion," "mutation," and "inspiration;"[12] so that the totality leads to the desired consequence.

As for "contraction," it is mentioned in the hadith that the Dajjāl wants to kill the believing Muslim a second time but God does not give him the power to do this, despite the fact of his genuine motive to kill and the soundness of his instruments.[13] As for "expansion,"

---

[11] That is, there is no evil which occurs in the sense of nature being defied but there can be evil in these latter two cases of relatively less desirable situations coming into existence.

[12] *Qabḍ, basṭ, iḥāla,* and *ilhām*. *Qabḍ* and *basṭ* are terms used in Sufi theories of a person's spiritual state implying respectively depression, i.e., contraction of the heart, or expansiveness and elation. *Iḥāla* is the mutation of the earthly processes of causality. *Ilhām* is the intuitive or instinctive inspiration which God bestows on all creatures.

[13] According to this hadith the Dajjāl or Antichrist figure would kill one of the believing Muslims and restore him to life, but would be unable to kill him a second time due to changes in his body. *Mishkāt*, p. 1148. Bukhārī Fitan 28 Khan trans. IX:186.

it is exemplified in God's causing a stream to gush forth for Job, may peace be upon him, through his stamping upon the earth,[14] although it is not usual for stamping on the earth to lead to water gushing forth. God empowered certain of His sincere worshippers participating in the Jihād to perform deeds which the mind would not imagine possible for that number of physical bodies, nor even on the part of several times that many. "Mutation" is demonstrated by His turning fire into cool air for Abraham, may peace be upon him.[15] The example of "inspiration" is shown in the tale of the breaking of the ship and the raising of the wall and the killing of the young man,[16] as well as in the revelation of the holy books and the divine laws to the prophets, may peace be upon them. Inspiration is sometimes given to the one affected and sometimes to another, for the former's sake.

The great Qur'ān has explained the types of divine management in a way that cannot be surpassed.

---

[14] Qur'ān 38:41, 42. "Stamp on the earth with your foot, this is cool water to wash with and to drink."

[15] Qur'ān 21:69–70. According to the account of the Qur'ān, Abraham's people wanted to burn him to death after he had destroyed their idols.

[16] Qur'ān 18:66–83. This is a reference to the story of Moses and Khiḍr. Moses was shocked when his companion Khiḍr made a hole in a ship, killed a young man, and restored a wall without claiming recompense for it. Later Moses was shown that the ship would have been seized by a tyrant, the young man would have become a rebellious unbeliever, and that the wall safeguarded the property of two orphaned youths.

CHAPTER 2

The World of Images ('Ālam al-Mithāl)[1]

Be informed that many traditions of the Prophet indicate that a non-elemental world exists in which abstract meanings are represented by quasi-bodily forms corresponding to them in quality. There, things take on their materialization in some form before they are materialized on earth. Thus when they come into existence they are the same in a certain sense of sameness. Many of the things which are commonly assumed not to be bodily move and descend, although people cannot see them. The Prophet, may the peace and blessings of God be upon him, said:

"When God created ties of relationship they stood up and said, 'This is the place of taking refuge with You from being cut off.'"[2]

"The chapters of the Qur'ān 'The Cow' and 'The Family of 'Imrān' will come on the Day of Resurrection as if they are two clouds or two canopies or two flocks of birds in rows disputing on behalf of their people (those who recited them)."[3]

"Acts will come forward on the Day of Judgment. First prayer will come, then alms-giving (*zakāt*) will come, then fasting will come."[4]

"Good and evil are creatures which will be set up for people on the Day of Judgment. As for good, it will give good tidings to its folk, while evil will say 'Go away, go away, while they are not able to do anything except stick to it."[5]

"God the Exalted will resurrect the days on the Day of Judg-

---

[1] The World of Images or "Imaginal" World, to use the expression favored by the scholar Henry Corbin, who wrote most extensively on this concept in Islamic thought, was most developed in the works of Muslim thinkers, especially those who were mystically inclined.

[2] That is, relatives should visit and respect each other. *Mishkāt*, p. 1024. Bukhārī Tafsīr 47, Tauḥīd 35, Muslim, Ibn Ḥanbal.

[3] A hadith explaining the benefits of reciting the second and third chapters of the Qur'ān. *Mishkāt*, p. 449. Muslim Musāfirīn 252, 253, Dārimī, Ibn Ḥanbal.

[4] Ibn Ḥanbal II:362.

[5] Ibn Ḥanbal IV:391.

ment just as they are, but He will resurrect Friday radiant and luminous."[6]

The Prophet further said:

"The world shall be brought forth on the Day of Judgment in the form of an old woman with gray-streaked hair, bluish teeth, and an ugly face."[7]

(In another hadith report) he said, "Do you see what I see? I see civil strife falling in the midst of your homes like drops of rain."[8] In the hadith of the Night Journey[9] he said, "There were four rivers, two concealed and two apparent. I asked, 'What does this mean, O Gabriel?' He answered, 'As for the concealed ones, they are in the Garden; and as for the apparent, they are the Nile and the Euphrates.'"[10]

In the hadith concerning the prayer at the time of a solar eclipse he related, "The Garden and the Fire were depicted for me" and in a version "hanging suspended between me and the Qibla wall."[11] In this he related that he stretched forth his hand to pluck a bunch of grapes from the Garden, and that he drew back from the Fire and gasped because of its heat.[12] In Hell he saw a thief who had robbed pilgrims to Mecca and a woman who had tied up her cat so that it died of hunger, while in the Garden he saw a prostitute who had given water to a thirsty dog.[13] It is clear that this distance (between Muḥammad and the Qibla wall) would not hold Heaven and Hell embodied as they are generally understood. He also said, "Paradise is surrounded by undesirable things and Hell by craved

---

[6] *Kanz al-ʿUmmāl* III:284. Baihaqqī *Shuʿb al-Īmān*, al-Ḥākim al-Nīsāpūrī in *al-Mustadrak*.

[7] The hadith continues, "Man beholding her will say, 'Mercy on us Who is this?' The angels will answer, 'This is the world for whose sake you quarreled and fought and embittered one another's lives.'" A Sufi hadith not found in the standard collections.

[8] Bukhārī Madīna 8, Fitan 4.

[9] The *isra'* or Night Journey of the Prophet occurred before the emigration to Medina. In addition to being transported from the furthest mosque in Jerusalem through the seven heavens, the Prophet was shown aspects of Paradise and Hell which are recounted in the hadith.

[10] *Mishkāt*, p. 1266. Bukhārī Bad' al-khalq 6, manāqib al-anṣār 42, ashraba 12. Muslim, Nasā'ī, Ibn Ḥanbal.

[11] Bukhārī Fitan 15, Duʿāt 34, Muslim, Ibn Ḥanbal.

[12] *Mishkāt*, p. 309. Bukhārī Kusūf 9, Ādhān 91, Nikāh 88, Muslim, Nasā'ī, Tirmidhī, Ibn Ḥanbal.

[13] See hadiths cited in *Mishkāt*, p. 404. Bukhārī Bad' al-khalq 16, Khan trans. IV:338, Ibn Ḥanbal. The story of the prostitute is found translated in Siddiqi trans. Muslim II:429–430.

things,"[14] then Gabriel ordered him to look at them. He also said, "Affliction descends and prayer counteracts it."[15]

The Prophet said, "God created human reason (*al-ʿaql*) and said to it, 'Come forward.' and it came forward, then He said to it, 'Run away.' and it ran away."[16] In another hadith he said, "These are two books from the Lord of the worlds."[17] In another, "Death will be brought like a ram and sacrificed between Paradise and Hell."[18]

There is also God's saying, may He be exalted, "We sent our spirit unto her (Mary) and it appeared to her in the form of a well-built man."[19] It is completely detailed in the hadith reports that Gabriel used to appear to the Prophet, and that he was visible to him and that he would speak to him, although the rest of the people could not see him.

It is further reported in the hadiths that the grave will be extended seventy cubits by seventy or contracted until it presses the ribs of the buried person together, and that the angels descend to the entombed and buried person and question him and that his deeds will appear before him, and that the angels descend to the dying person with either silk or sackcloth in their hands, and that they beat the buried person with a hammer of iron and he utters a cry which can be heard from the East to the West.[20]

The Prophet, may the peace and blessings of God be upon him, said, "An unbeliever will be afflicted by ninety-nine vipers in the grave which will bite him and sting him until the Hour comes," and "When the corpse (of the believer) is put in the grave the setting sun is made to appear to him; and he sits rubbing his eyes and says, 'Let me pray.'"[21]

---

[14] Bukhārī Riqāq 28, Muslim Jannat 1, Abū Dāwūd, Tirmidhī, Nasā'ī, Dārimī, Muwaṭṭa', Ibn Ḥanbal.

[15] Not found in the standard hadith collections.

[16] Ibn Athīr, *Jāmi' al-Uṣūl fī Aḥādīth al-Rasūl* IV (Beirut: Dār al-Fikr, 1970), 18, #1992.

[17] This refers to a hadith in which the Prophet came out one day with a book in each hand. In his right hand was a book containing the names of the People of Paradise and in his left hand a book of the future inmates of Hell. *Mishkāt* p. 108. Tirmidhī III:304 6667, Qadar 8, Ibn Ḥanbal.

[18] This will occur on the last day when the people of Paradise have entered Paradise and the people of the Fire have entered the Fire. Bukhārī Tafsīr 19:1 (variant), Muslim Jannat 40, Tirmidhī, Dārimī, Ibn Ḥanbal.

[19] Qur'ān 19:10.

[20] *Mishkāt*, p. 36. As in Bukhārī Janā'iz 67.

[21] That is, a person who prayed in this life, experiencing the darkness of the grave, will expect to pray the sunset prayer. *Mishkāt*, p. 38. Transmitted by Ibn Ḥanbal 2, 311.

It is frequently mentioned in the hadiths that God (on Judgment Day) manifests Himself in many forms for the people who are awaiting their judgment and that the Prophet, may the peace and blessings of God be upon him, will go in to his Lord while He is on His Throne[22] and that God will speak to the sons of Adam orally, and so on with innumerable other instances.

The one who studies these hadiths may adopts one of three attitudes. He may confirm their literal meaning, and then he is forced to affirm the existence of the world whose nature we have mentioned (the World of Images). This is the alternative required by the method of the People of the Hadith. Al-Suyūṭī,[23] may God have mercy upon him, pointed this out, and I accept and follow this. Or secondly, one may say that these events are perceived by the sense perception of the viewer and manifest to him in his sight although they do not exist in the phenomenal world. He thus takes a similar position to what ʿAbd Allāh ibn Masʿūd[24] said in the interpretation of the Qurʾānic verse, "Today the heaven will bring forth clear smoke,"[25] i.e., that a famine had assailed them and one of them had looked toward the sky and had seen something resembling smoke because of his hunger.

It was related from Ibn al-Mājishūn[26] that all of the hadiths concerning the movement and the vision of God on the Day of Assembly meant that the perception of His creatures would change and they would see Him descending, manifesting Himself, conversing with His creatures, and addressing them without it affecting His greatness, nor would He move; in order that they would know "that God is able to do anything."[27]

Or (thirdly) he may make them symbolizations for conveying other meanings, and I don't think that any people of the truth[28] stop short at the third position. Imām Ghazzālī's discussion of the punishment of the grave described these three levels of interpretation when he said,

---

[22] As in a hadith reported in Ibn Ḥanbal 1, 296.

[23] Al-Suyūṭī, Abū Faḍl. d. 1505. A prolific and versatile scholar and writer particularly known for Qur'ān commentary.

[24] ʿAbd Allāh ibn Masʿūd. d. 652/3. One of the earliest Companions and a specialist in interpretive Qur'ān commentary.

[25] Qur'ān 44:10.

[26] ʿAbd al-Mālik ibn Mājishūn, d. 212/827, one of four Medinan disciples of Mālik.

[27] A phrase often asserted in the Qur'ān, for example 2:20, 106, 109, 148, 259, 284; 3:29, 165, 185. etc.

[28] *Ahl al-Ḥaqq*—followers of the Ashʿarite school of theology.

The symbols of these reports have correct literal meanings and concealed inner meanings but these are clear to those who possess spiritual insight. The man to whom their inner meanings do not reveal themselves must not deny their literal truth, rather the lowest level of faith is acceptance and agreement (that they are true).

Question: If you say, "We look at the unbeliever in his grave for a while and observe him and we don't see any of these things, so how can one accept something which conflicts with his sense experience."

Answer: Know that you have three positions in believing in this type of thing. One of them, the most evident, sound and safe, is that you believe that they (the vipers) exist and that they bite the corpse but that you do not witness it. For this (earthly) eye is not suitable for witnessing heavenly (*malakūtī*) matters; and all those things connected with the next world are from the World of Dominion (*Malakūt*).[29]

Do you not see how the Companions, may God be pleased with them, used to believe that Gabriel descended to the Prophet, may the peace and blessings of God be upon him? Although they did not see him, they believed that the Prophet saw him. If you do not believe in this then correcting the foundation of your belief in the angels and in revelation is the most important thing for you. If you do believe in this, then you hold it possible that the Prophet, may the peace and blessings of God be upon him, saw what the community did not see—so why do you not allow this in the case of the corpse? Just as an angel does not resemble men or animals, the snakes and scorpions which bite in the grave do not belong to the type of snakes in this world but are of another kind and are perceived by another sense.

The second position is that you should bear in mind the case of the sleeping person who may see a snake biting him in his dream, and he feels the pain so that perhaps you will observe him cry out

---

[29] *'Ālam al-malakūt* (the World of Dominion) is the plane or level of existence related to the World of Images and the highest angels. Shāh Walī Allāh, following Ibn 'Arabī's school, incorporated a descending schema of four worlds: *Lahūt, Jabarūt, Malakūt*, and *Nāsūt* in his cosmology. *Lahūt* would be the level of the divine existence, *Jabarūt* (the World of Domination) represents the level of God's divine command (*amr*) and determination, and *Malakūt* the level of the World of Images and the angels where the order of God takes shape before being translated into physical manifestation at the level of *Nāsūt* or human existence. Al-Ghazzālī had a similar conception of *Malakūt* although he sometimes placed the World of Domination (*Jabarūt*) between it and the level of human existence (*Nāsūt*).

and see the sweat on his forehead and he may be aroused from his place. He perceives all of this within his own mind yet he suffers as the one who is awake suffers while he witnesses it. You see his exterior quiet and you do not see around him any snake or scorpion, but the snake exists so far as he is concerned and punishment occurs, although it is not witnessed by you. Since the punishment is due to the pain of the biting, there is no difference between a serpent which is imagined or which is seen with one's own eyes.

The third position is that you know that the snake itself does not cause pain but rather what you experienced from it was the pain of its poison. Thus the poison itself is not the pain, but your torment lies in the effect which the poison has on you. If some comparable effect occurred without poison the punishment would still obtain. It would be impossible to explain this type of punishment except through linking it to the cause which usually leads to it. For example, if there had been created in man the pleasure of sexual intercourse without the actual experience of intercourse, then this could not be made known except by relating it to intercourse, so that the relationship would make it (the pleasure) known by the cause. The effect of the cause would be achieved even if the actual cause were not, for the cause intends its result, not itself. These destructive attributes (of a person) are transformed into scathing torments and pains in his soul at the time of death and their torments are like the stings of the vipers without these vipers actually existing.[30]

[30] End of quotation from Al-Ghazzālī found in *Iḥyā' 'Ulūm al-Dīn* (Beirut: Dār al-Ma'rifa, 1982), book 40, chapter 7 on "The Punishment in the Grave." When Ghazzālī asserted that these objects of Islamic theology could be perceived through another sense by those with spiritual insight he opened the door for later Muslim philosophies and mystics to speculate that these existed in a world of their own. Shihāb al-Dīn al-Suhrawardī (d. 1191) seems to have been the first to affirm this realm. See Fazlur Rahman, "Dream, Imagination and *'Ālam al-Mithāl*" in *The Dream in Human Societies*, ed. Gustave von Grunebaum (Los Angeles: University of California Press, 1966), 409–419, for a discussion of the World of Images in Islamic thought.

CHAPTER 3

The Highest Council[1]

God, may He be exalted, said:

> Those who carry the Throne and those around it extol the praises of their Lord and believe in Him, and ask his forgiveness for believers. Our Lord, Your mercy and knowledge encompass everything. Lord, forgive those who repented and followed your path, protect them from the punishment of Hell and make them enter Paradise, the gardens of Eden which were promised to them by You, with those who are righteous from among their fathers and their spouses and their progeny, for Thou art the Great, the Wise. Deliver them from evil. To those whom You deliver from evil on that day, you have shown mercy, and that is the great triumph.[2]

And His Prophet, may the peace and blessings of God be upon him, said,

> When God decrees an order in the heavens, the angels beat their wings in submission to His speech with a sound like the clattering of an iron chain on smooth rocks. When their hearts are delivered from awe they ask each other, "What did your Lord say?" They reply, "The truth, and He is the Exalted, the Great."[3]

And there is the report,

> When He decides on a decree, the bearers of the throne glorify Him, then the inhabitants of the upper heaven, who follow them, until the glorification of the inhabitants of the heaven reaches the world. Then those after the bearers of the throne ask, "What did your Lord say?" They inform them of what He ordered and the people of the heavens tell each other until the news reaches the people of this sphere.[4]

---

[1] *Al-mala' al-a'lā*. The Highest Council or the Highest Host. The term is found in Qur'ān 37:8, 38:69. This group contains both angels and developed human souls, according to Shāh Walī Allāh.

[2] Qur'ān 40:7-9.

[3] *Mishkāt*, p. 959. Transmitted by Bukhārī Tafsīr 15, I; 34, I, Tirmidhī, Ibn Mājah, Dārimī.

[4] *Mishkāt*, p. 959. Transmitted by Muslim Salām 124, Tirmidhī, Ibn Ḥanbal.

On another occasion, the Prophet, may the peace and blessings of God be upon him, said,

> I arose during the night and performed the ablution and prayed my share of prayer, but I felt drowsy in my prayer so that I fell asleep. My Lord approached, Blessed and Exalted, in the best of forms and said, "O Muḥammad." I said, "I am at Your service, my Lord." The Lord said, "What are the Highest Council disputing about?"[5] I answered, "I don't know." He said it three times, then I saw Him putting His palm between my shoulders so that I felt the coldness of His fingertips through to my breast. Everything was revealed to me and I understood. He said, "Muḥammad." I said, "I am at Your service, O Lord." He said, "About what are the highest angels disputing?" I answered, "Concerning the atonements for sins." He said, "What are they?" I replied, "Walking on foot to the congregational prayers, sitting in the mosques after prayers, and properly completing the ritual ablutions, even in difficult conditions." He said, "Of what do they speak then?' I replied, "About the ranks (of pious actions)." He said, "And what are they?" I said, "The distribution of food, gentleness of speech and prayer at night while people are sleeping."[6]

The Prophet, may the peace and blessings of God be upon him, said,

> When God loves a person he calls Gabriel and says, "I love so-and-so, so love him." And Gabriel loves him, then he proclaims it in the heavens saying, "God loves so-and-so, so love him." Then the people of the heaven love him, therefore success on earth is spread out before him. If He hates a person, He calls Gabriel and says, "I hate so-and-so, so hate him." And Gabriel hates him and proclaims to the people of heaven, "God hates so-and-so, so hate him." Therefore they hate him so that hatred is spread out before him on the earth.[7]

The Prophet, may the peace and blessings of God be upon him, said,

> The angels pray for one of you as long as he remains seated at the end of his ritual prayer, saying, "O God, have mercy upon him, O God, forgive him, O God be gracious to him," if he does not damage it or break it (his ritual purity).[8]

---

[5] A reference to Qur'ān 38:69.
[6] *Mishkāt*, p. 151. Ibn Ḥanbal 5:243.
[7] *Mishkāt*, p. 1040. Muslim transmitted it.
[8] *Mishkāt*, p. 143. Bukhārī Ṣalāt 87, Buyū' 49.

And he, may the peace and blessings of God be upon him, also said,

> There is no day when people get up that two angels do not descend and one of them says, "O God, give to the charitable one a good replacement," and the other says, "O God, give the miser ruination."[9]

Be informed that it has been abundantly reported in the sharīʿa that God, may He be exalted, has worshippers who are the best among the angels, and the nearest to His presence. They unceasingly pray for the one who reforms and purifies himself and strives to reform the people, so that their prayer is the cause for the descent of blessing upon them. They curse the one who rebels against God and tries to spread evil so that their curse is the cause for the existence of grief and remorse in the soul of the perpetrator. Their invocations cause intuitions in the breasts of the Lower Council,[10] so that they become angry with this evildoer and they inflict harm upon him, either in this world or when the garment of his earthly body is removed by natural death. The highest angels are ambassadors and intermediaries between God and His worshippers and they inspire good in the breasts of human beings; that is, they are a cause for the emergence of good impulses in them through one sort of causality.

They have gatherings, however and wherever God wills. In view of this they are designated as "the highest companions" or "the highest circle" or "the Highest Council." The spirits of the most excellent and worthy humans enter among them and join them, as in God's saying, "O contented soul, return to your Lord satisfied and satisfying; enter among My servants and enter My Garden."[11]

And the Prophet, may the peace and blessings of God be upon him, said, "I saw Jaʿfar ibn Abī Ṭālib as an angel with wings flying in paradise with the angels."[12]

From here (the circle of the highest angels), the divine decree (*qaḍāʾ*) descends and the command (*amr*) is specified which is indicated by the following saying, "During the Night of Power every

---

[9] *Mishkāt*, p. 394. Bukhārī Zakāt 27, Muslim, Ibn Ḥanbal.

[10] The lower ranks of the angels who are near the earth and the lower reaches of the heavens.

[11] Qur'ān 89:27–30.

[12] The brother of ʿAlī ibn Abī Ṭālib. He died a martyr in battle. *Mishkāt*, p. 1354. Tirmidhī Manāqib 29. Variant (doesn't have "with wings"), V:320 #3852.

wise command is decided,"[13] and here the religious laws are determined in some aspect.

The Highest Council is composed of three categories. The first constitutes those whom God has designated as having the regulation of the good depending on them so that He created luminous bodies at the level of the fire of Moses[14] and He exhaled into them noble souls. The second are those who arose out of the occurrence of a mixture of the subtle vapors of the elements, requiring the emanation of the lofty souls which strongly reject the contamination of the bestial. The third group are those human souls whose origin is near the Highest Council. They continue to perform acts of salvation which keep them attached to the angels until the garments of their bodies are thrown off and then they become incorporated into them and counted among them.

The task of the Highest Council is to contemplate their Maker with a devoted contemplation which cannot be distracted by their turning to anything else. This is the sense of His saying, may He be exalted, "They extol the praises of their Lord, believing in Him."[15]

They are instructed by their Lord to condone the righteous order and condemn whatever opposes it, so that this condoning knocks at the doors of the Divine Generosity, and this is the meaning of His saying, "They ask for forgiveness for those who believed."[16]

The lights of the excellent among them are gathered together and interpenetrate each other at the level of the Spirit which the Prophet has described, may the peace and blessings of God be upon him, as having many faces and tongues.[17] There they become gathered like a single thing, and it is called the Holy Enclave (*ḥaẓīrat*

---

[13] Qur'ān 44:4. This Qur'ānic verse refers to the *Lailat al-Qadar*, or Night of Power when God's commands are established. "Amr" also refers in Sufi thought to the immaterial "World of God's Command" (*'ālam al-amr*) which includes the World of Images from which the divine decrees are later materialized in the World of Creation (*'ālam al-khalq*).

[14] The fire which Moses saw in the burning bush on Mount Sinai. The relationship of some angels to this fire is mentioned in Shāh Walī Allāh's *Lamaḥāt*. (Hyderabad: Shāh Walī, Allāh Academy, n.d.), Lamḥa 57.

[15] Qur'ān 40:7.

[16] Qur'ān 40:7.

[17] "*Min kathrat al-alsina wa-l-wujūh*." A possible reference to a Sufi hadith quoted by al-Ghazzālī, "Allāh has an angel who has 70,000 faces with 70,000 tongues in each face; with all of them he praises God." This would refer to the "Spirit" in contrast to the rest of the angels. *Mishkāt al-Anwār*, p. 52. Many other hadiths mention the number of 70,000 angels.

al-quds).[18] Sometimes in the Holy Enclave a consensus is reached to establish a means of saving human beings from the disasters of this life or the next world through the perfection of the purest man of the age and through causing his command to be implemented among the people. This (consensus of the Holy Enclave) requires inspirations in the hearts of those people prepared to follow him and to constitute a community brought out for humankind,[19] and it requires the representation in his heart of branches of knowledge which will make the people righteous and guide them, either by revelation, dreams, or a voice from the Unseen; and members of the Holy Enclave appear to him and speak with him directly. This consensus necessarily requires the victory of his companions and their drawing nearer to all good, and the cursing of those who impede the path of God and their being brought near to all pain.

This is one of the principles of prophethood and the continuing consensus of this council of angels is called "support by the Holy Spirit."[20] These blessings bear fruit such as is not normally met and these are known as miracles.

Below these ones (the Highest Council) are souls (the Lowest Council) whose emanation required the coming into being of a balanced combination among the subtle vapors. Bliss does not reach them to the extent of the first (angels), and their fulfillment consists of being free to await whatever filters through to them from Above. When something is showered down, in accordance with the capacity of the receiver and the effectiveness of the doer, they are dispatched to carry out these orders, just as birds and animals are triggered into movement by instinctual motivations. In this they are oblivious to what concerns themselves and subsist through what has been revealed to them from Above. They can influence the hearts of men and animals so that their wills and their inner

[18] The Enclave or Fold (*ḥaẓīrat*) of the Holy is used by Shāh Walī Allāh to indicate that area of the heavens where the highest angels, joined by a few select human souls, convene to assist the future course of human affairs. The term appears with this usage in Ghazzālī's *Mishkāt al-Anwār* together with other aspects of his angelology adopted by Walī Allāh such as the discussion of the Highest and Lower Councils of angels. According to Ghazzālī, this place is called the "Enclave" (*ḥaẓīrat*) of the Holy because nothing foreign to it can enter. *Mishkāt al-Anwār* Arabic text, p. 66. The term first appears in a hadith transmitted by Ibn Ḥanbal. *Musnad* V:257.

[19] An echo of Qur'ān 3:110.

[20] Such as in Qur'ān 2:87, 2:253, "We gave Jesus, son of Mary, the clear signs and supported him by the Holy Spirit." Other prophets are also supported in this manner.

promptings are transformed into something compatible with carrying out the desired object, and they influence certain natural phenomena by augmenting their movements or transformations. For example, a great stone may roll away and some powerful angel may help increase its motion so that it rolls along the ground with greater velocity than normal. Sometimes a fisherman may cast a net in the river and armies of angels will inspire instincts in certain fish to plunge into the net, and in some others to flee, and they will contract one rope of the net and loosen another, and they do not know why they do that, but they follow what they are inspired to do. Sometimes two armies may be warring with each other, and then angels come and create visions of courage and perseverance in the hearts of one side through inner voices or images as the situation requires. They may reveal the ruse for victory and aid in the archery and similar things.[21] In the hearts of the other side they may inspire the opposite of these traits; for as God decides an affair, thus it is done. Sometimes what is sent down is the paining of a human soul or its happiness, thus the angels will make the fullest efforts and try every way possible to effect this. Opposing these (angels) is another group possessed of fickleness, inconstancy, and thoughts contrary to the good whose origination was made necessary by the decaying of dark vapors, and these are the devils who unceasingly strive against the efforts of the angels, and God knows better.

[21] As in the account in the Qur'ān 4:42 of the battle of Badr.

CHAPTER 4

God's Customary Way of Acting (Sunna) Which is Indicated in His, may He be Exalted, Saying "You Will Never Find any Alteration in the Sunna of God"[1]

Be informed that certain of the acts of God are somehow structured according to potentialities which have been placed in the world. Both reason and textual revelation attest to this. The Prophet, may the peace and blessings of God be upon him, said, "God created man from a handful (of dirt) which He grasped from the whole of the earth, so that the sons of Adam arose in accordance with the nature of this earth; the red, the white, and the black, and some in between; and among them the soft and the rugged, the impure and the pure."[2]

ʿAbd Allāh ibn Salām asked him, "What makes the child take after his father or his mother?" He answered, "If the fluid of the male precedes the fluid of the female the child takes after him;, and if that of the mother precedes, she is resembled."[3]

I do not see anyone doubting that killing depends on striking with a sword or taking poison, and that the creation of the child in the womb follows the ejaculation of sperm, and that the creation of seeds and trees follows upon sowing and planting and watering.

Due to this capacity (*istiṭāʿa*) religious duties were imposed, and people were commanded, forbidden and permitted in what they did. Among these structuring forces are the properties of the elements and their natural constituents, the determinants which God put in every specific form, and included are the states of the World of Images and the existence decreed there, before earthly existence. Also among these structuring forces are the prayers of the Highest Council who concentrate their fullest zeal on those who refine their souls and make efforts to reform people, and (their negative

---

[1] Qur'ān 33:62, 35:43, 48:32.

[2] *Mishkāt*, p. 28. Abū Dāwūd IV:222 #4693 Sunna 16, Tirmidhī, Ibn Ḥanbal.

[3] *Mishkāt*, p. 1272. Bukhārī Manâqib al-Anṣār 51, Tafsīr 2:6, Ibn Ḥanbal.

invocations) on the one who opposes this. The structuring forces also include the religious laws ordained for human beings which establish obligation and prohibition, for they are the basis of the rewarding of the obedient and the punishment of the disobedient. Among them (the structuring forces) is that God decrees a thing and this sets in motion something else because it entails it according to God's law, and the breaching of the system of entailment is not pleasing (to God). The root of this is found in his saying, may the peace and blessings of God be upon him, "If God decrees that a person will die somewhere on the earth, a need will arise for him to go there."[4] All of these matters are spoken of in the prophetic traditions and necessarily confirmed by reason.

Be informed that if the causes through which His decree is structured are in conflict with each other according to what is the usual custom so that their requirements cannot possibly all be fulfilled, then the wisdom of God in this case takes into account the thing which results in the most absolute good. This is what is referred to by the scale in the Prophet's saying, "In His hand is the scale, (with which) He raises the share and lowers it"[5] and by the expression "mode" in His, may He be exalted, saying, "Every day He is in a mode."[6]

Therefore preference is sometimes given to the case of whichever causes are stronger, and sometimes to the case of whichever structuring effects are more beneficial. In the matter of the precedence of the domain of material creation (*khalq*) over the domain of divine ordering (*tadbīr*) and other matters of this sort; we, even though our knowledge falls short of comprehending the causes and knowing which of the conflicting aspects is worthier, certainly know that only the most worthy thing will come into being. The one who is certain of the truth of what we said will be relieved of many difficulties.

As for the dispositions of the planets, from their effects arise necessary results such as the alternation of summer and winter and the lengthening and shortening of the day in proportion to the po-

---

[4] *Mishkāt*, p. 29. Tirmidhī III:307 #6635 Qadar 11.

[5] Some interpreters say this refers to the determination of a person's share of sustenance, others say it refers to the scale weighing a person's deeds. *Mishkāt*, p. 26. Bukhārī Tafsīr 11:7, Tauḥīd 19, Khan trans. VI:168, Muslim Ibn Mājah, Ibn Ḥanbal.

[6] Qur'ān 55:29.

sitions of the sun, and the alternation of the ebbing and rising of the tides in correspondence to the moon's positions. In the hadith it was related, "When the star[7] rises the scourge will be lifted;," that is, in accordance with the laws of nature. However, the attribution of poverty, wealth, sterility, fertility, and the rest of what happens to humans to the movements of the planets is something not affirmed in the divine law. The Prophet, may the peace and blessings of God be upon him, forbade investigating this, saying, "the one who seeks knowledge of a branch of Astrology seeks knowledge of some branch of sorcery."[8] He emphasized this more strongly in the case of saying, "We have been given rain when such and such (a star) sets."[9] I do not hold that the sharī'a explicitly states that God, may He be exalted, does not create properties in the stars which generate events by effecting changes in the atmosphere surrounding people and this sort of thing.

You are well informed that the Prophet, may the peace and blessings of God be upon him, forbade divination as being information from the jinn, and he disassociated himself from anyone who went to a fortune-teller and believed him. Then, when the Prophet was asked about fortune-tellers he stated that when the angels are descending through the clouds and mentioning the matter decreed in Heaven, the devils eavesdrop; and reveal it to the fortune-tellers, so that the fortune-tellers tell one hundred false things along with that one.[10]

God, may He be exalted, said, "O you who believe, don't be like the unbelievers who said about their brothers, who traveled or went to war in the path of God, 'If they had remained (here) with us they would not have died or been killed.'"[11]

The Prophet said, may the peace and blessings of God be upon him, "Not one of you will be brought into paradise by his act,"[12]

---

[7] The Pleiades. Ibn Ḥanbal II:341.

[8] *Mishkāt*, p. 959. Abū Dāwūd IV:16 #3905, Ṭibb 2. Ibn Ḥanbal.

[9] "The one who said, 'We have been given rain when such and such a star sets' is the one who disbelieves in me and believes in the star." *Mishkāt*, p. 959. Transmitted by Bukhārī Adhān 156, Istiqsā 280, Maghāzī 35, Muslim, Abū Dāwūd, Nasā'ī, Tirmidhī, Dārimī, Muwaṭṭa', Ibn Ḥanbal.

[10] *Mishkāt*, p. 958. Bukhārī Bad' al-Khalq 6, Khan trans. VI:291, Tafsīr 15,1; 34, 1, etc., Muslim, Ibn Mājah, Ibn Ḥanbal.

[11] Qur'ān 3:156.

[12] That is, God's grace is also necessary.

and "You are only a person who can give kind treatment (to the ill) but God is the physician who cures."[13]

In sum, the prohibition (of fortune-telling and so on by the divine law) bears many benefits, and God knows better.

[13] The Prophet spoke these words to a doctor who offered to treat him. Hadith from *Musnad Ibn Ḥanbal*. IV:163.

CHAPTER 5

The True Nature of the Spirit (Rūḥ)

God, may He be exalted, said, "They ask you about the spirit. Say, the spirit comes from the order of my Lord and you have been given very little knowledge of it."[1]

Al-A'mash[2] read according to the reading transmitted by Ibn Mas'ūd, "They were only given a little bit of knowledge." It can be seen from this that the speech was addressed to the Jews who asked about the spirit, and this verse does not categorically state that no one from the Muslim community, on whom God has mercy, has knowledge about the true nature of the spirit, as is thought (by some), nor does it mean that knowledge of everything on which the religious law is silent is definitely impossible. Rather much of what is not mentioned in the sharī'a concerns sophisticated matters not suitable to be taken up by the general community even if it is possible for certain ones among them.

Be informed that the first thing to be comprehended about the spirit's true nature is that it is the source of life in the animal which is alive due to a breathing of the spirit into it and dies when it is separated from it. Then, if one ponders deeply it will be disclosed that in the body there is a subtle vapor which is produced in the heart from a quintessence of the humors which carry the faculties of perception, movement, and the distribution of food according to the rules of medicine. Experience reveals that every one of the states of this vapor, whether fine or thick or pure or turbid, has a particular effect on the faculties and the functions that proceed from these faculties.

The disease which may strike each member and the generation of the vapor related to that member corrupts this vapor and disturbs

---

[1] Qur'ān 17:85.

[2] Al-A'mash, Abū Muḥammad Sulaymān (d. 765), a traditionalist and Qur'ān reader who followed the tradition of Ibn Mas'ūd. His reading of the Qur'ān is one of the fourteen canonical ones.

its functions. The coming into being of this vapor causes life and its dissolution causes death, therefore this is equated with the spirit, at first glance, and after careful thought it is seen to be the lowest level of the spirit (*rūḥ*).[3] It corresponds in the body to the perfume of the rose or the fire in the coal. Then, if one reflects carefully it becomes disclosed also that this spirit (the vapor) is the vehicle of the real spirit and the material for attaching it (to the body). That is, we see that a child grows and become older while the humors of his body and the spirit arising from these humors change more than a thousand times. He becomes smaller at times, and bigger at other times, and he becomes dark at times and white at others, and he is ignorant at times and knowledgeable at others, and so on, with other changing attributes, while the individual remains himself. And if there is an objection raised concerning certain of these points then we may assume that these changes occur while the child is still himself, or we may say that we are not absolutely certain that these characteristics remain in one state, but we are certain that the child remains himself and that he is something other than the changing characteristics. The thing which makes him himself is not this spirit (the pneuma), neither this body, nor these individuating characteristics which you recognize and see without thinking twice. Rather the spirit is in reality a unique separate essence (*ḥaqīqa fardāniyya*)[4] and a luminous point. Its state is exalted above the condition of these changing and multifarious states, some of which are substances and some accidents. The spirit is found the same in the young as in the old and with the black as well as in the white and likewise with other opposites.

It has a special connection with the airy spirit[5] in the first place,

---

[3] In his detailed discussion of the levels of the spirit in *Alṭāf al-Quds*, (Gujrānwalā: Madrasa Nuṣrat al-'Ulūm, 1964), 20–27, Walī Allāh describes three levels of the spirit: 1) The pneuma (*nasama*), also called the airy spirit (*al-rūḥ al-hawā'ī*) which arises from the subtle vapors of the elements obtained through digestion. 2) The rational soul (*al-nafs al-nāṭiqa*) which is the specific form of the person which makes each person an individual. 3) The angelic spirit (*al-rūḥ al-malakūtī*) or Divine Spirit (*al-rūḥ al-ilāhī*), which is identified with the form of the person pre-existing in the World of Images. At resurrection, this form, which has been affected by the acts of the person throughout his life, unites with him. In this chapter of *Ḥujjat Allāh al-Bāligha*, Shāh Walī Allāh first explains how the rational soul becomes connected with the pneuma or nasama and then explains how the spirit remains in the afterlife to effect reward or punishment.

[4] In this instance "spirit" refers to the rational soul (*al-nafs al-nāṭiqa*).

[5] *Al-Rūḥ al-Hawā'ī*. This is the same as the pneuma, the most subtle essence of a person's humors.

and with the body secondly, in so far as the body is a substratum (*maṭiyya*)[6] of the pneuma (*nasama*). It (the rational soul) is like a small peep hole from the World of the Holy through which descends to the pneuma everything which it is ready to receive. The alterations in the changing things only take effect through earthly dispositions, in the way that the (same) light of the sun bleaches the robe and darkens the man doing the bleaching.

To us it has been confirmed by true direct intuition, that death is the separation of the pneuma from the body due to the loss of the body's capacity to generate it, not due to the separation of the Holy Spirit (*al-rūḥ al-qudsī*)[7] from the pneuma. When the pneuma decomposes during fatal illnesses it is necessary according to God's wisdom that the part of the pneuma remains to an extent which enables the connection of the Divine Spirit with it.[8] In like manner, if you suck air out of a bottle, the air will be pulled out until it reaches a limit after which it will not come out, so that you are then unable to suck in or the bottle will break. This is nothing other than a secret arising from the nature of air and likewise there is a secret in the pneuma and a limit of it which cannot be transgressed.

When a man dies the pneuma undergoes a regeneration and the emanation of the Divine Spirit generates in it a faculty, in what remains of the sensus communis,[9] which is enough to permit adequate hearing, sight, and speech through help from the World of Images. By this I mean the faculty intermediary between the non-material and the tangible which is dispersed through the spheres like a single thing. The pneuma may be prepared at the time of death to be clothed in the garb of light or darkness through help from the World of Images, and it is from there that the wonders of the intermediary world of the *barzakh*[10] appear.

Then, when the trumpet is blown,[11] that is, when a general emanation comes from the Creator of the forms—like the emanation

---

[6] This word (*maṭiyya*) literally means a riding animal and refers to how a higher form can attach to and guide a lower form.

[7] The highest level of the Spirit.

[8] The highest level of the Spirit.

[9] The sensus communis (*ḥiss mushtarik*) is the internal faculty of perception which receives all of the forms imprinted on the five senses.

[10] *Barzakh* refers to an isthmus or space between the material or earthly world and the heavenly or spiritual realm, and also the period between death and resurrection.

[11] That is, the trumpet which shall blow on Judgment Day, referred to in Qur'ān 27:87, 78:18, etc.

which proceeded from Him at the beginning of creation when the spirits were exhaled into the bodies and the world of elements was established; then the emanation of the Divine Spirit necessitates that the pneuma take on bodily clothing or clothing intermediary between the world of bodies and the World of Images.[12] All of what the truthful and trusted Prophet informed us about will actually come to pass, may the greatest of blessings and the most auspicious of greetings be sent to him.

Since the pneuma is an intermediate screen (*barzakh*) between the Divine Spirit (*al-rūḥ al-ilāhī*) and the earthly body, it is necessary that it have one aspect of the former and one of the latter. The aspect inclined toward the (world of the) holy is the angelic and that inclined to the earth is the animalistic. Let us restrict our account of the truth of the Spirit to these premises so that they will be accepted in this science, and its ramifications must be pursued before the veil over a science higher than this can be lifted, and God knows better.

[12] This is in explanation of the bodily Resurrection.

CHAPTER 6

The Inner Dimension of the Imposition of Religious Obligations

God, may He be Exalted, said, "We offered the Trust to the heavens and the earth and the mountains and they refused to bear it and shrunk from it, but man took it up, indeed he was unjust and ignorant. Therefore He will punish the hypocrites, male and female, and the polytheists, male and female, and forgive the faithful men and women, for God is forgiving and merciful."[1]

Al-Ghazzālī, al-Baiḍāwī and others have pointed out that what is meant by "the Trust" (*amāna*) is the assumption of the commitment to obey God's commands which exposed them[2] to the risk of reward and punishment on the basis of obedience and disobedience. By the Trust's being offered to them (the heavens, earth, etc.) their capacity was taken into consideration, and by their refusal (of the Trust), is meant natural refusal based on the absence of suitability and capacity. By man's taking it up is meant his fitness and capacity for it.

I hold that on this point there is His, may He be Exalted, speech, "he was unjust and ignorant" which functions as an explanation; for the unjust person is the one who is not just, while he should be just; and the ignorant person is the one who does not have knowledge when he should have. Beings other than man are either knowing and just, untouched by sin and ignorance, such as the angels are, or they are neither just and nor knowing, neither is it their place to acquire these qualities; such as is the case for animals. The only one suited for the imposition of religious obligations and having the capacity for it is the one who has the potential of perfection and not the actuality. The letter "l" prefixed to His saying "He will punish" (*liyuʿadhdhiba*) is the "l" of consequence,[3]

---

[1] Qur'ān 33:72–73.

[2] The heavens, earth, and so on.

[3] According to the rules of Arabic grammar.

as if He said that the result of taking up the Trust would be punishment or reward. If you desire that the truth of the matter be disclosed to you, then you must imagine the condition of the angels in their transcendence. They are neither disturbed by any state arising from the deficiency of the animal force such as hunger, fear, thirst, or sorrow; nor by its excesses such as lust, anger and vanity, and they are not preoccupied with nutrition and growth and what pertains to these. They remain exclusively occupied with waiting for what shall come to them from Above them. Then, when a command filters down on them from Above resolving the establishment of a desired system or pleasure or anger at a certain thing, they become totally filled with it and obey it, and they are dispatched to undertake whatever it requires. In this they are oblivious to their own volition and subsist through the purpose of what is above them.

Then imagine the state of the beasts, stained by the sensual forms, continuously infatuated with natural needs, totally absorbed in them. They are not provoked to do anything except by animalistic provocations traceable to a concrete benefit, they abandon themselves solely to the demands of their nature.

Thus you should know that the Creator, may his state be exalted, deposited in man through His splendid wisdom two forces. One is an angelic force which branches out from the emanation of the spirit particularized for man into the natural spirit (*al-rūḥ al-ṭabīʿiyya*). This natural spirit pervades the body, and it receives that emanation and is subject to it. The second is an animalistic force (*quwwa bahīmiyya*) branching out from the animating soul (*al-nafs al-ḥayawāniyya*) common to all animals, which is given shape by the faculties inhering in the natural spirit. It is self-sufficient and the human spirit submits to it and obeys its orders. Then you should know that competition and contention occurs between these forces. The former pulls toward the high and the latter toward the low. Whenever the animal force becomes apparent and its effects dominate, the angelic side is concealed, and likewise the reverse.

The Creator, may He be Exalted, has a concern for every system and a generosity toward everything which original and acquired capacities ask of Him. Thus, if someone acquires animalistic attitudes, God helps him in this and makes what is appropriate for them easy for him, and if he acquires angelic attitudes, God aids in them and makes what is appropriate for them easy for him. For

God, may He be Glorified and Exalted, has said, "As for the one who gives (to the needy) and fears God and affirms the good, We make these things easy for him, while the one who is miserly and deems himself independent, and denies goodness. We shall make this easy for him."[4] And He said, "Both of them, these and those, are aided by the gifts of your Lord, for the giving of your Lord is not restricted."[5]

Each force (the angelic and animalistic) has its own pleasure and pain. Pleasure is the perception of what is suitable for it and pain is the perception of what is not suitable for it. How similar is the condition of man to the condition of someone who used a painkiller in his body, so that he experienced no pain in being scorched by fire until its effect wore off, and when he returned to his normal state bestowed by nature he experienced such severe pain. This condition of man is also similar to that of the rose, which, according to what the doctors say, contains three forces: an earthly force apparent when it is crushed and made into powder, a watery force seen when it is squeezed and drunk, and an airy force apparent when it is smelt.[6]

Thus it is clear that religious obligation is one of the requirements of the species and that man asks his Lord through the means of his capacity that God will enjoin him with what is appropriate for the angelic and that He will reward him according to that, and that He will forbid his engrossment in the animalistic and punish him for that, and God knows best.

---

[4] Qur'ān 92:5–10.

[5] Qur'ān 17:20. "These" and "those" refer to persons who desire this life and persons who desire the Hereafter.

[6] That is, the human being is made up of elements derived from various components and levels of being.

CHAPTER 7

How Religious Obligation is Derived from What is Divinely Decreed

You should know that God has signs in His creation which guide the one who considers them to the fact that God has the conclusive argument (*al-ḥujjat al-bāligha*) in His imposing on man the religious obligations of the divine laws.

Look at the trees, their leaves, their blossoms, and their fruit and at all of their characteristics, visual, gustatory, or otherwise. He made for every species leaves in a particular form, blossoms with a particular color, and fruits with a particular taste. Due to these things it can be known that this individual is from such and such a species. All of these follow the specific form and are bound up with it, coming from where the specific form comes. God, may He be Exalted, decreed, for example, that this material would be a date palm, involving His specifying decree that its fruit should be a certain type and its leaves a certain type.

Some of the properties of the species are perceived by any intelligent person and others are such that only the most clever of all can perceive them, such as the effect of the ruby which brings happiness and courage to the one who wears it. Some of the properties (of the species) are generalized to every individual and others are only found in certain ones according to the capacity of the matter. For example, the myrobalan fruit[1] relieves constipation in the one who grasps it in his hand, and it is not for you to ask why the fruit of the palm has this property. This is an invalid question since the existence of the properties of essences in those essences cannot be inquired into by "why."

Then look at the types of animals, you will find that every species has a form and a natural constitution, just as is found among trees. Along with that you will find that they have voluntary movements, natural instincts and congenital traits which distinguish each

[1] *Ahlīlaj*—yellow myrobalan, a tropical fruit with laxative properties.

species. The pasture animals graze on grass and ruminate, while the horse, the donkey, and the mule graze but do not ruminate. Beasts of prey eat meat and birds fly through the air and fish swim in the water. Every species of animal makes a sound different than the sound of another and has a different way of mating and nursing its young than others, and the explanation of this will be very lengthy. He only inspires a species with knowledge appropriate for its temperament and through which that species will prosper.

All of these instincts trickle down to the species from their Maker through the tiny aperture of the specific form. An example of this is the designs on the flowers and the flavor of fruits in their being enmeshed with the specific form.

Among the determinants (*aḥkām*) of the species are those generalized to all individuals as well as those only present in certain ones according to the capacity of the matter and the confluence of the causes, although the basis of the capacity is found in all. Thus one leader arises among bees; and a (certain) parrot, after practice and effort, learns to imitate the sounds of human speech. Then look at the species of man, and you will find that he exhibits traits which you find among trees, or what you might find among types of animals such as coughing, stretching, burping, excretion, and suckling after birth. Along with these are found properties which distinguish man from the rest of animals such as rationality, understanding speech, and the derivation of acquired sciences from the ordering of axiomatic principles or from experience, induction, and intuition. Humans are concerned with matters which their reason judges commendable although they did not find these out through sense perceptions or imagination, such as the refinement of the soul, and the subjugation of regions under their control. Therefore all peoples have coinciding ideas on the principles of these matters, even the inhabitants of the remote mountain heights. This is due to nothing other than a secret arising from the basis of the specific form, and this secret is that the disposition of man requires that his intelligence dominate his heart and that his heart dominate his lower soul.[2]

---

[2] "*Nafs*" in this sense refers to the "Lower Soul" which represents the physical functions of the person and the lower or base types of desires such as greed and anger.

Look how God has designed and regulated every species, and at His instructing it, and His grace upon it. Since the plants do not perceive or move, He made roots for them which absorb matter composed of water, air, and fine particles of earth and they distribute this among the limbs and other parts in the proportions decreed by the specific form. Because the animals perceive and move at their own volition He did not make roots for them to suck matter up from the earth but rather He inspired them to forage for seeds, grass, and water in the most likely places and He inspired them with all the knowledge necessary to do useful things. God put in those species not created from the earth as worms are,[3] the faculties of reproduction and created in the female a moisture which is used to nourish the foetus and then changed it into pure milk and revealed sucking the breast and swallowing the milk. He created a liquid in the chicken which is used to produce the egg. When the egg has been laid the chicken is afflicted by dryness and the emptiness of its belly causes it to enter a stage of affection in which it abandons associating with others of its kind and prefers to incubate the egg which formerly filled its belly.

Among the traits of the dove is that affection between the male and the female which is seen in the female laying the eggs and the male taking incubating them upon himself. Then she takes pity on her chicks and He made the combination of her compassion and her extracted moisture produce vomiting and pushing seeds and water into the bellies of her chicks, and the male imitates her in this due to his affection. He created in the chicks a moist temperament so that their moisture could be transformed into feathers which enable them to fly.

Since man, with his sense perception, movement, and receptivity to the revelations of natural disposition and natural sciences, has an intellect and produces the acquired sciences, He inspired him how to cultivate, plant, and engage in trade and social transactions. He made some of them masters by temperament and opportunity, and some slaves by temperament and opportunity, and He made them the kings and the subjects. He made the wise man, speaking about the divine wisdom, and natural, mathematical, and practical sciences, and He made the stupid person who is not guided to that (wisdom) except through some kind of imitation. Therefore

[3] At this time it was believed that worms spontaneously emerged from the earth.

you see the nations of men, whether rural or urban, having identical ideas about things. All of this is an explanation of the outer properties and outer regulations connected to man's animal force and the supports of his livelihood (*irtifāqāt*). Now turn to his angelic force.

You should know that man is not like the other species of animals, rather his understanding is nobler than theirs. Among the sciences in which most individuals' ideas coincide, except for the one whose material defies the determinants of his species, is the inquiry into the cause for his coming into being and his being reared, and his being alerted to the evidence that there is an Orderer in the world who is the One who created him and sustains him. His supplication before his Creator and his Ruler through his concentrated resolve and knowledge is in accordance with how he and the members of his species are forever and eternally humbling themselves before Him (God) in their own way. This is reflected in His saying, may He be Exalted, "Have you not seen that whoever is in the heavens or on the earth, as well as the sun and the moon and the stars and the mountains and the trees and the animals and many men are prostrating before God; and punishment is decreed against many?"[4]

Isn't it the case that every part of the tree, its branches, its leaves, and its blossoms, eternally and forever implores in supplication to the vegetative soul directing the tree. If each part had a mind it would praise the vegetative soul with a praising different from that of the other parts. If it had understanding, then the form of supplication of its own state (*takaffuf ḥālī*)[5] would become imprinted in its knowledge and it would become a supplication with concentrated resolve (*himma*).

Know from this that man, since he has an intelligent mind, has imprinted on his soul an intellectual supplication commensurate with the supplication of his (physical) state. An additional human property is that there are those among the human species who can arrive at the source of rational knowledge[6] and who learn from it

---

[4] Qur'ān 22:18. Punishment is decreed against many for not prostrating.

[5] This is an "evolutionary" concept in Shāh Walī Allāh's thought whereby when a species reaches the limit of its current state, it implores God to expand its capacity or form. For example, in *Lamaḥāt*, p. 50, (no. 43), he describes the prototypes of the species which implore by means of their condition (*ḥāl*), not their speech (*maqāl*), that an exemplar of each will remain on the earth.

[6] Here Shāh Walī Allāh is referring to the Prophet's receiving direct knowledge from the Active Intellect.

through revelation, intuition, or dreams. The others may detect in this perfect one the effects of guidance and blessing so they follow what he orders and prohibits. There is no human individual who lacks the capacity to attain the Unseen through a dream which he has, a vision that he sees, a voice from the Unseen which he hears, or an intuition which he recognizes. However, among them are the perfect and the deficient, and the deficient needs the perfect one. He (the perfect one/Prophet) has attributes whose degree far surpasses the bounds of the bestial attributes, such as humbleness, purity, justice, and magnanimity;[7] and such as the manifestation of the glimmerings of Jabarūt and Malakūt[8] in response to his prayer, and the rest of the miracles, states, and stages.[9]

There are very many things which distinguish man from the rest of the animals but the totality and foundation of the matter resides in two qualities.

1) One of them is the preponderance of the intellectual traits, which has two branches:

a) a branch exercising skill in deriving the supports of civilization (*irtīfāqāt*) for the benefit (*maṣlaḥa*) of the human order and the discovery of its fine points, and

b) a branch readied for knowledge about the Unseen emanated by way of divine inspiration.

2) The second trait is the proficiency of the practical faculty (*al-quwwa al-ʿamalīyya*) which also has two branches.

a) One branch is "swallowing acts" by way of the "esophagus" of his choosing them and willing them. Animals do some actions by choice while the actions do not penetrate the basis of their souls and their souls are not tinted by the spirits of these acts. The acts only cling to them through the faculties inherent in the airy spirit (pneuma) so that it is easy for them to produce other acts like them.

Humans perform actions and the acts disappear while their spirits are extracted from them and the soul "swallows" them, so that there appears in the soul either light or darkness. The saying of

---

[7] These four qualities or cardinal virtues are later discussed as those which lead to the unique happiness of human beings. See Chapter 32.

[8] Jabarūt and Malakūt are two higher worlds or planes associated respectively with the archetypal forms in the mind of God and the planes of the angels.

[9] The states (*aḥwāl*) and stages (*maqāmāt*) refer to temporary or permanent levels of spiritual attainment according the mystics.

the divine law that the condition for being held accountable for actions is that they be voluntary, is like the doctor saying that the condition for being harmed by poison or being helped by the antidote is that either of these two enters the esophagus and then descends to the stomach. An indication of what we have said concerning the human soul "swallowing" the spirits of actions is that all the human communities agree concerning the performance of spiritual practices and acts of worship, due to spiritual recognition of the lights of these things through intuitive awareness. They agree as well on restraining sins and forbidden certain actions, since they see the seriousness of all of those things due to intuitive awareness (*wijdān*).

b) The other branch (of the practical force) consists of the resplendent states and stages of a sort not found among the animals such as the love of God and trusting in Him. You should know that the balance of the human temperament accords with what was given by the specific form. This is not effected except through branches of knowledge which only the purest one among them attains and then is imitated by the others, and through a divine law comprehending knowledge about God, the regulations of the arts of civilized life, principles that examine voluntary actions and divide them into five categories: the required, the recommended, the neutral, the reprehensible, and the forbidden,[10] and premises which clarify the stages of mystical progress (*iḥsān*). Therefore it was necessary according to God's wisdom and mercy that He prepare in the Unseen World of His holiness the provision of the intellectual faculty which the purest one among them may attain so that he can receive knowledge from There.[11] The rest of the people will be led by him in the same way that you see a leader among the species of the bees, who organizes the rest of the individuals. If not for this receiving (of Divine Knowledge) which occurs through an intermediary or directly, man's preordained perfection would not be effected.

Just as the reflective person, if he sees that a species of animal

---

[10] These are the five categories of actions, according to the Islamic legal system.

[11] According to Ibn Sīnā (1037) the intelligible forms which the human rational faculty receives are not produced by abstraction from matter but rather emanate directly from the Active Intelligence. The special feature of the Prophetic intellect is that it can immediately grasp universal truth from the Active Intellect. Fazlur Rahman, *Prophecy in Islam* (London: George Allen and Unwin, 1958), 15, 36.

requires grass to live, will be sure that God has arranged for this animal a pasture in which grass abounds, likewise the one who reflects on God's design will be sure that in the Unseen World, there is a group of sciences through which reason can make up for its deficiencies and fulfill the perfection ordained for it.

Among these sciences is the knowledge of the unity of God and His Attributes. It is necessary that this be explained in a manner which human reason comprehends naturally, and not so convoluted as to be only understandable by the exceptional person. The explanation of this gnosis (*maʿrifa*) is what is indicated by the Prophet's saying, "Glory to God," and by his praising Him, for He confirmed about Himself attributes that people know and use among themselves such as life, hearing, seeing, power, will, speech, anger, wrath, sovereignty, and self-sufficiency. He confirmed along with these that "there is nothing like unto Him"[12] in these attributes, for He is living in a way different from our life, seeing in a way different from our seeing, powerful in a way unlike our power, willing with a will unlike ours, speaking but with speech unlike ours, and so on. Then he explained the absence of equivalence by things which are impossibly remote from us. For example that it is said that He knows the number of the raindrops and the grains of sand of the deserts, or the number of leaves on the trees or the number of the breaths of animals, or that He sees the crawling ant in the dark of night and hears what is whispered beneath the blankets in houses whose gates are locked, and so on.

Among these sciences are the science of acts of worship, the science of the supports of civilization (*irtifāqāt*), and the science of disputation. That is, when errors of judgment arise among the baser souls which contradict the truth, how to solve these problems. Among them is the science of recalling to memory the blessings and the days of God,[13] and the science of the events in the intermediary state[14] and the Day of Assembly.[15]

God, may He be Exalted and Blessed, looked in pre-eternity at the human species and at man's capacity which he would bequeath to his descendants in the species. He took into consideration his

---

[12] Qur'ān 42:11.

[13] The afflictions which God sent upon previous nations.

[14] The "Barzakh," between this life and the Day of Resurrection.

[15] The elements of Islamic eschatology dealing with the assembling of souls before God for judgment.

angelic potentiality and the regulation which would encourage it through sciences (*'ulūm*) explained in accordance with his capacity. Then all of these sciences were represented in the Unseen of the Unseen (*ghaib al-ghaib*),[16] delimited and computed, and this representation is what the Ash'arīs refer to as his" psychic speech,"[17] This is something other than (the attributes of Divine) knowledge, will, and power.

When the time for the creation of the angels came, God knew that the best interest of human individuals would not be fulfilled except through noble souls whose relationship to the species of man would be like the relationship of the intellectual faculty in one of us to his lower soul. Thus He created them through the word "Be!"[18] with pure solicitude for human individuals and put into their breasts a shadow of these sciences ('ulūm) which has been delimited and computed in Unseenness of His Unseen, so that they appeared in spiritual form (as angels). They are referred to by His, may He be Exalted and Blessed, speech, "Those who bear the Throne and those who are around it."[19]

When certain of the celestial conjunctions took place marking a change in the empires and religions, God decreed another spiritual existence for these branches of knowledge and they became explained in detail according to what was suitable for these conjunctions.[20] This is indicated in His saying, may He be Exalted, "We made it descend on a blessed night, We were ever a Warner. On that night every matter of wisdom was decided."[21]

Then the wisdom of God awaited the existence of a pure man, prepared for the revelation. He had ordained the exaltedness of his character and the elevation of his station, so that when this man came into existence, He chose him for Himself, singling him out

---

[16] A reference to the World of Images and the Realm of the angels.

[17] "*Kalām nafsī*" which is God's own uncreated speech to Himself from which divine scriptures descend. This refers to the dispute between the Ash'arites and the Mu'tazilites as to whether the Qur'ān was eternal or created. The Ash'arite position is that the Qur'ān as an aspect of God's attribute of speech or His psychic speech, is eternal, but that the individual texts and words are created.

[18] As referred to in the Qur'ān, "He says to it (a thing) 'Be' and it is." Verses 2:117, 3:47, 6:73, etc.

[19] Qur'ān 40:7.

[20] Thus Shāh Walī Allāh explains a relationship between certain periods of time, developments in the world, and the revelation of divine laws. This is referred to in his work *Lamaḥāt*. Lamḥa 59.

[21] Qur'ān 44:2–4.

as a means for fulfilling His will. He revealed His book to him and made obeying him incumbent upon His worshippers. This is God's saying to Moses, may peace be upon him, "I made you for Myself."[22]

The fixing of these branches of knowledge in the Unseen of the Unseen was only required by His providence toward the species. Nothing asked God to emanate the souls of the Highest Council except the preparedness of the species, and nothing beseeches, asking for this particular divine law at the time of the celestial conjunctions,[23] except the conditions of the species, and "God has the conclusive argument."[24]

If it were said, "Wherefore is it incumbent upon man to pray and why is it necessary for him to obey the Prophet, and why are adultery and theft forbidden to him?" Then the answer is, "The former are required of him and the latter forbidden for him for the same reason that livestock must graze upon grass and are prohibited from eating meat; and that the predators must be carnivorous and not graze upon grass; and for the same reason that bees must follow their leader; except for the fact that the animals receive their branches of knowledge inspired by innate instinct and man merits obtaining his knowledge by acquisition and speculation, or through revelation and imitation."

---

[22] Qur'ān 20:41.

[23] Understood as indications of changes in the prevailing material and spiritual circumstances of the time.

[24] Qur'ān 6:149.

CHAPTER 8

Why Religious Obligation Necessitates Requital

You should know that people are requited on the basis of their actions, if good then with good, and if bad then with bad, from four aspects.

1) One of them is the natural requirement of the specific form. Just as when livestock feed on grass and predators eat meat, both remain in good health, and if livestock were to eat meat and the predators, grass, their health would be spoiled—so it is with man. If he performs acts whose spirits are submission to God, purity, magnanimity, and justice;[1] his angelic nature thrives. If he performs acts whose spirits are opposed to these qualities, then his angelic nature is harmed. At the time when he is relieved of the weight of the body, he feels the suitability and incompatibility (of his actions) as one of us would feel the pain of burning.

2) The second of them is from the direction of the Highest Council. Just as one of us has perceptual faculties placed in his brain with which he feels whatever his foot treads on, whether a live coal or a piece of ice, likewise there are angelic attendants with the form of humanity[2] represented in Malakūt. The providence of God brought them into being together with the human species because the human species could not thrive without them, just as one of us would not be sound without the faculties of perception.

Whenever an individual human performs an act which would lead to salvation these angels emit beams of delight and happiness, and whenever he commits a destructive act, beams of repugnance and anger are emitted by them. These beams penetrate the soul of that person causing him joy or gloom, or they may penetrate the souls of certain angels or persons so that inspiration will occur for them to love him and do good to him, or conversely to

---

[1] The four cardinal virtues, according to Shāh Walī Allāh, discussed in detail in Chapter 32.

[2] By the form of humanity (*ṣūrat al-insān*), Shāh Walī Allāh is referring to the ideal prototype or model for the human species.

despise him and cause him evil. This is similar to what you see when one of us steps on a hot coal and his perceptual faculties feel the pain of burning, then beams emerge from them which affect the heart, so that it becomes sorrowful, and the disposition, so that he develops fever. The effect of these angels on us is like the effects of perceptions in our bodies. If one of us expects pain or humiliation, the muscle beneath his shoulder blade will tremble, his color will become yellow, and his body will become weak, and perhaps his carnal desires will cease and his urine will turn red, or perhaps he will urinate or defecate out of intense fear. All of this reflects the influence of the perceptual faculties on the physical nature of man, their impression on it and their domination of it. Likewise instinctive inspirations and natural states shower down on humans and the souls of the lower angels from the angels which are the guardians of human beings. All of the human individuals are in the position of the natural faculties of these angels which are in the position of perceptual faculties for them.[3] Just as these beams descend to what is below, likewise a coloration ascends to the Holy Enclave from the angels. This prepares for the emanation of an attitude which will be known as mercy, pleasure, anger, and cursing; just as placing water near a flame prepares for its being heated up, and premises lead to the conclusion, and the preparation of petitionary prayer leads to the response. The renewal of states[4] is thus accomplished in Jabarūt in this manner, so that anger may occur and then repentance, and mercy, then vengeance.

God, may He be Exalted, has said, "Verily, God does not change the state of a people until they change what is in themselves."[5] The Prophet, may the peace and blessings of God be upon him, informed us in many hadith reports that the angels bring the deeds of men up to God, may He be Exalted, and that God asks them, "In what state did you leave my servants?;"[6] and that the deeds of the day are brought to Him before the deeds of the night.[7]

---

[3] The analogy is being made between angels overseeing and guiding humans, and the perceptual faculties guiding the physical functions of the body.

[4] That is, this alteration or renewal (*tajaddud*) is accomplished in God's attitude to people and nations as a response to their actions which is then transmitted through the angels.

[5] Qur'ān 13:11.

[6] They reply, "We left them while they were praying, and we came to them while they were praying." *Mishkāt*, p. 127, transmitted by Bukhārī Bad' al-khalq 6, Nasā'ī, Muwaṭṭa'.

[7] As in a hadith cited in Muslim Īmān 295.

He also, may the peace and blessings of God be upon him, indicated a sort of mediation of the angels between humans and the light of God set up in the middle of the Holy Enclave.[8]

3) The third aspect (of the requital for acts) are the requirements of the divine law which has been decreed for humans. The astronomer knows that when the stars form one of their celestial aspects, a spirituality mixed from their forces is obtained that is represented in a part of the celestial sphere, then when this is transmitted to the earth by the transmitter of the celestial decrees, that is, the moon, their (humans') thoughts are transformed in accordance with that spiritual force. Similarly the one who has realized God[9] knows that when a certain time comes, which in the divine law is called "the blessed night" on which "every wise matter will be decided,"[10] a spiritual energy is generated in Malakūt (the angelic plane) which is composed of the determinants of the human species combined with the requirements of this time. (At this time) inspirations filter down from there upon the most intelligent of God's creatures and on the souls beneath him in intelligence, through his mediation. Thus the rest of the people are inspired to accept these inspirations, approve of them and to support their protector and forsake their opponent. The lower angels are inspired to do good to the one who obeys them and evil to the one who disobeys them. Therefore there ascends from them a coloration[11] to the Highest Council and the Holy Enclave, and pleasure or rage results there.

4) The fourth (aspect of requital) is that when the Prophet, may the peace and blessings of God be upon him, was sent to people, God intended through sending him grace for them and their drawing nearer to good. He made it incumbent upon them to obey him; and the knowledge which He revealed to him became concretized and represented and it was mixed with the zeal of this Prophet, his prayer, and God's decree that he should triumph and then this was confirmed and realized.

As for requital from the first two aspects,[12] it is an original nature

---

[8] This light refers to the Greatest Theophany (*al-tajallī al-aʿẓam*) of God. Shāh Walī Allāh speaks of the Holy Enclave as surrounding this light like rays surrounding a jewel. See *al-Tafhīmāt al-Ilāhiyya*, I Tafhīm 25. *Saṭaʿāt* (Hyderabad: Shāh Walī Allāh Academy, 1964) #28, 46. *Lamaḥāt*, #49–55.

[9] *al-ʿārif billāh*

[10] The Night of Power. Qur'ān 44:3–4.

[11] Coloration (lawn), here in the sense of the emission of a certain influence.

[12] I.e., from the requirements of the specific form and due to the angels.

(*fiṭra*) according to which God created man, and "there is never a change in the original nature created by God."[13] This is nothing other than the essentials of piety and sin and their general axioms, not their ramifications and extremities.

This original nature (*fiṭra*)[14] is that religion (*dīn*) which does not differ according to the differing of eras and all of the prophets agree on it. As He said, Blessed and Exalted, "This, your community, is one community (umma)."[15]

The Prophet said, may the peace and blessings of God be upon him, "The prophets are all step-brothers, sons of the same father by different mothers."[16] Blame and being held accountable is justified according to this (natural or constitutional factor) before the missions of the prophets and after them also.

As for requital from the third aspect (according to the divine law), it changes with the variations of the eras, and it is linked to the sending of messengers and prophets. This is referred to in his saying, may peace and blessings be upon him, "Indeed an allegory for me and the message with which God has sent me is that of a man who came to a people and said, 'O people, I saw the army (of the enemy) with my own eyes and I am a simple warner so save yourselves, save yourselves!' Thus a group of his people obeyed him and set out at nightfall, and they decamped at their leisure and were saved; and a group did not believe him and they stayed in their place, and the army fell upon them in the morning and destroyed and exterminated them." Likewise is (the case of) the one who obeys me and follows me and what I have brought, and he who disobeys me and disbelieves in what I have brought from God."[17]

As for requital from the fourth aspect, it only applies after the sending of the prophets, the clarification of the obscure and the correct propagation of the faith.

"Let he who perished (on that day) perish by a clear proof, and let he who survived, survive by a clear proof."[18]

---

[13] Qur'ān 30:30.

[14] That is, the natural constitution of the entire species.

[15] Qur'ān 21:92.

[16] *Mishkāt*, p. 1225. Bukhārī Anbiyā' 48, Muslim, Abū Dāwūd.

[17] *Mishkāt*, p. 40. Bukhārī Riqāq, 26, I'tiṣām 2, Muslim. This hadith is also cited in the introduction to the present work, p. 17.

[18] Qur'ān 8:42.

CHAPTER 9

The Variation in the Natural Dispositions of People Leading to Differences in their Virtues, Actions, and the Degrees of their Perfection

The basic principle of this is the saying which was reported from the Prophet, may the peace and blessings of God be upon him, "If you hear that some mountain had moved from its place, believe it; but if you hear that a man's nature has changed, don't believe it, for he remains true to his inborn disposition."[1] He also said, "Lo, the children of Adam were created at various ranks and among them are those who were born as believers."[2] He then cited this hadith in its entirety mentioning their ranks in anger and readiness to pay debts, and saying, "People are mines, like mines of gold and silver."[3]

And God, may He be Exalted, said, "Say, everyone acts according to his own manner,"[4] that is, the manner according to which he was constituted.

If you want to discover what God has disclosed to me regarding this topic and made me understand about the meaning of these hadiths, then know that the angelic force is created in man with two aspects. One of them is the aspect appropriate for the Highest Council whose task is becoming colored by the knowledge of the Divine Names and Attributes, the gnosis of the intricacies of Jabarūt, comprehensively learning about an order, and focusing their resolve in seeking to bring this order into existence.[5]

---

[1] *Mishkāt*, p. 33, transmitted by Ibn Ḥanbal VI: 443.

[2] "... live as believers, and die as believers. Some are born unbelievers, live as unbelievers, and die as unbelievers. Some are born believers, live as believers, but die as unbelievers. Some are born unbelievers, live as unbelievers, but die as believers. Thus, some are quick in anger but slow to repent. Some are slow to anger and quick to repent, and so on." Tirmidhī Fitan 26, Ibn Ḥanbal III: 19.

[3] That is, they are of different types and capacities.

[4] Qur'ān 17:84.

[5] The order which leads human beings to their specific perfection and the universal beneficial purpose.

The second is the aspect related to the Lower Council (of angels) whose mode is springing forth at a motivation which showers down on them from above them, neither comprehending it, nor with focused resolve, nor with gnosis and illumination, and they reject the stains of the animalistic.

In a similar manner, the animal force is created with two aspects. One of them is coarse, harsh animosity, such as the attitude of the sturdy stallion who develops due to plentiful nourishment and who is given appropriate care. His body is splendid and powerful, his voice strong, and his courage is great. He possesses effective resolve and great pride, strong anger, jealousy, and abundant lust. He is competitive in domination and appearance, and his heart is brave.

The second is weak, flimsy animality like the attitude of the animal which is gelded and born prematurely, brought up in a drought and given inappropriate care. His body is humble and weak, his voice feeble, his courage weak, and his heart cowardly. He is without resolve, and lacks competitiveness in domination and appearance.

Each of the two forces has an innate disposition which specifies one of the two aspects, and acquired experience (*kasb*) reinforces, strengthens, and aids this. The combining of the two forces in people also occurs in two (different) ways.

1) Sometimes they combine in competition with one another (*tajādhub*) wherein each one passionately strives to seek its requirements, desiring its ultimate goals, and seeking its innate habits, so that inevitably competition occurs between them. If one predominates, the influences of the other are diminished, and vice versa.

2) At other times they come together in (a relationship of) inner integration (*iṣṭilāḥ*) in which the angelic descends from pursuing its absolute standard to what is close to it in terms of rationality, generosity of the lower soul, continence, preferring the general good to his own benefit, concern for the future (life) rather than being content with the present, and loving cleanliness in everything with which he is involved.

The animalistic side correspondingly advances from seeking its absolute standard toward that which not far from the comprehensive outlook (*al-ra'y al-kullī*)[6] and does not conflict with it, and

[6] A term used by Shāh Walī Allāh to indicate a motivation to act for the universal good as opposed to particular, personal interest.

thus the two forces are reconciled and a temperament is obtained which is free from conflicts.

Each of the classes, the angelic, and the animalistic; and their mixture, has two sides and a middle,[7] and levels closer to one of the sides or to the middle; so that in this way infinite divisions arise. Nevertheless, there are eight chief categories classifiable by their particular properties, and the other categories can be recognized through recognizing them. These are obtained by (first) dividing what was joined by competition into four (types):

1. A high angelic with a strong animalistic.
2. A high angelic with a weak animalistic.
3. A lower angelic with a strong animalistic.
4. A lower angelic with a weak animalistic.

Combinations through inner integration (*iṣṭilāḥ*) also fall into four sub-categories along these lines. Every category has an invariable determinant, and the person who succeeds in recognizing their determinants will be spared many confusions. We will mention here those which we require in this book.[8]

The people most in need of doing arduous spiritual practices are those whose animal force is strong, especially those whose aspects are in a relationship of mutual contention. The one who is most suited to achieve perfection is the one whose angelic side is of the "high" type, but the one whose two aspects are integrated is the best among them in action and proper deportment. The one whose two aspects are in mutual contention is the greatest among them in knowledge, if he manages to free himself from the bonds of the animalistic, and he does not concern himself greatly with the decorous performance of acts. The one who most consistently avoids challenging matters is the one whose animal side is the weakest, while the one with the high angelic aspect abandons everything to be free to devote his attention to God. The one whose angelic side is of the lower sort, if he is able to detach himself, will devote himself to the next life, and if not he will abandon everything out of laziness and laxity.

---

[7] In other words; higher, lower, and medium levels.

[8] The persons representing these various combinations are discussed in more detail in the author's work, *Hama'āt* (Hyderabad. Sindh: Shāh Walī Allāh Academy, 1964), 96–107.

The one who most energetically plunges into challenging matters is the one with the strongest animal side, while the one (who combines this) with "high angelic" traits is most fit for leadership and the type of things connected with the comprehensive outlook (*al-ra'y al-kullī*). The one with a "lower angelic" component is the one who rushes most readily into battle and who bears heavy burdens.

The person of competing aspects, if he inclines toward the lower, will become preoccupied with worldly affairs alone, and if he progresses toward the higher he will become preoccupied with religious matters only; refining the lower self and purifying it. The man composed of integrated elements works with the higher and lower together and desires both simultaneously. The one whose angelic side is the highest of all is the one sent for the leadership of religion and the world simultaneously. He subsists[9] through what God wills and works on His behalf in all of the universal order, such as in the Caliphate and the leadership of the religion. These ones are the prophets and their heirs,[10] the prominent ones among the people and their Sultans. The leaders among them and those who should be followed in matters concerned with the religion of Allah are those whose two sides are integrated and whose angelic nature is of the high type. The ones most obedient to them are those whose components are integrated and who have a preponderance of the lower angelic side, for they can apprehend the divine ordinances through their figurative expressions and forms.

The furthest from them are the people of competing sides because either they are engrossed in the turbidities of nature, and thus do not carry out the rightly guided practice, or they may overcome it (the animalistic).[11] If they are people with high angelic sides they can grasp the spirits of the divine revelations while they are indulgent with regard to their (formal) expressions. Most of

---

[9] The use of the term "subsist" (*baqā'*) here, has a special reference to the Prophetic consciousness and mode of activity in which the state of *baqā'*—i.e., remaining active in the world while subsisting through the divine, is stressed. This is in contrast to annihilation of self (*fanā'*), generally associated with the function of sainthood.

[10] Heirs in the sense of working for the preservation and propagation of their message.

[11] In *Hama'āt*, pp. 102 ff. Shāh Walī Allāh says that it is those whose two aspects are in a state of competition who have the potential to become prophets, zealous devotees and heroes, if both of these sides are strong.

their concentrated resolve is directed toward gnostic realization of the subtleties of Jabarūt and becoming colored by its hue. If they fall short of that, then they concentrate on spiritual practices and prayers, and they will be amazed by the scintillations of the angelic in terms of mystical disclosure, insight into what is in people's hearts, the answering of prayers, and this type of thing. They will not grasp the divine ordinances in the core of their being except through ruses which subdue the physical nature and earn the lights.[12]

These are principles which my Lord has bestowed on me. Whoever masters them well will find the states (*aḥwāl*) of the People of God revealed to him, as well as the extent of their perfection and the goal of their symbolic expressions about themselves, and he will be able to interpret the levels of their spiritual itinerary.[13]

"These things are given to us and to the people through God's grace but most of the people are not grateful."[14]

---

[12] In other words, they only come to realize the inner meanings, acquire the lights of acts of obedience, and perform them with eagerness rather than out of habit; once they have subdued their animal side.

[13] Understanding how the aspects of which a person is composed naturally interact and how they may be influenced by spiritual practices is an important aspect of Sufi teaching. The states (*aḥwāl*) of the Sufis involve a succession of spiritual experiences which lead to the permanent acquisition of the stages (*maqāmāt*) of progressive spiritual development. The spiritual path or itinerary (*sulūk*) incorporates the practices and theory of development according to Sufi teaching. See Shāh Walī Allāh, *Alṭāf al-Quds*, Persian with Urdu translation by 'Abd al-Ḥamīd Swātī (Gujrānwalā: Madrasa Nuṣrat al-'Ulūm, 1964). English translation, *The Sacred Knowledge*. G.H. Jalbani and D. Pendelberry. (London: Octagon Press, 1984), for further details on this aspect of his ideas.

[14] Qur'ān 12:38.

CHAPTER 10

The Causes of the Thoughts Which Motivate Actions

Know that the thoughts which persons finds themselves having, on the basis of which they are motivated to act, must necessarily have causes, just as is the way (sunna) of God, may He be exalted, with the rest of temporally originated events. Investigation and experience show that among these causes, and in fact the greatest among them, is the natural disposition of humans according to which they were created, as the Prophet, may the peace and blessings of God be upon him, informed in the hadith which we previously related.[1]

Among them is a person's natural temperament which is liable to change due to the attendant regime of food, drink and so on, Thus the hungry person seeks out food and the thirsty wants water and the lustful desires women. Many a person eats food which increases lust, so that he is attracted to women and his mental dialogue with himself concerns them, and this in turn stimulates him to carry out many actions. Many a man eats strong food whereby his heart is hardened and he is made bold enough to kill, and he becomes quick to anger at things which do not anger others.

If these two types of person were to train their lower souls by means of fasting and getting up to pray at night, or they were to become old, or they were struck by a serious illness, then many of their habits would change and their hearts would become gentle and their souls chaste and continent. Therefore you see the difference between old men and youths, and the Prophet, may the peace and blessings of God be upon him, permitted the aged man to kiss (his wife) while fasting, but did not allow the young man this.[2]

Among these causes are customs and familiar habits, for when someone often associates himself with a thing, the attitudes and

---

[1] This hadith was cited in the previous chapter, "If you hear that some mountain has moved from its place, believe it . . ."

[2] *Mishkāt*, p. 427, Abū Dāwūd II:312 #2387 ṣaum 35, Muwaṭṭa', Ibn Mājah.

shapes associated with that thing become consolidated in the tablet of his soul and many of his thoughts incline to that thing.

Among the causes of certain thoughts arising is that the rational soul at certain times extricates itself from the bonds of the animalistic so that it manages to grasp from the Realm of the Highest Council something which facilitates (its obtaining) an illuminated attitude. Sometimes this attitude falls under the category of intimacy and contentment, and sometimes it would consist of the determination to undertake some (good) action.

Yet another cause of thoughts motivating actions is that certain despicable souls are influenced by the devils and take on certain of their tones, and sometimes these attitudes require thoughts and actions.

You should know that the case of dreams is like that of thoughts except that the soul is totally detached from them, so that their images and forms are represented to it.[3]

Muḥammad ibn Sīrīn[4] said: "There are three sorts of dreams. The speech of the soul (to itself), the devils' frightenings, and good tidings from God."[5]

---

[3] Thus the dreamer usually has no sense of thinking or imagining his dreams, but experiences them represented as if they are happening to him, in contrast to thoughts.

[4] Ibn Sīrīn (d. 728), a famous hadith transmitter and interpreter of dreams.

[5] Bukhārī, Ibn Mājah, Ibn Ḥanbal.

CHAPTER 11

The Adhering of Acts to the Soul and Its Being Held Accountable for Them

God, may He be Exalted, has said, "We have made a sign on the neck of every man and on the Day of Judgment we will bring out a book and present it to him opened. Read your book, and on this day you by yourself are a sufficient account taken against yourself."[1]

The Prophet, may the peace and blessings of God be upon him, said, narrating from his Blessed and Exalted Lord (in explanation of this verse), "Indeed these are your acts which I will count for or against you, then I will requite you for them. The one who finds good should praise God, may He be exalted, and the one who finds something other than that should only blame himself."[2]

He also said, "The lower soul wishes and desires while the sexual organ confirms that or denies it."[3]

You should know that the acts which a person definitely intends and the ethical behaviors which are firmly grounded in him originate at the root of the rational soul and return to it, and then they cling to it and are counted for or against it. As for origination from it, it is due to what you have learned to the effect that there are categories of the angelic and animalistic and their combination, and every category has its own determinant. The predominance of the natural temperament and its being influenced by the angels and the devils and so on are among causes which can only operate according to the way the natural disposition permits so that whatever

---

[1] Qur'ān 17:13–14.

[2] Muslim Birr 55, "*innamā hiya a'mālukum iḥṣīkum lakum.*" Abū Dāwūd, Tirmidhī, Ibn Mājah, Ibn Ḥanbal.

[3] The entire hadith is, "It is predetermined for the son of Adam as to the amount of adultery he should commit. Now, the adultery of the eye is a (lustful) look, the adultery of the tongue is talk; the soul wishes and desires while the sexual organ only confirms that or denies it." *Mishkāt al-Maṣābiḥ*, p. 25. Bukhārī Istidhān 121, Qadar 9, Muslim, Abū Dāwūd, Ibn Ḥanbal.

is consistent will occur. Therefore this goes back to the root of the soul, with or without other means.[4]

Have you not seen how the effeminate man is created, so that even early in life his constitution is delicate. The knowledgeable person is able to deduce from this that when he matures according to his temperament, it is necessary that he will adopt the habits of women, dress himself up with their clothes and adornments, and assume their behaviors. Similarly, the doctor understands that a child, if he grows up according to his temperament, and as long as he doesn't meet with any unforeseen accident, will be either lively and strong, or weak and humble.

As for virtues going back to the rational soul, it is because when some person carries out an action, the habituation of his soul to it increases, and it is easier for this action to originate from the soul, as he will need no deliberation and no taking up a (fresh) motivation. Inevitably the soul will be influenced by this and accept its tone. Unavoidably each one of these similar acts participates in this influence even if it is minute and its role is obscure. This is the import of his speech, may peace and blessings be upon him, "Temptations strike the heart in the way that a reed mat is woven stick by stick. Any heart which is influenced by them becomes marked with a black dot, and any heart which rejects them becomes marked by a white dot. The result is two hearts; one like a white stone, and no calamity or test can harm it 'as long as the heavens and earth endure;'[5] while the other heart becomes black and clouded, like an inverted jug, which neither knows good (*maʿrūf*) or evil (*munkar*) but only what it absorbs of its own desire."[6]

As for acts clinging tenaciously to the soul, this is because initially a soul is created as prime matter, free from everything with which it later becomes tinged. Thereafter it continues to emerge from potentiality into actuality day by day, and every subsequent one of its states is prepared for it, and all of the preparations constitute an ordered sequence, and later states cannot precede the earlier ones. Associated with the attitude of the soul existing now is the decree of everything which preceded, although this may be hidden from the soul due to its preoccupation with what is external

---

[4] "Soul" is used here in the sense of the rational soul which bestows the individual form. The "other means" refers to external events which would affect the temperament.

[5] Qur'ān 11:107–108.

[6] *Mishkāt*, p. 1120. Muslim Īmān 231, Ibn Ḥanbal

to it. Indeed, unless whatever supports the faculty which produces these acts from the soul fades away, as we mentioned in the case of the old man and the sick man,[7] or an attitude from Above assails it, changing its regime, such as the change cited in the case mentioned. As God, may He be Exalted said, "The good deeds will wipe out the evil ones,"[8] and He said, "If you associate others with God, then all your acts will be futile."[9]

As for the soul being held accountable, its secret, according to what I have learned through mystical experience (*dhauq*), is that in the high realm (the World of Images) a form for every human manifests what the higher order has bestowed on him, and that which appeared in the account of the Primordial Covenant is a ramification of this.[10]

When this person comes into existence this form is congruent with him and is merged with him. When he performs a (good) action this form rejoices due to this act with an involuntary natural happiness.

Perhaps the soul will appear in the afterlife such that its actions will be counted for or against it from Above, for example through the reading of the scrolls. Perhaps it will appear with its acts clinging to its limbs, and this would be the (meaning) of the hands and feet "speaking."[11] Therefore, the form of every act clearly expresses its consequence in this world and the next. Sometimes the angels hesitate to represent it so that God, may He be exalted, says, "Write the act as it is."

Al-Ghazzālī said,

> Everything which God, may He be Exalted, decreed from the beginning of the creation of the world until its end is written and confirmed in a created thing which God, may He be Exalted, made. This is sometimes referred to as "the Tablet," and sometimes as the

---

[7] *Mishkāt*, p. 427. Mentioned in the previous chapter, "When they become old or are struck by serious illness, their habits change."

[8] Qur'ān 11:114.

[9] Qur'ān 39:65.

[10] The Primordial Covenant (*Mīthāq*) refers to the occasion in pre-eternity when all of the human souls to ever come into being acknowledged Allāh as their Lord. Thus, Shāh Walī Allāh associates the pre-existing form of the soul in the World of Images with this covenant mentioned in Qur'ān 7:172.

[11] These are all elements of Islamic eschatology. The hands and feet speaking (i.e. bearing witness) is referred to in Qur'ān 24:24. The reading of the scrolls is mentioned in Qur'ān 81:10 "When the scrolls are laid open."

> "Clear Book," and sometimes as the "Clear Register" as recorded in the Qur'ān. Everything which has happened or will happen in the world is written on the Tablet and engraved on it with an engraving not witnessed by this eye. Don't suppose that this Tablet is made of wood, or iron, or bone; or that the book is made of paper or leaves. Rather it must be definitely understood that the Tablet of God, may He be exalted, is not similar to the tablet of men, nor is the Book of God, may He be exalted, like the book of men, just as His essence and attributes are not like the essence and attributes of men. Rather if you are seeking for an example which will bring this closer to your comprehension, then take the fixing of the determined things on the preserved Tablet as analogous to the words and letter of the Qur'ān in the brain and the heart of the one who has memorized it. These words are written on his heart as if he were reading them off, looking at them. If you were to examine his brain piece by piece you would not find a single letter of this script. In this fashion you must understand the engraving of the Preserved Tablet with everything which God, may He be Exalted, has decreed and determined.[12]

Frequently the soul remembers the good or evil which it did, and anticipates requital for this, and this becomes another aspect strengthening a person's action, and God knows better.

---

[12] *Iḥyā' 'Ulūm al-Dīn*, IV (Beirut: Dār al-Ma'rifa, 1982), 504–505. The Tablet (*al-lauḥ*) is referred to in Qur'ān 85:22; the Clear Book (*al-kitāb al-mubīn*) in Qur'ān 5:15, 6:59; the Clear Register (*al-imām al-mubīn*) in Qur'ān, verse 36:12.

CHAPTER 12

The Connection of Actions to Psychological Attitudes

You should know that actions are the outer manifestations of psychological attitudes,[1] and illustrations of them, which provide ways of grasping them. They are associated with them according to natural convention. That is, the majority of people agree that actions express psychological attitudes due to a natural cause which is provided by the specific form. This is because when motivation seeks to produce an act and the soul complies, it (the soul) becomes expanded and joyful, and if it refrains from doing it, it becomes contracted and depressed.

When this act is carried out, its origin, whether angelic or animalistic, becomes independent and stronger; while its opposite is deflected and becomes weaker. This is what is referred to in the Prophet's, may the peace and blessings of God be upon him, saying, "The lower soul wishes and desires and the sexual organ either confirms that or denies it."[2]

You will not see any person who does not have acts and attitudes which reflect on him and express who he is, and their form is represented in order to disclose him. If a man were to describe someone else as courageous and was asked to illustrate this, he could only explain this by his mighty endeavors. If he were to describe someone else as generous, he would explain this by his freely spending dirhams and dinars. If a man wanted to summon up for himself the form of courage and generosity, he would be forced to imagine these actions, unless it should occur remarkably that "the original nature (*fiṭra*), according to which God, may He be exalted, created people"[3] had changed.

If someone wants to acquire a trait which he doesn't have, there is no way for him to do this except by being in situations where it

---

[1] *Al-hai'āt al-nafsāniyya* are attitudes of the soul (psyche), in that sense translated as "psychological attitudes."

[2] This hadith was cited in the previous chapter.

[3] Qur'ān 30:30.

is most likely to be found, taking upon himself actions associated with it, and keeping in mind the attainments of those strongest in this trait. Therefore acts are the precise matters which are intended at set times and they may be seen, shown, spoken of, and have influence, and they come under ability (*qudra*) and choice (*ikhtiyār*), and it is possible to blame and punish according to them.

Furthermore, souls are not equal in the way their acts and acquired traits (*malakāt*) are accounted.[4]

1) Among them there are powerful souls whose traits are more readily represented than their acts, so that only virtues are directly considered among their perfection; still acts are represented for them because they are the receptacles and forms of the virtues. Thus the accounting of acts is weaker than the accounting of virtues. An analogy to this is the way that symbols of the intended meaning take shape in dreams, such as the example of the seal upon the mouths and the sexual organs.[5]

2) Among them are weak souls whose actions are reckoned to be the same as their basic perfection due to the lack of independence of their psychological attitudes. They are only represented as fading into their acts, thus the acts themselves are accounted for or against them. This group comprises the majority of people and they are greatly in need of having proper timing established,[6] and therefore great care has been given to (setting) actions in the divine ordinances. Moreover, many acts become established with the Highest Council and their approval or disapproval is directed to them immediately without regard to the psychological attitudes from which they emerge.

Thus the performance of righteous actions on their part is analogous to receiving inspiration from the Highest Council in drawing nearer to them, imitating them, and acquiring their lights, and the case of perpetrating evil acts on their part is the opposite of this.

The establishment of actions with the Highest Council has several aspects.

---

[4] "Malaka" is an acquired trait, the Latin "habitus."

[5] Al-Ghazzālī cites this example in *Iḥyā' 'Ulūm al-Dīn* IV:505 and in *Mishkāt al-Anwār* based on Ibn Sīrīn's famous example of a dream interpretation in which the "seal" represents the *mu'adhdhin's* call in Ramaḍān which ends eating and sexual activity for the day.

[6] In performing their ritual obligations.

1) Among them are that they (the Highest Council) are instructed by their Creator that the order of mankind will only flourish through people's executing certain acts and refraining from (other) acts, thus these acts are represented among them and are sent down in the divine laws from There.

2) Another aspect is that as humans have carried out and persisted in acts, these have become transferred to the Highest Council, and the Council's approval or condemnation has been directed to them. As centuries and ages have passed along this line, the forms of actions have become established among them.[7]

In summary, at that time such actions have efficacy like the influence of the effective litanies and incantations passed down from the pious ancestors, by means of their form and quality, and God knows better.

[7] Thus the Highest Council is here depicted as a cumulative reservoir of attitudes to human actions. Centuries later the approved actions are sent down from there as part of the religious laws.

CHAPTER 13

The Causes for Requital

You should know that although there are a multiplicity of causes for requital, they all go back to two basic ones.

1) One of them is that the soul, on account of its angelic force, feels that an act or quality acquired by it is unsuitable for it, and therefore remorse, grief, and pain will become embodied in it. Sometimes this requires the representation of events during sleep or while awake including those causing pain, insult, and threat. Many a soul is disposed to receive inspirations of what is a sin, thus he is addressed by means of the angels appearing to him (to inform him of his bad deeds); just like other knowledge which he is predisposed to have. Concerning this principle we have the indication of His, may He be exalted, saying, "Yes, whoever acquires evil will be surrounded by his sin, and these are the People of the Fire and they will remain in it for ever."[1]

2) The second cause is the attention of the Holy Enclave directed towards human beings. According to the Highest Council attitudes, actions, and virtues are either pleasing or loathsome. Therefore they urgently ask their Lord to bless the people of the former, and to punish the people of the latter. He answers their prayer and their concentrated thought forms (*himam*) surround humans so that the form of satisfaction or reproach filters down upon them just as the rest of knowledge filters down. Thus pain-causing or blessing-bestowing events become embodied (for them) and the Highest Council appear either threatening them or being delighted with them.

Sometimes the soul is affected by their wrath so this presents itself in the form of fainting or illness. Sometimes their concentrated resolve filters down influencing insignificant accidents, such as sudden thoughts, and so on; so that the lower angels or other people are inspired to do good or evil to a person.

---

[1] Qur'ān 2:81.

Sometimes one of the (good or bad) things which he has acquired translates into goodness or harm, and matters which result in being blessed or punished are manifested. Still the unequivocal truth is that God, may He be Exalted, had Divine Providence for people on the day when He created the heavens and the earth which required that human individuals are not neglected nor do they serve no end; and (which required) that He take them to task for what they do. However, due to the subtlety of grasping this we have made the invocation of the angels a sign for it, and God knows better.

This basic point is what has been indicated in His saying, may He be Exalted, "Those who disbelieve and die as unbelievers are cursed by God, may He be Exalted, and by the angels and men altogether, and they are eternally in this state. Punishment will not be reduced for them nor will they be granted a respite."[2]

The two basic principles (for requital)[3] form combinations. Many remarkable forms result from their combinations, according to the predisposition of the soul and the action. However the first principle is stronger with respect to the acts and morals which better the soul or corrupt it, and the purest and strongest souls are most susceptible to this. The second operates more intensely with respect to acts and morals which oppose the universal beneficial purposes and which are incompatible with the soundness of the human order, and the souls most susceptible to this are the weaker and more loathsome ones.

Each one of the two causes has an impediment which may postpone its decree for a time. The first one (the soul's angelic force feeling pain) is impeded by the weakness of the angelic force and the strength of the animalistic, so that the soul seems to be only an animalistic one unresponsive to the paining of its angelic component. Then when that soul divests itself of the animalistic clothing and its support diminishes, and the angelic lights begin to sparkle; it gradually comes to be pained or given pleasure.

The second (the attention of the Holy Enclave) is impeded by a confluence of causes to oppose its decree, until when the (soul's) appointed hour comes, which God, may He be exalted, has decreed,

---

[2] Qur'ān 2:161–162.

[3] The paining of the person's angelic force and the invocations of the angels in response to his actions.

punishment flows out like a flood. This is what is meant by His, may He be Exalted, saying, "Every people has an appointed hour, when their time comes, they cannot delay it by a single hour nor advance it by an hour."[4]

[4] Qur'ān 10:49.

BOOK II

THE SECOND INVESTIGATION: THE MANNER OF REQUITAL DURING LIFE AND AFTER DEATH

CHAPTER 14

Punishment for Acts in This Life

God, may He be Exalted, said, "Those misfortunes which strike you are the ones which your own hands earned, and He forgives many things,"[1] and "If they had acted according to the Torah and the Gospel and what was revealed to them by their Lord, they would have been sustained from above and from beneath their feet."[2]

There is also what God, may He be Exalted, said in the story of the owners of the garden when they denied alms for the poor.[3]

The Prophet, may the peace and blessings of God be upon him, interpreted His, may He be Exalted, sayings, "Whether you show what is in your hearts or conceal it, God, may He be Exalted, will bring you to account for it,"[4] and "Whoever does evil will be requited for it,"[5] explaining,

> These constitute God's punishment of His servant by means of afflicting him with calamity and fever, even to articles which he puts in his shirt sleeve and loses. Then he becomes agitated when he loses them. (This is) so that the servant may emerge from his sins like the red gold ingot emerges from the furnace.[6]

You should know that the angelic force has an emergence after being concealed in the animalistic, and a disengagement after being entangled with it. Sometimes this is due to natural death, for at that time its support from nourishment no longer comes, its material is dissolved without any replacement, and transitory states like hunger, satiety, and anger do not stir the soul, so that the color of the World of the Holy (*'ālam al-quds*)[7] filters down upon it.

---

[1] Qr'ān 42:30.
[2] Qur'ān 5:66.
[3] Their garden then was destroyed. Qur'ān 68:17–33.
[4] Qur'ān 2:284.
[5] Qur'ān 4:123.
[6] *Mishkāt*, p. 325. Tirmidhī IV:289 #4075 Tafsīr Sūra 2, Ibn Ḥanbal VI:218 reported it.
[7] The realm of the angels.

Sometimes (the two forces are disengaged) due to "voluntary" death in which one continues to curb his animal side by spiritual exercises and by continuously focusing attention on the World of the Holy, so that he experiences some glimmering of the angelic.

Everything experiences happiness and expansiveness through actions and attitudes which suit it, and depression and constriction consequent to acts and attitudes which conflict with it. Every pain and pleasure has a bodily shape by which it is embodied, so that the embodiment of the burning mixture (of humors) is goading drives (in the body), and the form of the pain from the heat of yellow bile is worry and disgust and that a person should see fires and flames in his dreams. The form of paining by phlegm is severe cold and that a person will see water and ice in his dreams.

When the angelic force predominates, either while awake or sleeping, embodiments of affection and joy will appear if a person has taken on cleanliness and humility and the other things befitting the angelic. Their opposites are embodied in the form of properties opposing the balance (of the temperament) and events which include insult and threat. Anger appears in the form of a rending beast of prey, and miserliness in the form of a biting snake.

The general rule with regard to the external punishments (of this world) in that they occur within the context of causes. The man who comprehends these causes and is able to conceptualize the order from which they arise, knows definitely that God does not leave the disobedient unpunished in this world as part of His guardianship of that system. Thus when causes are impartial regarding a person's being rewarded or punished, he is rewarded because of his righteous acts or punished because of his profligate acts. If the causes are massed toward causing him pain while he has been righteous, and if their being restrained due to the opposing force of his righteousness is not repugnant, then his acts are used to repel the calamity or to mitigate it. If they are massed to bestow favor on him while he has been a profligate, then his acts are used to cut off his favor and he is like one who has resisted the causes of blessing. Those causes which accord with his acts combine to reinforce them more patently.

Sometimes the decree of the order (of created things in this world) is more compelling than the decree of the acts, so that the immoral person gradually advances while the righteous one apparently experiences hardship. The righteous person takes advantage of the straitened circumstances to defeat his animalistic side, and he understands this and is content, like the one who willingly drinks bitter medicine. This is the meaning of the Prophet's, may the peace and blessings of God be upon him, saying, "The believer is like the supple stalk of the plant, which the wind blows to the ground and then back again upright until his appointed hour comes; and the hypocrite is like the pine tree standing firmly which the wind does not affect until all at once it snaps,"[8] and of his saying, "No Muslim is afflicted by an injury, be it sickness or some such thing, but that God, may He be Exalted, makes his sins drop off (by means of this) just as a tree sheds its leaves."[9]

There is many a land where obedience to the devil prevails and its people have come to resemble animalistic souls so that some of their punishments will be reserved until an appointed time. And this is His, may He be Exalted, saying,

> We did not send any prophet, to a town unless we afflicted its inhabitants with injury and harm so that they would be humbled (before God). Then we changed the evil to good, so that they became prosperous and said, calamity and adversity previously affected our forefathers.[10] We took them by surprise and they did not realize it. If the people of the towns had believed and been pious, we would have opened the blessings of the heaven and earth to them, but they denied the truth and We held them responsible according to what they had earned.[11]

In summary, the way things are in this life can be exemplified by the case of the master, who is not free (due to other obligations) to requite; then when the day of Judgment comes it will seem that he has become free to do this. This is what is indicated by His, may He be Exalted, saying, "We shall have time for you, O you two dependents."[12]

Furthermore, requital is sometimes effected in the soul of the person through the emanation of joy and contentment, or of depression and terror. At other times it occurs bodily, like the onset

[8] *Mishkāt*, p. 323. Bukhārī Riqāq, 26, I'tiṣām 2, Muslim.

[9] *Mishkāt* p. 323, Bukhārī Marḍā 3,16, Muslim, Ibn Mājah, Dārimī, Ibn Ḥanbal.

[10] I.e., they thought that the calamity and ample circumstances had no connection with their actions.

[11] Qur'ān 7:93–97.

[12] Qur'ān 55:31. The "two dependents" are interpreted as mankind and the jinn, whom God will be free to requite at judgment.

of diseases provoked by an attack of sorrow or fear. An example is the Prophet's, may the peace and blessings of God be upon him, falling down unconscious before his prophetic call, when he exposed his nakedness.[13]

Sometimes punishment (in this life) strikes one's property or one's family. At times people, angels, and animals are intuitively motivated to do good or evil to the person, and at other times he is brought nearer to good or bad situations through his own intuitions or changed ideas.

The one who understands what we have just said and puts everything in its proper place will be relieved of many confusions. An example is the problem of the (apparent) conflict among those prophetic reports which say that piety is the cause for greater sustenance from God, may He be Exalted, and sin the cause of deprivation; and those hadiths which indicate that the sinful are more quick to succeed in this world while the people who experience the most tribulation are the virtuous ones and the best, and so on, and God knows better.

---

[13] This refers to an incident which occurred while the Prophet was helping to repair the Ka'ba before he received the call to be a prophet. There is a hadith narrated by Jabīr ibn 'Abdullāh: "When the Ka'ba was being built the Prophet and 'Abbās went to bring stones. Al-'Abbās said to the Prophet, 'Take off your waist sheet and put it on your neck.' (When he did this) he fell to the ground with his eyes open towards the sky and said, 'Give me my waist sheet.' Then he covered himself with it." *Ṣaḥīḥ al-Bukhārī*, translated by Muhammad Muhsin Khan II:381. According to some commentators, the need to wrap his sheet around his neck arose since the skin on his shoulders had been scraped off by carrying the stones. Also, people in Pre-Islamic times used to perform the circumambulations with their bodies uncovered.

CHAPTER 15

The True Nature of Death

You should know that every form, whether mineral, plant, animal or human has a substratum (*maṭiyya*)[1] different from the substratum of the others, and each has a basic perfection, different from the perfection of the others, even though there may be an apparent resemblance.

The elements,[2] when they are divided into little pieces and mix together in varying greater and lesser proportions, form binary mixtures (composed of two elements) such as vapor, dust, smoke, mud, plowed land, charcoal, flame, and sparks; and tertiary combinations such as yeasted mud and water moss; and quaternary compounds similar to what we have mentioned. These things have properties composed of the properties of their constituents, and there are no new properties in them other than that.[3] They are called "things existing in the atmosphere."

Then, the mineral (form) comes and supervenes upon that mixture, making it a substratum for itself. It makes it take on its specific properties, and it preserves the mixture (of this matter).

Then comes the vegetal form; and it takes the body preserved from the mixture as a substratum, and it makes it into a faculty for transforming the parts of the elements and the things existing in the atmosphere into its own composition in order to bring its expected perfection into actuality.

Then comes the animal form, and takes the airy spirit which carries the faculties of nutrition and growth as a substratum, and it effects the control of its limbs through perception and volition in

[1] *Maṭiyya* literally means "riding animal" which refers to the material to which the form attaches in this life.

[2] Fire, air, earth and water.

[3] According to this theory, two types of composite things may exist, chemical (*kīmiyya*) mixtures or those constituted of a blend of elements (*mizājī*) in which the properties of the original elements remain and no new ones are produced, i.e., salt plus water produces salty water. Based on note in Urdu translation by M. Haqqānī, p. 68.

order to accomplish what it desires, and in order to withdraw from what it fears.

Then the human (form) comes and takes the pneuma (*nasama*) which flows through the body as a substratum. It (the human form) pursues the virtues which are the sources of motivations (to good) and aversions (from bad), so that they are embellished and their regulation is improved and it makes them a place to display that which it receives from Above. Although the matter may be ambiguous at first consideration, still careful reflection connects each of the effects to its source and classifies every form detached from its substratum.

Every form needs matter wherein may it inhere and it must be only that matter which is suitable for it. The situation of the form is like that of a human figure molded from wax; it is not possible for the molded form to exist except in the wax. Whoever says that the rational soul particularized for the person completely departs from matter at the time of death, has perpetrated a falsehood. Certainly the rational soul possesses matter both essentially, in the form of the pneuma (nasama), and accidentally, in the form of the earthly body. When a man dies his soul is not harmed by the cessation of the earthly matter and it remains lodged in the matter of the pneuma. It would be like the case of a master scribe who loves writing; if his hand is cut off, still his acquired aptitude for writing remains as it was. Similar is the case of the one who is fond of walking if his legs were to be cut off, or the one who hears and sees and then goes deaf or blind.

You should know that there are acts and attitudes which man carries out due to his own inner motivation of his heart, so that if he were left to himself he would be driven to that and would be restrained from what opposes it. Among his acts are those which he performs so as to conform with his fellows or those performed due to an extrinsic accident of hunger, thirst or such things. As long as it has not become an unshakable habit, once the accidental factor vanishes the motivation will cease. Many a one who is devoted to the love of a person or poetry or to some other thing, is forced to conform with his nation in dress and fashion. If he were left alone to make his own choice, and its style changed; he wouldn't feel any qualm in his heart. Many a person loves fashion for its own sake and even if left to his own choice, he would not permit himself to abandon it.

There are among people those who are by nature awakened and who intuitively comprehend the principle which synthesizes the multiplicities. Such a person's heart grasps the cause behind the effects and the character trait behind the deeds. There are also those who are by nature drowsy and who remain more preoccupied with the multiplicity rather than the unity, with the deeds rather than the traits, and with the embodiments rather than the spirits.

You should know that when a man dies his earthly body dissolves while his rational soul remains connected with the pneuma and devoted to its concerns. It discards those things which were necessary for worldly life but which lacked a heartfelt motivation, while those things which it held to at its core continue to subsist in it. At this time the angelic becomes prominent while the animalistic weakens, and there filters down on the soul from Above a certainty about the Holy Enclave and about what is accounted for it (the soul) there. At that time the angelic side will either be hurt or delighted.

You should know that the angelic, while it is submerged in the animalistic and mixed with it, must yield to it to some extent, and be somewhat influenced by it. However, the most harmful thing is that extremely incompatible attitudes should become embodied in it; while the most beneficial thing is that extremely appropriate attitudes should become embodied in it. Among the most unsuitable things are:

1) That a person becomes strongly attached to wealth and family, not being convinced that behind them is a (higher) goal.

2) That he strongly clings to the lower attitudes at their very root, and other such things which consolidate his being on the side opposed to magnanimity (*samāḥa*).

3) His involving himself in impure things and being arrogant before God, neither acknowledging Him and not submitting to Him even for a day, and other such things which consolidate that he is on the side opposed to inner development (*iḥsān*).

4) That he gives insufficient attention to the Holy Enclave in their working for the truth, and he gives insufficient reverence to God's command, the mission of the Prophets, and establishing the order which is pleasing to God, for then he will be afflicted with hatred and curses from above.

Among the things appropriate (for the angelic) are:

1) carrying out of acts attuned to purity and submission to the Creator,
2) the constant remembrance of the state of the angels,
3) beliefs which detach the soul from satisfaction with worldly life,
4) that a person be magnanimous and forbearing,
5) that the invocations of the Highest Council and their special concern for the order pleasing to God will incline favorably to him, and God knows better.

CHAPTER 16

Variation of the States of People in the Intermediary World

You should know that people in the intermediary world[1] are at various levels which are innumerable, but which fall into four broad classes.

1) The first type are those who are awake (in the intermediary state), and these ones are punished or rewarded by these same inappropriate or appropriate things.[2] The situation of this type is what is indicated in His, may He be Exalted, saying, "A soul may say, alas, for how careless I was regarding God, may He be Exalted. For I was among the mockers."[3]

I saw a group of God's people whose souls had become like a big pot filled with still water which the wind did not stir. At midday the light of the sun struck it and it became like a patch of light. That light is either the light of pleasing actions, the light of remembrance (*yād dāsht*),[4] or the light of mercy.[5]

2) Another group are those near in origin to the angels, but these are the people of natural sleep,[6] for they are afflicted by dreaming. Dreaming for us consists of the representation of information stored in the sensus communis.[7] A hint of the waking state prevents us from being submerged in these representations and from forgetting that they are imaginary, but when a person goes to sleep, he does not doubt that these forms are exactly what they appear to be.

Perhaps someone of choleric temperament.[8] will dream that he is

---

[1] *Barzakh* is the world between death and the Last Judgment.

[2] That is, suitable or opposed to their angelic side, see previous chapter.

[3] Qur'ān 39:56.

[4] A Sufi term of Persian origin which refers to keeping the mind concentrated on God at all times.

[5] These saintly persons are awakened to the light which becomes reflected in their souls through performing pious actions and spiritual practices.

[6] The Cairo edition has "natural light," which seems to be an error.

[7] In the Islamic theory of perception, such as that of Ibn Sīnā, the sensus communis was the place where memories of sensory perceptions were stored.

[8] According to the theory of the four humors, one or the other tends to dominate the person's temperament. The choleric person is dominated by yellow bile and therefore

in a dry forest on a summer's day while a hot wind is blowing. Meanwhile, he finds himself suddenly surrounded by fire on all sides, and he tries to flee but finds no place to escape. Then the fire burns him so that he suffers severe pain.

The phlegmatic[9] one dreams of a winter night, and of a cold river and a freezing wind, and waves buffeting his boat so that he tries to flee but finds no way to do so. Then finally he is drowned and suffers great pain.

If you make an investigation among people, you won't find anyone who hasn't experienced for himself the depiction of events linked with pleasant things or painful ones as associated with these representations and with the soul of the dreamer. The one thus afflicted (in the intermediary world) is in a dream except that this is a dream from which one does not wake up until Judgment Day. The one dreaming does not know during his dream that these are not external things and that the pain and pleasures do not exist in the external world. If he does not awake, he will not become aware of this secret, so perhaps it is more correct and precise to call this intermediary world an outer world, rather than a dream.[10] Possibly the savage person will see (in the intermediary world) that a wild beast is mauling him and the miser will see that vipers and scorpions are biting him, and the state of the cessation of higher knowledge will be represented by the two angels who ask, "Who is your Lord, and what is your religion and what do you say about the Prophet, may the peace and blessings of God be upon him?"[11]

3) There is a (third) group of whom both the angelic and animal sides are weak. (After death) they become attached to the lower angels either because of causes of natural disposition which preclude their angelic side from becoming too immersed in the animalistic, submissive to it, or influenced by it; or due to causes which they have acquired such as their maintaining purity through heartfelt motivation, which makes it possible for their souls to receive inspirations and the angelic glimmerings.

For example, sometimes people are created in the masculine form

---

is easy to anger. Hence the imagery which corresponds to the element "fire" suggested here.

[9] One dominated by the humor phlegm, associated with the element "water".

[10] As in the hadith, "Men are asleep, when they die they awaken."

[11] Which occurs in the grave according to Islamic eschatology based on hadith reports, for example, *Mishkāt*, p. 36. Abū Dāwūd IV:239 #4751 Sunna 24, Tirmidhī, Ibn Ḥanbal.

but are effeminate in temperament and incline to feminine attitudes. In childhood the passions of the male do not differ from those of the female, for the important things then are the longing for food or drink and the love of play. Thus such a boy follows what his being characterized as a male decrees and he avoids what this would preclude him from doing, in terms of choosing feminine apparel. Then when he grows up he reverts to his natural disposition and does not care what is said about him. He is free to choose women's dress, and to habituate himself to their ways; and the desire to be a passive homosexual overcomes him. He does as women do, adopting their speech mannerisms and giving himself a female name, so that by this (behavior) he completely leaves the domain of maleness. Likewise is the person who in his earthly life may be occupied with desire for food, drink and sex and other requirements of nature and habit, but he is near in origin to the lower angels and strongly drawn to them, so that when he dies the bonds are broken and he re-turns to his true nature, joining the angels and becoming one of them, being inspired just as they are and working for the same purposes which they do. (This is shown) in the hadith, "I saw Jaʿfar ibn Abī Ṭālib as an angel with wings flying in Paradise with the angels."[12]

Sometimes such people concern themselves (after death) with promoting the word of God and the victory of His party and sometimes they send inspirations of good to human beings. Sometimes one of them may have a very strong longing for a corporeal form arising from the root of his disposition, so that this knocks at the door of the World of Images and a force from it becomes mixed with the airy pneuma and it becomes like the body of light.[13] Another one of them may crave food, and so on, so that he is aided in what he desired thus satisfying his longing. This is referred to in His, may He be Exalted, saying, "Don't think that those who are killed in the path of God, have died; rather they are alive in the presence of their Lord, sustained and happy with what Allah has granted to them of His Grace."[14]

---

[12] *Mishkāt*, p. 1354. Tirmidhī V:320 #3852 Manāqib 29, transmitted a variant (doesn't have "with wings"). Jaʿfar's limbs were cut off at the battle of Muʾta and the wings were thus provided as replacements for them.

[13] The airy pneuma (*al-nasama al-hawāʾiyya*) refers to the subtle part of the pneuma or the lower part of the spirit connecting the bodily and spiritual components of the person.

[14] Qurʾān 3:169–170.

In contrast to them there is a group who by natural disposition are close to the devils such that their temperaments are corrupt, requiring ideas which contradict the truth, and which are incompatible with the comprehensive outlook and remote from good moral standards. They envelop themselves in vile qualities and corrupt thoughts and follow the whisperings of the devils, and (God's) curse surrounds them. Thus, when they die they join the devils and are clothed in robes of darkness and certain vile pleasures whereby they satisfy some of their desire are represented for them.

The first ones are rewarded by the joyfulness which arises in their souls and the second are punished by depression and sadness, just as the effeminate man knows that being effeminate is the worst condition for a male, but is not able to renounce it.

4) A (fourth) type are those of the inner integration[15] (*iṣṭilāḥ*) of a strong animal force and a weak angelic force. Most people are of this type and most of their actions predominantly follow an animalistic pattern which predispose them to being occupied and absorbed in the body. Death is not the complete disengagement of their souls from their bodies, rather the soul separates from managing but not from imagining. Due to this the soul thinks, in so far as no conflicting possibility comes to mind, that it is the body itself, to the point that if the body is crushed or cut the soul is sure that this happened to it. The distinguishing mark of these people is that they deeply believe that their spirits are the same as their bodies or are accidents contingent on them, even if they give lip-service to a traditional explanation or a conventional one which is different than that.

When these ones die a weak glimmer shines upon them and a faint vision appears to them like the one which appears to people who do spiritual exercises in this world. Sometimes things are given shape in imaginary forms and other times in externalized[16] forms from the World of Images, such as may appear to those who do spiritual practices.

If they were associated with angelic actions, then knowledge of the fortunate news will be imparted in the shapes of beautiful angels bearing silk in their hands, and through gracious words and attitudes, and heaven's door will open allowing the fragrances of paradise to waft outward.[17]

---

[15] See Chapter 9.
[16] The Cairo text is missing the word "*khārijiyya*."
[17] As reported in the hadith, for example, Tirmidhī Janā'iz 70. *Mishkāt*, p. 36.

If they enveloped themselves with acts opposed to the angelic or which disposed them to be cursed, the knowledge of this will be imparted in the shapes of black-faced angels with rough words and attitudes, just as anger may be conveyed by the image of a predatory animal and cowardice in the form of a hare.

There (in the world of *barzakh*) there are angelic souls whose capacity causes them to be put in charge of these sorts of fellow inhabitants of that place, and they are commanded to punish them or to make their experience pleasant. The afflicted people see them with their own eyes even if the people of this world do not.

Be assured that the world of the grave is nothing other than the remnant of this world, but there this knowledge filter down from behind a veil. The determinants (*aḥkām*) of each soul are manifest to each individual on his own there, unlike the events of the Day of Assembly.[18] The events of the Day of Assembly appear to the souls once they have ceased to exist through the determinants particularized for separate individuals while they remain subsisting through the form of the (whole) human species, and God knows better.

---

[18] The assembly (*ḥashr*), when the souls will be resurrected and gathered before God to receive the final judgment.

CHAPTER 17

Concerning Some of the Secrets of the Events of the Day of Assembly

You should know that the human[1] spirits have a place of gathering to which they are drawn with an attraction like that of iron to a magnet. This is the Holy Enclave (*ḥaẓīra al-quds*), the location where the souls who are free from their bodily garb gather with the highest spirit (*al-rūḥ al-aʿẓam*)[2] which the Prophet, may the peace and blessings of God be upon him, has described as having many faces, tongues, and languages. Indeed this is an embodiment of the form of the human species in the World of Images or the Dhikr,[3] whichever you prefer to call it. This is the place of its (the human spirit's) annihilation (*fanāʾ*) from those of its firm determinants arising from individualistic properties. It is the place of its subsistence (*baqāʾ*) through determinants arising from the human species, or where aspects of the species become predominant.

To elaborate, human individuals have determinants which distinguish certain ones from others as well as traits shared by all which are identical in all of them. Inevitably these (common things) are characteristics of the species as indicated in the Prophet's, may the peace and blessings of God be upon him, saying, "Every child is created according to the original nature (*fiṭra*)."[4]

Every species is particularized by two sorts of determinants. The first are the external, i.e., the physiognomy such as color, shape, size, and voice. That is, an individual will be found to have a form given by the species unless he is premature due to some recalcitrance of the material component. He must thus necessarily take

---

[1] Cairo text is missing the word "*bashariyya*."

[2] This association of the highest spirit with the ideal form of the human species or the cosmic person (*al-insān al-kulliyya*) is present in Sufi emanationist theory following Ibn ʿArabī's school.

[3] The term "*dhikr*" refers to the "Preserved Tablet," a symbol derived from the Qur'ān 85:22 and in Sufi terminology equivalent to the World of Images.

[4] *Mishkāt*, p. 26. Bukhārī Janā'iz 2, 80. Tafsīr 30,1; Qadar 3, Muslim, Ibn Ḥanbal.

form according to it (the specific form) and be in conformity with it. Humans have an erect stance, speech, and have bare skin; while horses are curved in stature, neigh, and are covered with hair, and so on with more things which are never lacking in individuals when their temperaments are sound. The second type of determinants are the internal determinants such as comprehension, finding sustenance, and the capacity to deal with situations which may happen to them.

Thus every species has a certain divine law (sharīʿa). Haven't you seen how the bee is inspired by God, may He be Exalted, to track the plants and eat from their blossoms, then how to make a home in which all the members of its species may gather, and then how to collect the honey there.[5] He inspired the sparrow instinctively that the male should desire the female, and then to build a nest, and that the two should incubate the egg, and then feed the young birds with their beaks. When the young birds begin to fly, He teaches them where to find water and where there are seeds and teaches them how to differentiate enemies and friends. He teaches them how to flee from the cat and the fowler, and how to compete with other members of their kind to effect benefit or ward off harm. Doesn't sound nature provide the idea that these determinants go back to the requirement (*iqtiḍā'*) of the specific form?

You should know that the happiness of individuals is plentiful and perfect if the determinants of the species are firmly established in them and their matter does not resist them. Therefore individuals of the species differ in terms of the happiness or distress reckoned for that species. As long as individuals stick to the pattern defined by their species, they experience no pain. However, their nature may become altered due to some incidental causes such as a tumor. This is what is referred to by the Prophet, may the peace and blessings of God be upon him, saying, "Then his parents make him a Jew or a Christian or a Magian."[6]

You should know that the human spirits are attracted to this place of gathering,[7] sometimes because of insight and (spiritual) resolve and sometimes due to the embodiment of its influences in

---

[5] Cf. Qur'ān 16:68.

[6] This is the conclusion of the hadith previously cited, "Every child is created according to an original nature (*fiṭra*) but then his parents make a Jew, a Christian, or a Magian." *Mishkāt*, p. 26. Transmitted by Bukhārī and Muslim. The import is that surrounding circumstances affect the original nature of the person.

[7] The Holy Enclave.

them, giving paining or comfort. As for attraction due to insight, there is no one who manages to divest himself of some of the animalistic contaminations but that his soul makes a link with it (the Holy Enclave) and something from there is disclosed to it (the soul). This is the import of the Prophet's saying, may the peace and blessings of God be upon him, "Adam and Moses met in the presence of their Lord,"[8] and it has been reported from the Prophet, may the peace and blessings of God be upon him, through many lines of transmission that the spirits of the righteous will assemble before the highest spirit (*al-rūḥ al-aʿẓam*).[9]

As for the other types of attraction, be informed that the gathering of the physical bodies and the reinstatement of the spirits to them[10] is not a new life, but is only the completion of the preceding creation, analogous to indigestion due to over-eating. If this (continuity) were not so, then these (spirits) would not be the original ones and would not be punished for what they had done.

You should know that many of the externally materialized things are like dreams in embodying meanings in the forms appropriate for them. An example is when the angels appeared to David, may peace be upon him, in the form of two adversaries and posed the test to him; and he realized that this symbolized his excesses (in desiring to marry) the wife of Uriah and he asked for forgiveness and turned repentantly to God, may He be Exalted.[11] When two cups of wine and milk were offered to the Prophet, may the peace and blessings of God be upon him, and he chose the milk, this signified the offering of either the original nature (*fiṭra*)[12] or carnal appetites to his people, and the choosing of the sound original nature on the part of the rightly guided among them.

---

[8] The Cairo/Ṣābiq text has "*ijtamaʿa*" which may be a variant of the hadith with "*iḥtajja*". Bukhārī Anbiyā' 31, Tauḥīd 37, Muslim, Abū Dāwūd, Tirmidhī, Ibn Mājah, Ibn Ḥanbal. *Mishkāt*, p. 23.

[9] This refers to Shāh Walī Allāh's theory that the highest human spirits unite with the angels to influence the course of the world through human destiny.

[10] At the Day of Assembly before the Last Judgment.

[11] All of these details do not appear in the Qur'ānic account of David. Shāh Walī Allāh refers to this aspect of David's story in his *Ta'wīl al-Aḥādīth* ed. G.M. al-Qāsimī (Hyderabad. Sindh: Shāh Walī Allāh Academy, 1966). English translations by G.H. Jalbani, (Hyderabad, Sindh: 1972) and J.M.S. Baljon, (abridged) *A Mystical Interpretation of Prophetic Tales by an Indian Muslim: Shāh Walī Allāh of Delhi's Ta'wīl al-Aḥādīth*. (Leiden: E.J. Brill. 1973), pp. 45–46. The allegory of the two angels appears in Qur'ān 38:22–25.

[12] That is, Islam. This hadith may be found in *Mishkāt*, p. 1224. Bukhārī Tafsīr 1, 3; Ashraba 1, 12, Nasā'ī, Dārimī, Ibn Ḥanbal.

Likewise when the Prophet, may the peace and blessings of God be upon him, was sitting with Abū Bakr and ʿUmar at the bench beside the well and ʿUthmān sat separately from them, this signified what God, may He be Exalted, has decreed concerning the arrangement of their graves and their burial according to what Saʿīd ibn al-Musayyab interpreted,[13] and let it suffice to mention it; and most of the events of the Day of Assembly are of this nature.

You should know that the attachment of the rational soul to the pneuma is very firm in most people. With respect to those sciences (of the higher worlds) which are very unfamiliar to it, the rational soul can be compared to the person born blind who does not represent to himself colors or lights at all. The soul has no desire to obtain that knowledge except after many years and extended stretches of time spent among embodiments and representations.

When the souls are first resurrected they are requited with an easy or difficult accounting, or by passing over the Bridge in safety or being lacerated (by it); or by each one following his leader and reaching salvation, or being destroyed; or by the speaking of the hands and feet; or by the reading of the scrolls; or by the manifestation of what one was miserly with, and his having to carry it on his back or be branded with it.[14] In summary, the forms and embodiments of what happens to the souls accord with what the properties of the specific form define for them.

Whichever man has a stronger lower soul or a more extensive pneuma will experience the representations related to the Day of Assembly more completely and abundantly. Therefore the Prophet, may the peace and blessings of God be upon him, related that the major part of the punishment of his community would occur in their graves. There,[15] the things represented for the souls would be equivalences in their sight, such as the guidance extended in the mission of the Prophet, may the peace and blessings of God be upon him, being represented by a cistern;[16] and their acts which will be counted for or against them being represented by the scales

---

[13] As in the hadith narrated in Bukhārī. Faḍā'il al-Ṣaḥāba 5, Khan trans. V:19.

[14] These are all elements of Islamic eschatology; the weighing of acts, the passing over the Bridge which narrows according to the extent of the person's sinfulness, religious communities following their leaders, the hands and feet attesting to actions, people reading scrolls listing their deeds, etc.

[15] In the intermediary state.

[16] The cistern of Kauthar, from which the believers will drink on Judgment Day.

and so on; and blessing being represented by delicious and wholesome food, tasty drink, delightful maidens, fine garments, and a magnificent dwelling.

The going out from the turbidities of delirium to the blessing (of God) is through amazing stages as the Prophet, may the peace and blessings of God be upon him and his family, explained in the hadith about the man who was the last man to come out of the fires of Hell.[17] The souls automatically have some identical desires received from their species through which blessing is represented, and they have other desires besides this by which they are distinguished from one another. These are the saying of the Prophet, may the peace and blessings of God be upon him:

I entered Paradise and saw a young girl, tanned and with dark red lips, and I asked, "Who is that, O Gabriel?" He answered, "God knew the desire of Ja'far ibn Abī Ṭālib for tanned and red-lipped maidens and created her for him."[18]

And his saying,

> If God makes you enter the garden; you have only to wish to ride in it on a horse made from a red ruby, which will fly around with you wherever you want, and you will get this.[19]

And his saying,

> A man among the people of Paradise will ask his Lord for permission to plant crops. God, may He be Exalted, will say to him, 'Haven't you received all that you wanted?' He answered, 'Yes, but I would like to grow plants.' So he sowed seeds and the plants poked up tops and grew erect and ripened and instantly were like mountains. And God, may He be Exalted, said, 'Here you are, O son of Adam; indeed, nothing satisfies you.'"[20]

---

[17] *Mishkāt*, p. 1186–1187. Bukhārī Riqāq 51, Tauḥīd 24, Tirmidhī, Ibn Mājah, Ibn Ḥanbal. According to this hadith the last of the inhabitants of Hell will enter Paradise after passing through successive stages, having requested God to bring him gradually closer and closer.

[18] *Kanz al-'Ummāl* V:153. From Ja'far ibn Aḥmad in *Faḍā'il Ja'far ibn Abī Ṭālib*, and in *Ta'rīkh Rāfiī* from 'Abd Allāh ibn Ja'far.

[19] *Mishkāt*, p. 1202. Tirmidhī Ṣifāt al-janna 11, vol IV p. 87 #2666, Ibn Ḥanbal V:352.

[20] *Mishkāt*, p. 1205. Transmitted by Bukhārī Tauḥīd 38, Harath 20, Ibn Ḥanbal II:511–512.

The ultimate of these is the vision of the Lord of the Worlds and the appearance of the Master of theophanies in the garden of Kathīb.[21] After that there will occur something about which I will not speak, following the example of the lawgiver, may peace and blessings be upon him.

[21] Kathīb refers to hills of musk where some persons will be located at the time of resurrection.

BOOK III

THE THIRD INVESTIGATION: THE SUPPORTS OF CIVILIZATION (THE *IRTIFĀQĀT*)

CHAPTER 18

The Manner of Discovering the Irtifāqāt

You should know that the human being is such that all members of his species accord in their need for food, drink, sexual intercourse, protection from the sun and the rain, seeking warmth in the winter, and so on. It was due to the providence of God, may He be Exalted, for man that He naturally inspired him based on the requirement of his specific form how he could manage to accomplish these needs. Thus inevitably all individuals are equal in this except for the defective one whose substance is resistant.

He revealed to the bee how to eat fruits, how to make a dwelling in which the members of its species assemble, how to follow its leader, and then how to produce honey. He revealed to the sparrow how to forage for nourishing seeds, how to find water, how to flee from the cat and the fowler, how to struggle against the things which prevent it from obtaining its needs, and how the male should mate with the female at the time of mating and then make a nest in the mountains, and how they should share in incubating the eggs and how they should feed the chicks. Likewise every species has a law (*sharīʿa*) infused into the breasts of its individuals by means of the specific form. In like manner He inspired human beings how to accomplish these necessities although He added for them along with these things, three other things which necessitate that his specific form be higher than all others.

1) One of these is being motivated to do something due to a comprehensive outlook (*al-raʾy al-kullī*). Animals are only motivated to a sensorially perceived or an imagined objective due to a motivation arising from their physical nature, such as hunger, thirst, and lust. Sometimes humans devote themselves to a rational benefit that has no motivation in their physical nature, so that they may want to establish a just order in the town, to perfect their character and refine their soul, to be saved from the punishments of the next world, or to win a special place in peoples' hearts.

2) The second is that humans combine aesthetic sensibility (*ẓarāfa*)

with the supports of civilization. Animals only desire a thing to meet their requirements and to fulfill their needs, while a person at times want to be aesthetically and emotionally delighted beyond the basic need, and therefore he seeks a beautiful wife, delicious food, sumptuous clothing, and a magnificent dwelling.

3) The third is that there are found among them people of intelligence and awareness who discover the appropriate supports of civilization (*irtifāqāt*), while there are people who are motivated by the same concerns, but who have been unable to discover a solution. Thus when they see and hear what the wise men have discovered they accept it wholeheartedly, and firmly adhere to what they find to be in agreement with what they know.[1]

Many a person is hungry and thirsty and does not find food or water so that he is afflicted by severe distress until he finds them. Thus he seeks for a mode of development corresponding to this need but does not find a way. Then he chances to meet a wise man who had experienced the same difficulty, and has learned which seeds were edible and discovered how to sow them, water them, reap them, thresh them, winnow them, and store them until the time when they would be required. He has learned to dig wells for those far from springs and rivers and to make large jugs, water skins, and bowls, so that he has made this an inroad into the supports of civilization.

Then a person chewed seeds as they were but they could not be digested by his stomach and he fed on raw vegetables[2] and could not digest them. He wanted to find a solution for this but could not find a way to do it until he met a wise man who had discovered cooking, roasting, grinding, and baking so he adopted this as another inroad into the supports of civilization; and take this as an example for all of his needs. The reflective person has evidence in what we have mentioned of the origination of many beneficial things which previously had not been in existence in the countries. Generations passed in this way and they did not cease doing this until

---

[1] In his later work *al-Budūr al-Bāzigha*, Shāh Walī Allāh terms this third uniquely human trait "*takammul*," an inner urge to self perfection. *al-Budūr al-Bāzigha*. (Arabic text). (Hyderabad. Sindh: Shāh Walī Allāh Academy, 1970. English translations by J. M. S. Baljon. (Lahore: Ashraf, 1988) and G.H. Jalbani (Islamabad: Hijra Council, 1985).

[2] The text has fruits (*fawākih*) but in this context, vegetables seems to be the intended meaning.

a great deal of the revealed sciences supported by acquired experiences were collected and they became attached to them and they lived and died in accordance with them.

In summary, the situation of the necessary inspirations with these three things is like that of the breath. It is basically involuntary like the pulsation of the heart, but the choice of taking small or large breaths remains with the person.

These three things are not found equally in all people due to the variations in their temperaments and in their minds which cause the infusing of a comprehensive outlook (*ra'y kullī*), a love of the aesthetic, and the deduction of the supports of civilized life, and their adopting these as models. Due to their differences in applying themselves to deriving these, and this sort of causes, the supports of civilization (*irtifāqāt*) have two degrees.

The first is one which even the people of primitive societies possess, such as the Bedouin, the people who inhabit the mountain peaks, and the regions far from the sound climates, and this we will term the "First *Irtifāq*." The second is the one of the people of settled populations and the urban centers of the healthy regions where naturally people of superior virtues and wise men have arisen; since there are more groups of people, needs are more complex, and the types of experiences are greater. Thus abundant numbers of practices were discovered, and each person held to them firmly. The highest aspect of this degree is what the kings who are people of complete luxury put into effect, to whom the wise men from other nations come, so that they adopt sound practices from them, and this is what we shall term the "Second *Irtifāq*."

When the Second *Irtifāq* was completed this required a "Third *Irtifāq*," which is that when social transactions went on among them and avarice, envy, the postponement of paying debts, and denying the rights of one another, entered these relations. Thus there emerged among them disagreements and disputes, and there arose among them those who were overpowered by these destructive desires or naturally inclined to boldness in killing and plundering. They had supports of civilization of collective benefit but no one of them (alone) was able to carry them all out, or it was not easy for him, or he did not have the energy to do it. Thus people were forced to set up a king to judge among them with justice, restrain the rebellious one, oppose the insolent one, and collect from them taxes to spend as they should be spent.

The Third *Irtifāq* required a Fourth *Irtifāq*. This occurred because when each king was singled out in his city and wealth was collected for him, and champions gathered around him, greed, covetousness, resentment befell them (the kings) and they disputed among themselves, and battled one another, so they were forced to set up the caliph or to obey the one who had the authority of the great deputyship (*khilāfat*). I mean by the Caliph one who has obtained so much power that it is seen to be impossible that another man could dispossess him of his kingdom except after many gatherings and the spending of much wealth—an occurrence which during long periods of time only one may find possible.

Due to the variations in people and customs the caliphs differ, and the community of people who have more severe and violent natures is more in need of kings and caliphs than those who are the less rapacious and hateful. We want to inform you about the basic principles of these *Irtifāqāt*, and the contents of their divisions such as the minds of the righteous nations of superior virtues require. They have adopted them as an accepted practice and do not differ anywhere in it, so listen closely to what is recounted to you.

CHAPTER 19

The First Irtifāq *(Stage in the Supports of Civilization)*

A aspect of this is the language that expresses what is in people's minds. This is based on acts, attitudes, and bodies, which come to be associated with some sound through onomatopoeia, causal connection or something else. Then this sound is imitated as it is, and then used to derive forms[1] corresponding to varying meanings. Those things known by sight or arising from intuitive attitudes in the mind are represented by the first type (onomatopoeia), thus a sound is assigned which resembles them. Then languages were expanded through associations due to resemblance, proximity, or by the transferal of some relationship. There are other principles (of the first level) which you will find in certain of our discussions.

Among them are:

Agriculture, planting, digging wells, methods of cooking and making stews, and making pots and water skins.

Domesticating the animals and taming them to make use of their backs, meat, hides, hair, fur, milk, and their young.

(Taking) a dwelling for protection from heat and cold such as caves or huts and so on.

Clothing functioning as a covering which is made from animal skins or the leaves of trees or whatever their hands manufactured.

That a male is guided to select a mate and not share her with any one else, in order to alleviate his lust and perpetuate his descent, and to be assisted by her in his domestic needs and in raising and educating the children. Non-humans only select a female by chance, due to their being from the same litter and maturing together in each other's company, or for other such reasons.

That he is guided to tools without which cultivating, planting, digging wells, and domesticating animals would not be achievable, such as the pickax, bucket, plow, ropes, and so on.

---

[1] "*Ṣīgha*" means forms of speech, tenses, cases, etc.

That he is guided to exchanges and cooperation in certain matters.

That the wisest and strongest will subjugate the others and become the leader and chief by one means or another.

That there will be an accepted practice for settling their quarrels, controlling the oppressor, and warding off whoever wants to attack them.

Necessarily, there must be someone in every nation who discovers the methods for the supports of civilization as they relate to their situation and sets them out for others to follow. There will be among them ones who love beauty, luxury, and repose from whatever aspect; and those who pride themselves on their own courage, magnanimity, eloquence, cleverness, and so on; and ones who want their fame to spread or their rank to be higher.

God, may He be Exalted, blessed His worshippers in His Great Book by inspiring the branches of this first stage of civilization with His knowledge that the imposition of the religious duties in the Qur'ān extends to all types of people, and that only this level of the first stage of civilization encompasses all of them, and God knows better.

CHAPTER 20

The Art of the Proper Manner of Livelihood (The Second Irtifāq)

This is the science (*ḥikma*) which examines the manner of progressing beyond managing the needs previously established to the second degree (of the settled populations and towns). The basic principle in it is that the First *Irtifāq* should be tested by correct experience in every field. Then those attitudes should be chosen which are further from harm and closer to benefit, while those other than this should be abandoned. It should be tested according to those superior virtues to which the people of the most perfect temperaments are predisposed, so that whatever this necessitates and requires will be chosen while everything else will be abandoned. (Finally) it should be tested on the basis of friendly interaction among people and proper association with them, and other goals arising from the universal outlook.

The major issues of livelihood concern the proper manner of eating, drinking, walking, sitting, sleep, travel, defecating, sexual intercourse, clothing, dwellings, cleanliness, decoration, polite conversation, charms against disorders, foreseeing the onslaught of events, banquets celebrating births, marriage, festivals, or the return of a traveler, and other such things; as well as mourning over calamities, visiting the sick and the burial of the dead.

The people of sound temperaments who inhabit the populated lands agree that harmful food should not be eaten, such as an animal which dies of natural causes, rotten food, and animals which have unbalanced or immoderate natures. They like to put food in vessels, set them on tablecloths and other such customs, and to wash the face and hands before eating. They avoid frivolousness and gluttonous behaviors or ones which would cause a grudge to arise among the company. Also they agree not to drink foul-smelling water, nor to drink water by lapping it up or gulping it down.

They agree on preferring cleanliness, and keeping the body, garments, and dwelling free from two impurities: from the smell of defilement and from the dirt which is naturally generated such as the residue left on the tooth cleaning stick, the hair under the arms and in the pubic area, the grime on clothes, and household rubbish.

They also agree that a man should be well-dressed and that his clothes should be orderly and his hair and beard properly combed; and that a woman after marriage should adorn herself with makeup, jewelry, and such things. They agree that nakedness is a disgrace but clothing is a beautification, that showing the genital area is dishonor and the most complete clothing is that which conceals the entire body, and the clothing used to cover the genital area should be separate from that concealing the body.

People also are in agreement about (censuring) prediction through any means, whether by dream interpretation, astrology, omens, fortune telling from the flight of birds, sooth-saying, geomancy, and so on.

Everyone who was created with a sound temperament and good taste will necessarily choose to speak without using coarse or unpleasant words. The composition of his utterances will be firm and faultless, and his manner of speaking will make people listen to him and trust him, and this person sets the standard of eloquence.

In summary, in every topic there are issues collectively agreed on among the people of all countries, even if they are far from each other, and people after them have differed in laying down the principles of proper conduct. The naturalist laid them out according to what is approved by medical science, the astronomer according to the properties of the stars, and the spiritual person according to spiritual development (*iḥsān*), as you will find in detail in their books. Every nation has a style and set of manners which distinguish it, as required by variations in temperaments, habits and so on.

CHAPTER 21

The Management of the Household

This is the science examining the way to preserve the ties existing between the members of the household according to the second degree of the stages of the supports of civilization. In it are four departments: marriage, having children, being the master, and associating with one another. The basis of this is that the need for sexual intercourse necessitated relationship and companionship between man and woman, then affection for the children required their cooperation in raising them, and by nature the woman is the one of them more guided to bring up the children and the less intellectual of the two, the one less able to bear hardships, and the more totally modest and attached to the home. She is more meticulous in doing humble tasks and is the more completely obedient, while the man reasons more soundly, is more concerned with defending his honor, more enthusiastic in throwing himself into difficult tasks and more bent on domination, dispute, and so on. Thus the livelihood of the latter would not be complete without that of the former, and the former needs the latter.

The rivalry among men for women and their sense of jealousy concerning them requires that their relationship is not proper unless the man's exclusive possession of his wife is settled in the presence of witnesses. The desire of the man for the woman and the depending of her honor on her guardian[1] and his (the guardian's) defending her, necessitate the dowry, engagement, and her guardian's remaining disinterested (in her as a wife). If the guardian were allowed to desire the relative put under his protection, this would lead to great harm coming to her, since she would be prevented by him from marrying the one she wanted, and she would not have anyone who would demand marital rights on her behalf despite her great need for this. The offense against the ties of kinship through the quarrels of co-wives, and the like, in addition to

---

[1] Her father or brother, in most cases.

soundness of temperament require that a man should not be (sexually) attracted to the one who bore him, or towards the one who was born from him, or the one who is like a branch of the same tree with respect to him (i.e. his sister).

Modesty in mentioning the need for sexual intercourse required that it should be concealed within the highest aspiration expected of them as if that was the goal for which the couple came together. Politeness in publicizing (the marriage) and setting up the domestic framework as the highest aspiration required holding a banquet to which people are invited and where there is singing and drumming.

In summary, for many reasons, some which we have mentioned and others which we have omitted—being confident of the cleverness of the intelligent ones—marriage, according to the usual form, I mean marriage to those not forbidden due to close relationship; takes place in the presence of people, with the presentation of a dowry and an engagement, with equality in social status, with the guardian refraining from (desiring) the woman, and accompanied by a feast. "Men support the women"[2] and have responsibility for them in matters of livelihood, and women are helpful in the home, raise the children, and are obedient. These are binding practices and matters agreed on by all people, and an original nature according to which God, may He be exalted, created people, and neither the Arabs nor the non-Arabs disagree about this.

The spouses will not expend efforts to cooperate, in such a way that each one of them feels the other partner's harms and benefits as affecting himself or herself, unless they have committed themselves to continuing the marriage. Still there must inevitably be preserved a way to disengage if they do not accede to one another's wishes and come to terms with one another, and this is the most hateful of permitted things.[3] Therefore in the case of divorce one should pay attention to certain stipulations and to a waiting period. Likewise on the death of the husband, (there is a waiting period) out of respect for marriage in the minds of people, in order to render certain rights of continuity, out to fidelity to the time of companionship, and in order to avoid confusion about the lineage.

The need of children for parents, and their natural sympathy for their children require that the training of the children be such as

[2] Qur'ān 4:34.

[3] According to a hadith found in al-Dārimī, Ibn Mājah.

to benefit their original nature. The precedence of the parents over the children is necessary, for even when they grow up their parents' superiority in wisdom and experience continues, along with the requirement of sound morality that kindness should be exchanged for kindness. Honoring parents is a compulsory practice, since they suffered in bringing them up to an extent that needs no elaboration.

The variations in the capacities of human beings requires that among them are masters by nature, who are the most intelligent, independent in their livelihood, and who by nature possess strategy and luxury; and there are servants by nature who are foolish and subordinates who do whatever they are ordered to do.

The livelihood of one can only be achieved through the other, and cooperation in both the pleasant and the disagreeable can only occur if they reconcile themselves to continue this relationship. Now, other circumstances may have required that some of them became captives of others, thus when this took place, ownership (of slaves) became organized. Necessarily there must be a customary practice that each will hold himself to obey, and will be blamed for not upholding it, and there must continue to be a way to attain freedom, whether through money or without it.

Often it occurs by chance that needs and maladies may afflict a person, due to sickness or calamity, and he faces demands and requirements which he is unable to meet without aid from his fellows. All people are equal in this and therefore they need to establish affection among themselves and to make it last so that aiding the one who asks for help and assisting someone in trouble will be a customary practice among them, which they can be required to do and blamed for not doing.

Needs are at two degrees. One degree is not fulfilled until each considers the hurts and benefits of the other as if they were his own, and this will not happen unless each expends all he is capable of in the friendship of the other, and in spending on him and ensuring his inheritance rights. In sum, circumstances require on both sides that each take the profit with the loss. The people who are the most suitable for this degree are relatives, for their mutual love and companionship is a natural matter.

The other degree is fulfilled by less than this. Thus it is necessary that consolation and empathy for those struck by calamities be accepted practice among all, and that the bond of kinship should be firmer and stronger than anything else.

The majority of the issues in this field concern recognizing the causes requiring getting married and leaving a marriage, the proper way to marry, the qualities of the husband and wife, the responsibilities of the husband in dealing kindly with his wife and protecting the wife from dishonor and vices, the extent of the wife's duties of chastity, obedience to the husband, and exerting her capacity in household matters; the way to resolve cases of spousal mistreatment, the method of divorce, the mourning period of the woman who has lost her husband, the bringing up of the children, honoring the parents, the way of managing slaves and treating them kindly, the way for slaves to serve their masters, the way to manumit slaves, how to behave with relatives and neighbors, how to give charity to the poor of the locality and cooperate in alleviating the calamities which assail them, the proper behavior of the leader of the group and how he should monitor their condition, how to divide the legacy among the heirs, and how to preserve the lineages and the lines of descent.

Therefore you will not find any nation among humanity which does not have convictions about the principles of these topics and which does not strive to establish them, despite their varying religions and the distance between their countries, and God knows better.

CHAPTER 22

The Art of Economic Transactions (Muʿāmalāt)

This is the science examining the manner of establishing exchanges, cooperation, and means of earning, at the second stage of the supports of civilization. The basis of this is that when needs become numerous and excellence is sought in them, and these (needs) are aesthetic and emotional, it becomes impractical for everyone to undertake them. Some found food in excess of their needs but did not find water, and others had extra water but no food. Thus each desired what another had, so exchanging appeared to be the only solution. Thus this exchange arose due to necessity, so that they were forced to agree among themselves that each one would accept to take care of one need, and achieve mastery in it and make efforts to attain proficiency in the use of its tools, and to make it a means for achieving further needs through the medium of exchange, and this became a practice on which they agreed.

Since many people would desire a certain thing or dislike a certain thing, and could not find anyone to trade in it in that case, they were forced to provide a way to commission things and have them prepared in advance and they were forced to agree to set the convention of mineral substances which would last a long time as a means of exchange among them. The most suitable among these were gold and silver due to their small bulk, their homogeneity, their greatly beneficial effect on the human body, and since they could be used for adornment, so these two became the natural currency while other things were given conventional monetary values.

The foundations of the professions are agriculture, herding, and the gathering of permitted goods from land, sea, mines, plants, and animals; and industries such as carpentry, iron-smithing, weaving, and so on, which involve working with natural substances, in order that the desired level of societal development in achieved through them. Then trade also became a profession, and running the affairs of the city became a profession, and then engaging in all the things which people need became professions.

The more people became refined and devoted to pleasure and luxury, the more the branches of the professions became diversified. Each man would specialize in a profession for one of two reasons. Firstly, that it was appropriate to his strengths, such as the brave man finds military raiding appropriate, and the clever man with a good memory finds accounting suitable, and the very strong person would carry heavy burdens and perform difficult tasks. (The second reason) was due to coincidental circumstances, as the art of smithing seems easier for the son or neighbor of the smith than any other profession does to him, and easier for him than for another person. The resident of the seashore takes to catching fish as no other person does, and more than to any other profession. There remain those who are incapable of doing things in a sound way, and they stoop to occupations harmful to the city such as robbery, gambling, and begging.

Exchange is either property for property, which is barter; or property in exchange for usufruct, which is hire and lease. Since the organization of the city could not be accomplished except through good will and love among people, and good will often leads to giving without a return, or delaying payment, therefore gifts and loans evolved. This good will is also only accomplished through sympathy for the poor; and thus charity evolved. Material circumstances required that there be among them fools, competent ones, impoverished ones, and wealthy ones, those who disdained low-status work and those who did not, those whose needs were pressing and those who were free from need. Thus, no one's livelihood could be achieved without the cooperation of another, and there could be no cooperation except through contracts, (setting) conditions, and conventions according to a customary practice. In this way the branches of share-cropping, limited partnership (*muḍārabah*), hire and lease, partnership, and power of attorney came into being. Needs arose which led to borrowing and keeping in trust; then some attempted deception, lying and delaying of payments; so they were forced to institute witnessing, writing up documents, mortgaging, guarantee and bills of exchange.

The more their level of comfort increased, the more the types of cooperation diversified, and no community of people exists which does not practice these transactions and recognize justice from injustice, and God knows better.

CHAPTER 23

The Governing of the City-State (the Third Irtifāq*)*

This is the science examining the manner of maintaining the bond existing among the people of the city.[1] By the city I mean a group living in close proximity to one another who have dealings with one another and who dwell in separate houses. The basic principle of this is that the city is a single individual in the sense of this bond composed of many parts and a collective attitude. Every composite thing may be afflicted by disorders in its material or its form, and it may be struck by a "sickness," that is, a state which is less suitable for it than another with respect to its species; while "health" would refer to a state which would improve and embellish it.

Since the city comprises a great assembly (of people), it is not possible that all their views will all agree on preserving the just practice, and certain of them will not be able to rebuke others unless they are distinguished by rank, for this would lead to extensive fighting and killing. The city's affairs will not be put in order unless the majority of the influential people agree to obey a person who has his own supporters and might, and whoever is more greedy, violent, and prone to killing and anger, needs to be regulated more than others.

Among disorders (which may strike a city) are:

1) That a group of wicked people who have vigor and power may join together to follow their own desires and throw off the just practice, either out of desire for the wealth of people, such as highwaymen; or in order to do harm to people out of hostility, malice or a desire to dominate. In this instance people have to gather forces together and prepare to fight.

2) That an offender may kill, wound or strike a man, or harm his family, by molesting his wife or desiring his daughters or sisters

[1] "*Madīna*", which could also be appropriately taken as "state" rather than city, arising from the usage of the term in the Islamic intellectual traditions influenced by Plato's *Republic* which dealt with the ideal "city-state" of his time.

without any right; or his wealth by open force or secret theft, or that he may impugn his honor by linking him with a despicable matter for which he comes to be blamed or by addressing him rudely.

3) Acts which harm the city secretly such as black magic, surreptitiously administering poison, teaching the people wickedness, turning the subjects against their king, the servant against his master, and the wife against her husband.

4) Immoral habits, which lead to the disregard of the necessary supports of civilization. Examples are homosexuality, lesbianism, and bestiality, for these impede marriage or cast off the sound original nature, as in the case of the man who behaves effeminately, and the woman who acts like a man. They may lead to extensive disputes, such as in the case of many men going to the same woman without her being anyone's exclusive partner, or like addiction to wine.

5) Transactions harmful to the city such as gambling and compound interest, bribery, giving short weights and measures, cheating in commodities, adulterating goods, merchants who corner the market and raise prices, and hoarding crops.

6) Difficult controversies in which all (sides) are drawn in over some specious argument while the real nature of the situation is not uncovered. What is required in this case is adherence to clear evidence, oaths, documents, circumstantial evidence, and so on; referring this back to an accepted practice, openly declaring the cause for preferring one side over the other, recognizing the subterfuges of the antagonists, and so forth.

7) That the people of the city revert to nomadic live and suffice with the First *Irtifāq*, or that they move away to other cities or distribute themselves in professions in such a way that it harms the city. For example, most of them become merchants and agriculture declines, or most of them make their living through warfare, and so on. Indeed it is necessary that the agriculturalists be considered like the food, and the manufacturers, traders, and those protecting property be considered like the salt seasoning the food.

8) The spread of ferocious and harmful wild beasts and injurious vermin: efforts must be made to wipe them out.

Under the topic of effective protective measures is the construction of structures from which all benefit, such as walls, forts, fortifications, border garrisons, markets, and bridges. Other protections are digging wells, discovering springs and docking provisioned boats at the shores of the river. Among them are extending friendliness and kindness to the merchants of provisions and advising the people of the city to deal well with outsiders, for this will provide access to their coming often. The farmers should be admonished not to leave any land uncultivated, and the craftsmen to improve their work and to be proficient in it. The people of the city should be encouraged to acquire superior graces such as calligraphy, arithmetic, history, medicine and the proper methods of advancing knowledge.

Among protective measures is being aware of the news of the city, so as to recognize the immoral one from the moral, to know who is in need in order to help him, and to know the best workman in order to seek his services.

In this time there are two principle causes for the deterioration of cities. One of them is the depletion of the treasury, due to the fact that people have made a habit of earning a living by taking from it on the basis that they are soldiers or ulema who have a right to it, or that the kings have customarily given gifts to some persons, such as ascetics or poets; or through some other forms of parasitism. Thus, the main goal among them has become gaining a livelihood without doing any worthwhile work, and one group succeeds another in making unpleasant the lives of others and have become a burden on the city.

The second is the levying of heavy taxes on farmers, merchants and professional people and raising them until this leads to the obedient ones being ruined and destroyed, and the ones with enough strength rejecting this and rebelling.

Certainly the city is improved through collecting small taxes, and establishing defense to the necessary extent. The people of this era should remain aware of this, and God knows better.

CHAPTER 24

The Conduct of Kings

The king must be characterized by praiseworthy virtues, otherwise he becomes a burden upon the city. If he is not courageous and is weak in opposing the ones fighting to overthrow him, his subjects will only regard him with contempt. If he is not forbearing he will nearly destroy them through his strength, and if he is not wise he will not discover the beneficial way to manage things. He must be in full possession of his mental faculties, of mature age, free, male, having his senses of sight, hearing, and speech intact.[1] The people must agree on his nobleness and the nobility of his ancestors, they must have seen praiseworthy feats done by him and his ancestors, and they must know that he will not spare any effort in putting the realm in order.

This is all corroborated by reason and the people of all nations have concurred in this, despite their far-flung locations and variations in their religions, for they perceive that the benefit intended in appointing a king is only achieved through these things. If they see him overlooking any of this then they will think that he is not what is needed and they will despise him, and if they remain silent, they are harboring latent anger.

The king must establish a place of honor for himself in the hearts of his subjects and preserve it, and he must see that his good reputation is perpetuated through appropriate means. The one who seeks a high position must be endowed with superior leadership qualities, such as courage, wisdom, generosity, forgiveness for the wrong-doer, and the desire to promote the common good. He must deal with the people as the hunter deals with the wild beast. For the hunter goes to the forest, then studies the gazelle and contemplates the attitude appropriate for its characteristics and habits and adapts himself to that attitude. Therefore he shows himself to

---

[1] As stipulated in manuals of Islamic political theory regarding rulership such as al-Māwardī's (d. 1058) *Aḥkām al-Ṣulāniyya*.

it from afar, and he focuses his glance on its eyes and ears so that when he knows that it is being watchful he stays in his place as motionless as a rock. When he sees that it is not paying attention he stealthily crawls toward it. Sometimes he charms it with music and throws to it fine tidbits which it craves to eat, as if he were generous by nature and does not wish to ensnare it by means of this. Favors result in the love of the recipient and bonds of love are stronger than bonds of iron. Likewise the man who wishes to be outstanding among people should display the attitude which they like in clothing, speech, and manners. Then he should approach them humbly and should offer them advice and affection in a fashion which is not frivolous; while not displaying any evidence that this is done in order to ensnare them. He teaches them that no one else would be able to equal him in their regard, until he perceives that their hearts have become convinced of his superiority and pre-eminence, that their breasts are filled with loving and respecting him, and that their limbs are pervaded by humility and submissiveness; and he must maintain them in that state, and do nothing to turn them against him. If anything like that should happen he must make amends with kindness and favor, demonstrating that the common good required him to do what he did, and that he is working for them, not against them. In spite of this, the king must compel obedience to himself by taking vengeance on the one who rebels against him. Whenever he notices that some man is capable in war or collecting taxes or management then he should multiply his reward and raise his rank to display his favor to him. Whenever the king perceives treachery, opposition and disobedience from some man, his pay should be lowered, his rank reduced and he should shun him.

The king requires a more comfortable life than that of the people, yet he should not be too strict in assigning them tasks such as trying to cultivate wasteland, or protecting a very remote district, and so on. He should not punish anyone until it is confirmed by men of authority that this person deserves it and that the general good requires this.

The king must be intuitive about human nature so that he knows what is secreted in people's hearts, and he must be sagacious, thinking a thought simultaneously with you as if he saw it and heard it. He must not procrastinate, delaying what must be done until tomorrow, nor must he hesitate if he detects animosity from someone, to destroy his position and weaken his strength, and God knows better.

CHAPTER 25

The Policy towards Assistants

Since the king cannot establish all of these beneficial measures (*maṣāliḥ*) by himself he needs helpers to deal with every need. Among the conditions required of helpers are trustworthiness, ability to carry out orders, obedience to the king, and their good will towards him, publicly and in private. Anyone violating these conditions merits dismissal, so that if the king neglects to dismiss him he betrays the city and harms his own position. He must not take as assistants those whose dismissal would be difficult or who would have some claim over him due to familial relationship, and so on, so that their dismissal would be distasteful.

The king should discriminate among the ones who love him. There are those who love him out of fear or greed so that they seek to draw him to them through ruses, and there are those who love him for himself and perceive his gain as their gain and his harm as their harm; these are the sincere advisors.

Every man has a natural disposition according to which he was formed and a habit to which he is accustomed, and it is not desirable that the king expect more from someone than is the capacity of that person. The assistants may be a protection from the evil of opponents in analogy to the hands of the body which carry the weapons; or the assistants may be managers of the city analogous to the natural faculties of the human being, or they may be the counselors of the king in analogy to the mind and senses of a person. The king must ask every day what news they have, and know whatever improvement or negative development has occurred.

Since the king and his assistants perform a beneficial service to the city it is proper that their livelihood be provided by it. It is necessary that there be a just practice for collecting the *ʿushr* and *kharāj* taxes[1] which does not harm the people but suffices the need.

---

[1] The land taxes levied on various types of terrain. The *kharāj* is a land tax levied on non-Muslim subjects while the *ʿushr* is levied for public assistance on Muslim-held land.

It is not desirable that every person and property be taxed, and the kings of the nations of the East and West have agreed that taxes should be levied on the people of abundant wealth and amassed riches, and on wealth multiplying from productive property such as breeding animals, agriculture, and commercial activity; and if more than this is required, then on the tradesmen.

The king must have a policy toward his army, and the method of management here is like that of the skilled riding master with his horse, in that he knows the types of gaits; whether trotting, running, galloping, and others, and the bad habits such as obstinacy, and so on, and the things which train the horse by making a profound impression such as the goad, the crop, and the whip. Thus he observes the horse and whenever it does what displeases him or stops doing what he wants, he warns it through what its nature obeys, and through this he tames its impetuousness. He doesn't intend by this that its mind should become confused, nor that it should not understand why he strikes it, but rather he wants the image of what he is teaching it to be represented in its mind and fixed in its heart, and that the fear of punishment will become established in its mind. Once the desired action is achieved and the reprehensible one is avoided, he should not abandon the training until he sees that the desired pattern of behavior has become second nature and habitual for it, so that even without the whip, the horse will not be disposed to things which conflict with the desired goal. Likewise it is necessary for the trainer of the army that he know the desired method in taking action and refraining, and the things which will serve as a warning to them, and he must never be neglectful of any of these things

The number of assistants is not limited but is proportional to the needs of the city. Sometimes two assistants will be required for a task while sometimes one assistant will suffice for two tasks. However there are five principal helpers:

1) The judge (*qāḍī*), who must be a free person, male, mature, possessing his mental faculties intact, capable, knowing the practice of transactions and the artifices of the disputants in their litigation. He must be strict and forbearing, combining both at the same time. He must consider two aspects. One of them is knowing the true status of the case whether it is a contract, misdeed, or a matter contested between them. The second (aspect is) what each one of the claimants wants from the other, and which of the two

claims is more sound and preferable. He must also consider the way of examining information, for there may be evidence about which people have no doubt that demands an unequivocal judgment, and there may be evidence which is not so clear, requiring a judgment short of the former one.

2) The second is the commander of the warriors who must know all about war materiel and how to mold champions and heroes. He must recognize the potential usefulness of every man, the manner of preparing and training the army, the deployment of spies, and how to get information about the strategies of the opponents.

3) The third is the manager (chief of police) of the city, who must be experienced, having recognized the means to the good and the bad for the city. He must be both strict and forbearing and be one of those people who does not remain silent when he sees something which displeases him. He must select a deputy from each group who knows about its affairs, so that through him they can be regulated, and who will be held accountable for their situation.

4) The fourth is the revenue collector (*ʿāmil*) who must know the means of taxing wealth and distributing it to the ones entitled to it.

5) The fifth is the chamberlain (*wakīl*) who is responsible for the well-being of the king. For the king, with all of his other duties, is not given the opportunity to pay attention to the proper ordering of his own daily life.

CHAPTER 26

The Fourth Irtifāq *(the Caliphate)*

This is the science investigating the political strategy (*siyāsa*) of the rulers of the cities and their kings, and the way of preserving the bonds existing among the people of the regions. That is, when each king is independently selected in his city, property is collected as taxes for him and brave people gather around him. The fact that their (the kings') temperaments will vary and their capacities are different necessitates that injustice and the abandonment of the rightly guided custom will arise among them. Certain ones will desire to take over another's state and they will envy and fight with one another for petty reasons[1] like the desire for wealth or land, or due to envy or malice. When this spreads and increases among kings they are forced to appoint a Caliph. The Caliph is a person who has an army and equipment which make it clearly impossible for someone else to usurp his domain. This could only be imaginable after a general calamity, great strife, many gatherings, and the expenditure of huge sums of money which people would fall short of attaining, and which ordinarily are impossible.

Once the Caliph is established and the quality of life in his realm improves, and the haughty submit to him and the kings obey him, the blessing of God is accomplished and the land and the people become content. The Caliph must undertake combat to prevent harm befalling them from savage people who plunder their property, imprison their sons, and dishonor their wives.

This is the need which prompted the Children of Israel to say to one of their prophets,[2] "Send us a king so that we may fight in the path of God."[3]

---

[1] *Arā' juz'iyya*—literally, partial outlooks. The opposite of the comprehensive outlook (*al-ra'y al-kullī*).

[2] I.e., Samuel.

[3] Qur'ān 2:246.

In the beginning when persons of lustful or savage conduct turned to evil in their way of life and sowed corruption on the earth; God, may He be Exalted, inspired the prophets, either directly or through an intermediary, that He would take away their power and that He would kill those who could not ultimately be reformed, for they, in relation to the human species, are like a limb afflicted by gangrene. This is the thing indicated in His, may He be Exalted, saying, "Had God not driven back some people by means of others, hermits' cells and churches, oratories and mosques wherein the name of Allah is often mentioned, would have been pulled down,"[4] and His saying, "Fight against them until there is no more persecution."[5] The Caliph can not conceivably fight against tyrannical kings and eradicate their power except by means of monetary resources and raising forces. In order to do this he must know the required principles of each one of the following: war, truce, and imposing land and poll taxes.[6]

First he must well consider what is the goal of the fighting, whether the prevention of wrongdoing, the destruction of savage, foul people whom there is no hope of reforming, crushing people whose evil is of a somewhat lower order by wiping out their power, or crushing a group corrupting the earth through killing or imprisoning the leaders who were organizing them or by confiscating their wealth and land, or by getting the subjects to desist from supporting them. The Caliph should not rush into trying to accomplish a goal greater than his capacity, nor should he wish to acquire wealth through annihilating a righteous body of his supporters. He must win over the hearts of the people and understand the potential benefit in each one, and not depend on any person beyond his capacity. He must elevate and respect the leading class and the resourceful people. In goading them to battle he must both incite and intimidate, but first he must give attention to dividing the union of his opponents, blunting their strength, and striking fear in their hearts, so that under his power they appear to be helpless. When he triumphs in this, he should implement the intention concerning them which he had planned before the war. If

---

[4] Qur'ān 22:40. This is a verse giving permission to those who fight because they were wronged.

[5] Qur'ān 2:193.

[6] The kharāj is the land tax paid by non-Muslim subjects while the jizya is a head or poll tax levied on non-Muslims living in Muslim territories.

he fears that they will rebel another time he should impose on them a grueling land tax and a poll-tax which will extirpate them, and he should destroy their strongholds and render them unable to perform their (rebellious) acts.

Since the Caliph must ensure the soundness of a mixture derived from the combination of extremely incongruous elements, he must be alert and send spies into every quarter and effectively employ perspicacity about human character. When he sees that a faction is forming among his forces he must not delay in forming another such group which will never connive with the first. If he sees that a man solicits the Caliphate he must not hesitate to shield himself from his boldness and he must bring his power to an end and weaken his strength.

He must make obeying his orders and bearing good will towards him accepted practice among them. It is not sufficient that these be merely accepted, but rather there must be external signs of their acceptance to which his subjects should be obliged to adhere, such as their praying for him, acclaiming his glory in great assemblies, and conforming to a style and attitude ordained by the Caliph, such as the convention of having coins stamped with the name of the Caliph in our time, and God knows better.

CHAPTER 27

The Agreement of People on the Principles of the Irtifāqāt

You should know that no city in the inhabited climes and no people of the nations of balanced temperaments and virtuous morals is devoid of the supports of civilization, from the time of Adam, may peace be upon him, until Judgment Day. Their principles are agreed upon by all, century after century, and generation after generation. People will always severely repudiate anyone who disobeys them, and they consider them to be self-evident due to their great repute. Do not be dissuaded from what we have mentioned because of people's variations in the patterns and ramifications of the Irtifāqāt. Thus they agree, for example, on preventing the stench of the dead and concealing their private parts, but they differ in the forms. Some of them choose burial in the earth and some of them cremation. They agree on publicizing marriage and openly distinguishing it from fornication in the presence of witnesses but they differ in the forms. Thus, some choose witnessing, acceptance, and feasts, while others choose drumming and singing, and put on magnificent clothes which are only donned for important banquets. Likewise they agree on the punishment of adulterers and thieves, but differ in that some choose stoning and cutting off a hand while others choose severe beating, painful confinement, and crushing fines.

Don't let yourself be led to reject our view due to its being contradicted by two groups. One of them are the foolish who are like animals; the multitude has no doubt that their character is deficient and that their minds are defective. They can deduce the stupidity of these people from their failure to bind themselves by these restrictions.

The second group are the profligates. If what is in their hearts were carefully examined, it would become apparent that they believe in the *Irtifāqāt*; however, their passions have overcome them and they disobey the laws, testifying to their own profligacy. They fornicate with the daughters and sisters of the people, but if someone committed such acts with their own daughters or sisters, they would nearly explode with rage. They would definitely realize that what happens to other people is the same as what they experience, and that the harm of these things is a violation of the order of the city; it is only that lust has blinded them. Theft, extortion, and so on may be discoursed upon similarly.

It must not be thought that people have agreed on these things for no reason, in the same way that the people of the East and the West may agree on eating a certain type of food; such a view is a blatant error. Rather the original nature decrees that people, with all their varying temperaments, the distance between their countries and the various schools of thought and religions that they follow, can only agree on these things due to a consistency of the original nature derived from the form of the human species, and from commonly occurring needs which are identical for individuals of the species, and due to virtues which the well-being of the species requires in the temperaments of its members.

If a man were brought up in the wilderness far from civilization and did not learn its customs from anyone; still he would necessarily experience hunger, thirst, lust, and the inevitable desire for a woman, and necessarily, if their temperaments were sound, children would be produced. They would join with other families and have social relations with one another, and thus the First *Irtifāq* would be completely organized. Then when these people multiplied, people of superior virtues would inevitably be found among them and circumstances necessitating the emergence of the rest of the supports of civilization would arise, and God knows better.

CHAPTER 28

The Conventions Current Among People

You should know that in relationship to the supports of civilization conventions (*al-rusūm*) are analogous to the heart in the human body. The divine laws primarily and essentially consider these conventions, and they are what is discussed and referred to in the heavenly injunctions. There are causes due to which they arise, such as their being discovered by wise men, and such as the inspiration of God, may He be Exalted, which is sent into the hearts of those supported by the angelic light.[1] There are reasons why these conventions become disseminated among mankind; for example their being the practice of a great king before whom people submit, or they may be an elaboration of what people find in their hearts, so that they accept them on the basis of their inner testimony. There are reasons why people cling firmly to them such as their having experienced a punishment from the unseen world when they were neglected, or the occurrence of evil when they were ignored, or that the people of rightly-guided opinions condemn abandoning them, and so on. The insightful man may be able to verify this according to the vitality of practices or their dying out in many countries, in correspondence to what we have mentioned.

The current practices, if they are true in their basic premise, will preserve the sound supports of civilization and lead individuals to their theoretical and practical perfection. If it were not for these, most people would be like animals. How many men conduct marriage and other transactions in the desired manner, and if they were asked the reason for abiding by these conditions they would not find any answer other than its having been agreed upon by the people. At the very most a person might possess a general knowledge which he can not articulate, let alone being able to give a preliminary explanation of its social function (*irtifāq*). Thus such a person, if he did not follow customary practice, would almost be

[1] The light from the Holy Enclave which supports prophetic missions.

like the animals. However, these customary practices may also come to include invalid elements and people may come to be deceived in this regard. This occurs when a group takes charge who are dominated by limited perspectives rather than (considering) the comprehensive beneficial purposes. Thus they perform predatory acts such as highway robbery and extortion, and lustful acts such as homosexuality and men behaving effeminately women, or they take up harmful professions such as usury and giving short measure and weight. Or they may become accustomed to stylish clothing and feasting which tend towards extravagance and require extreme preoccupation with earning money. Alternatively, they may become involved in amusements which lead to the neglect of their present livelihood and of the next life, such as musical instruments, chess, hunting, acquiring pigeons and so on. They may impose crushing taxes on travelers, or a land tax which ruins their subjects, or they may be miserly and quarrelsome in their dealings amongst themselves. They approve of doing such things to the people, but do not like it done to be done to themselves, and no one can criticize them because of their power and their despotic authority. Then the sinful members of the community arise to follow in their footsteps and help them, making efforts to spread these things. A group of people then comes forth, in whose hearts no strong inclination to righteous acts has been created but who are neither inclined to their opposite, and they are led to adhere to such actions by what they see their leaders doing; and perhaps the righteous procedures may be unknown to them. There remains among the rest of the people a group whose original nature is sound and who do not mix with them, and they conceal their wrath, so that an evil practice becomes established and reinforced (due to their silence).

It is incumbent on people of comprehensive outlooks that they expend efforts in spreading the right, and in suppressing and obstructing the wrong. Sometimes this is only possible through disputations or fights; and all of this is counted among the best of pious actions.

Once a rightly-guided practice is established the people assent to it age after age, and live and die according to it, and their souls and sciences become engrained in it so that they consider it inseparable from the principles, present and absent. No one would want to deviate from it or rebel against it except a person whose soul was loathsome, whose reason was confused, whose lust had gained ascendancy and who was controlled by passion. Then, if he carries out this deviation a testimony to his depravity is secreted in his heart, and a veil is lowered between him and the universal beneficial interest. If his act is completed this veil becomes an illustration of his psychological sickness, and this will be an impediment in his religion.

When this rightly-guided practice is clearly established the Highest Council raise prayers for those who conform to it and against those who oppose it, and satisfaction is established in the Holy Enclave for the person who follows this practice, and wrath upon him opposes it. When the customary practices are of this type, they are counted as being part of "the original nature (*fiṭra*) according to which God, may He be Exalted, created human beings,"[2] and God knows better.

[2] Qur'ān 30:30.

BOOK IV

THE FOURTH INVESTIGATION: HUMAN FELICITY

CHAPTER 29

Concerning the True Nature of Felicity

You should know that the human being has a perfection required by the specific form, and perfection required by the position of the human species, in terms of the proximate and remote genus.[1] The (ultimate) felicity,[2] the lack of which is detrimental to humans, and which people of sound reason surely strive for, is the first of these two perfections.

This is because a person may customarily be praised for attributes which he shares with mineral bodies, such as length and great height, and if "felicity" (*sa'āda*) were these, then the mountains would be more complete in well-being. Or a person may be praised for properties which he shares with the plants, such as growing properly and emerging with beautiful designs and radiant forms, so that if well-being consisted of these, then the red anemone and the rose would have more complete perfection. Or a person may be praised for properties which he shares with the animals like great strength, loudness of voice, being very lustful, eating and drinking plentifully, and having abundant anger and envy; and if this were felicity, then the donkey would have more complete felicity.

Characteristics which are particular to humans, such as cultivated virtues, sound supports for civilization (*irtifāqāt*), elevated crafts, and high position—it immediately appears that these are the ultimate felicity of humans. Thus you see that in every nation of mankind those who have greater discernment and a more sound opinion desire their acquisition; and that they consider things other than

---

[1] The animal is the proximate genus and the inanimate the remote one.

[2] Felicity or happiness—"*sa'āda*"—is a term used in the tradition of philosophical ethics (*akhlāq*). Plato's theory of virtue and Aristotelian ethical theory had a strong influence on the Islamic tradition once these Greek works were translated into Arabic. The use of the term "felicity" or "well-being" here, and elsewhere in this text, should therefore be understood as the ultimate felicity which arises through the acquisition of virtues and the refinement of the soul. Islamic thinkers such as al-Ghazzālī and Shāh Walī Allāh understood this as being consistent with the Qur'ānic concept of human nature and the effect of actions on it in this life and the next.

these not to be praiseworthy attributes. However, the matter has still not been precisely stated, because the basis of these (praiseworthy attributes) is found among individual animals. Thus courage is based on anger, the love of revenge, being steadfast in adversity, and advancing into dangerous situations, and all of these are abundantly found in male beasts. However, it is not called "courage" until after the emanation of the rational soul refines these so that they become guided by the universal beneficial purpose and arise due to a reasoned motivation. Likewise the basis of the crafts is found among animals such as the sparrow, which builds nests. Indeed, many things that animals naturally make, man could not make even with great difficulty. However, on the contrary, the truth is that this felicity is incidental, and the true felicity is the guiding of the animalistic by the rational soul, the compliance of the passions with reason, the rational soul's vanquishing the animalistic, and reason being dominant over the passions. All of the other characteristics are negligible.

You should know that the matters involved in true felicity are of two categories.

1) One category comes under the topic of the emergence of the rational soul's emanation in daily life through the decree of innate disposition (*jibbila*). It is not possible to acquire the desired virtue through this category, but rather sometimes a dipping into these acts may take place due to their attractiveness, especially through a fragmentary concept, as is the case of the deficient person; in contrast to the desired perfection. An example would be the person who wants to acquire courage through provoking anger, conflict, and so on; or the one who wants to acquire eloquence through knowing Arabic poetry and oratory. Virtues (*akhlāq*) only emerge through contact with other members of the species, and the supports of civilization are only devised in response to coincidental needs, and the crafts are only accomplished through tools and material; and all of these things come to an end with the passing away of the life of this world. Thus, if the deficient person dies in this condition, and he was loathsome, he remains devoid of perfection. If the images of these (worldly) connections cling to his soul, then the harm to him is greater than the benefit.

2) The second category has as its spirit the attitude of submission on the part of the animalistic to the angelic force, so that the former comports itself according to the inspiration of the angelic

and is colored by its hue, while the angelic prevents it from taking on the colors of the world, and from having its base imprints stamped on it like the imprinting of the sealing ring on the wax. There is no way for this to occur except that the angelic (force) requires something through its essential nature and sends an inspiration about it to the animalistic, and suggests this to it, so that (in turn) the animalistic obeys it and does not rebel or withhold itself from it. Thus, it (the angelic) keeps on ordering and the animalistic obeying, and things continue in this way until the animalistic force becomes accustomed to this and is trained. Those things which the angelic decrees through its essential nature, and which the animalistic is compelled to do despite its reluctance are of the sort that make the former happy and the latter depressed, such as the imitation of Malakūt and the striving for Jabarūt, for they are a particularity of the angelic, and the animalistic is at the utmost distance from them. Or (this continues until) whatever the animalistic decrees, takes pleasure in, and craves for in its excesses, is abandoned.

This second category is called acts of worship and religious exercises,[3] and these participate in the acquisition of the desired virtues which are missing. The investigation of this point goes back to the fact that true felicity is not ensnared except through acts of worship. Therefore the universal beneficial purpose calls to human individuals from the aperture of the specific form[4] and insistently orders them to carry out the improvement of the attributes which are a second perfection for it,[5] to the extent which they are needed. It also commands people to make the ultimate goal of their concentrated attention and the target of their sights the refinement of the soul, and its adornment with attitudes which make it resemble what is above it among the Highest Council, and to prepare it for the revelation of the realities of Jabarūt and Malakūt. (It commands them) to make the animalistic submissive to the angelic and obedient to it, and to make it a showcase for the manifestation of its commands.

When human individuals are sound according to their species, and their matter is capable of manifesting the properties of the species, completely and abundantly, they yearn for this (ultimate)

[3] Acts of worship (*ʿibādāt*) reinforce the requirements of the angelic, while religious exercises (*riyāḍāt*) train the animalistic.

[4] As mentioned in Chapter 5.

[5] The development of the supports of civilization, great skills, virtues, and so on.

felicity and are attracted to it as iron is attracted to a magnet. This is the innate character (*khalq*) according to which God, may He be Exalted, created man, and the original nature, according to which God, may He be Exalted, constituted human beings. Therefore there has been no nation possessing balanced temperaments among humankind which does not have among it a group of great people who give importance to the perfection of this innate character and consider it to be the ultimate felicity. The kings and wise men, and those beneath them, see them as attaining what is exalted above all the felicities of this world, and as being joined to the angels and affiliated with their circle so that they (these people) seek their blessing and kiss their hands and feet. Thus how can it be that the Arabs and the non-Arabs, despite the differences of their customs and religions and the distances between their cities and countries, agree on one thing with a unity of species—except due to the affinity of the original nature? How could this not be so, once you have recognized that the angelic force is present in the basic human original nature, and once you have recognized the superior and greatest people among them, and God knows better.

CHAPTER 30

The Differences among People in Felicity

You should know that courage and the rest of the virtues are found in varying degrees in human individuals.

Among them there is the person who is totally lacking in it, whose attainment of it is forever hopeless because of the establishment of the contrary form at the base of his nature. Examples are the effeminate one or the weak-hearted person with respect to courage.

Among them is the person deficient in courage, in whom it may be hopefully (be developed) after his practicing actions, sayings, and attitudes appropriate for it, learning such things from courageous people, and remembering the stories of those most perfect in it, i.e. what happened to them in the past, for they remained firm in adversities and confronted dangerous situations.

Among them is the person having the basis of the virtue created in his nature but who is constantly prone to errors, so that if he is ordered to restrain himself from them, this is very oppressive for him and he becomes silent in anger. If he is ordered to do something compatible with his nature it would be like sulfur to which fire is set, for its combustion would not be delayed.

Among them is the person in whom the virtue is created completely and abundantly. He will necessarily rush to accomplish its requirements, and if he is most strongly induced to be cowardly, for example, still he will not accede. Bringing forth acts consistent with this virtue and attitudes suitable to it will be easy for him by nature, without any convention or inducement. This person is the leader in this virtue and needs no teacher at all, and those who are inferior to him in this virtue should follow his example and firmly grasp his ways. They must hold themselves to the imitation of his attitudes and bear in mind his experiences in order that they may bring forth the perfection of this virtue anticipated in them according to that which has been decreed in their case.

Likewise human beings differ with respect to the virtue upon which their felicity depends.

Among them is the person totally deprived of it whose becoming righteous cannot be hoped for, like the one whom Khiḍr[1] killed, who was by nature an unbeliever. This is what is indicated in God's, may He be Exalted, saying, "Blind, deaf and dumb, they will not repent."[2]

Among them is the deficient one in whom this (virtue) may be anticipated after arduous exercises and protracted actions which punish his lower soul. Such a person needs the incitement of the call of the prophets and the customary practices handed down from them, and these ones are the majority of the people. It is for them that the sending of prophets is primarily and essentially intended.

Among them is person in whom the virtue was placed in a general sense but who also makes mistakes. Thus, he must have an exemplar in most cases in order to elaborate and set out the attitudes which are appropriate for the virtue. Concerning these ones God, may He be exalted, says "Whose oil would almost glow forth (of itself) though no fire touched it."[3] These people are called the "Ones who excel."[4]

Among types of people are the prophets who were able without a previous exemplar or inducement to bring forth the perfection of this virtue and choose the attitudes suitable for it and the means for making the one lacking it acquire it, to make it persist in someone in whom it is already present, and to perfect the person who is deficient in it. Thus while following the requirements of their natural disposition, they systematized the codes of behavior, which people bear in mind and take up as a statute. Indeed, when ironworking, trade and their likes are not grasped by most people except through practices transmitted from their forefathers; what is one to think about these noble pursuits to which only those given success by God are guided? Under this topic it is necessary to realize the compelling need for the prophets, and the need to follow their practices and to study their reports (*aḥādīth*), and God knows better.

---

[1] Qur'ān 18:75ff. Khiḍr is alluded to in the Qur'ānic account of Moses where he appears as his mysterious teacher. Part of Moses' training was not to question Khiḍr's actions which include killing a young man. Later it is disclosed that the man was by nature an unbeliever who would have committed very evil actions. This is elaborated in hadith reports, see for example, *Mishkāt*, p. 1222. Bukhārī Anbiyā' 23, Khan trans. IV:405.

[2] Qur'ān 2:18.

[3] Qur'ān 24:35.

[4] "*Al-Sābiqūn*" are "those who excel in good deeds" mentioned in Qur'ān 56:10. In Shāh Walī Allāh's terminology these are people whose angelic side is strongest. See Chapter 69 of the present work.

CHAPTER 31

The Divisions among People in their Manner of Obtaining This Felicity

You should know that this (highest) felicity is obtained in two ways.

1) One of them resembles sloughing off the animalistic nature, and this is done by adhering to stratagems which bring about a suspension of the rules of nature, a quenching of its heat, an extinguishing of the flame of its sciences and conditions. This is effected by turning with complete attention to what lies beyond Jabarūt, by the soul accepting knowledge completely divorced from space and time and pleasure different from the familiar pleasures in every way, until he no longer mixes with other people, nor desires what they desire, nor fears what they fear, for he is in a distant region and a place far removed from them.

This is the felicity which the Theosophs[1] among the philosophers and the *Majdhūbs*[2] among the Sufis crave. A very few of them attain the utmost peak of its summit and the rest remain ardently longing for it with their sights set on it, their glances aspiring to it, holding themselves to the formal imitation of its attitudes.

2) The second (of these ways to obtain felicity) is in the line of reforming the animalistic side and correcting its deviation while retaining the connection to its essential nature. This is effected when one tries to have the animalistic imitate the acts, attitudes, memories, and so on, of the rational soul. For example, as the mute tries to convey the speech of people by his gestures, and the one who is portraying psychological states of fear and shame (imitates) by using visible attitudes associated with these states; and as the woman who has lost a child expresses grief with words and laments which

---

[1] "*Muta'allihūn*"—those philosophers such as the *Ishrāqīs* who follow the illuminationist philosophy of Shihābuddīn Suhrawardī (d. 1191) or the *Wujūdīs* who follow Ibn al-'Arabī's (d. 1240) doctrine of the Unity of Being.

[2] *Majdhūbs* are the Sufis who are uncontrollably "drawn or attracted" toward the Divine presence.

no one can hear without becoming sorrowful and having the image of grief arise in him.

Since the divine management in the world is based on the choice of what is ever more proximate and ever more simple, and the concern is for the betterment of those following the same course as the generality of the individuals of the species rather than the exceptional and extraordinary ones, and the best interests of the two worlds are regulated so as not to disturb the order of anything in them; the grace and mercy of God required that He send the prophets in the first place and essentially to establish the second way; to summon to it and urge people on in it. The first way is indicated through entailed allusions and implied intimations and nothing else, and "God has the convincing argument."[3]

The elaboration of this is that the first way of attaining felicity only arises among people having a strong contention of the two sides (*tajādhub*)[4] and they are few; and through performing severe religious exercises and having a strong detachment (from the world), and very few do this. Their models are a group who neglect their own livelihood and for whom the world does not matter. This first way is only achieved through the advance preparations of a righteous group from the second (path).[5] This group (of mutually conflicting sides) cannot avoid neglecting one of the two felicities; either bettering the supports of civilization in this world or reforming the lower soul for the next life. Thus, if most people took up the first way, the world would go to ruin, and if it were imposed on them it would be the imposition of an impossible thing, since the supports of civilization have become second nature to people.

The models of the second way are the people who are instructed (*mufahhamūn*)[6] and they are the reformers who take on the governing of religion and the world at the same time. Their summons is accepted and their precedents are followed, and encompassed within them is the perfection of those who were equipped with inner integration (*al-muṣṭaliḥīn*) among "the Ones who Excel in good works" (*sābiqūn*), and the People of the Right Side (*aṣḥāb al-yamīn*), and

---

[3] Qur'ān 6:149.

[4] That is, a pulling in opposite directions of the angelic and the animalistic sides.

[5] That is, the prophets preceded the ascetics.

[6] They are instructed by God and the angels, for further details of this concept refer to Chapter 55.

these are the majority of people.[7] The intelligent, the foolish, the busy or the idle person is capable of following this second way, and there is no hardship in this. This way suffices the servant of God in straightening out his lower soul and preventing its deviation, and in protecting him from the dreaded torments of the Next World, since every soul has angelic actions whose presence makes it happy, or by whose absence it is pained. As for the principle of stripping away (the animalistic side), the regenerations of the grave and the resurrection will happen to the soul in such a way that the person does not understand their nature, even after some time. Poem:

> Time will reveal to you that of which you were ignorant,
> and someone will bring you news, whom you had not commissioned to do so.[8]

In sum, comprehending and deeply penetrating the aspects of the good is at the level of something impossible for most people, and simple ignorance is not harmful, and God knows better.

---

[7] According to Shāh Walī Allāh the ones of the right side (*ashāb al-yamīn*), Qur'ān 56:27, are people who receive the forms but not the spirits of things, since both their angelic and animalistic sides are weak. These types of persons are discussed in Chapter 69 of the present work.

[8] Lines by the pre-Islamic poet, Ṭarafah ibn al-ʿAbd, d. c. 564 C.E. Author of a Muʿallaqa of which these lines are the last verses. *Sharḥ Dīwān Ṭarafah ibn al-ʿAbd* (Beirut: Dār Maktaba al-Ḥayāt, 1983), 28.

CHAPTER 32

The Principles to Which the Attainment of the Second Way Goes Back

You should know that there are very many ways to obtain felicity by the second way, but God in His grace made me understand that they go back to four qualities (*khiṣāl*) which the animalistic side takes on when the rational soul dominates it, forcing it to do to what is suitable for it. These are the states of man most similar to the attributes of the Highest Council, and a preparation for becoming associated with them and joining their circle. He made me understand that He sent the prophets to summon and urge on to them, and that the divine laws are an elaboration of this and are based upon it.

1) One of the qualities is purity (*ṭahāra*). The true nature of this is that when a person's original nature is sound, his temperament healthy, and his heart free from the lower states which distract it from contemplation, and then he becomes soiled with impurities and needs to urinate or defecate or has recently had sexual intercourse or what leads up to it; his soul becomes constricted, he is struck by depression and sadness, and he finds himself in a great stupor. Therefore when he unburdens himself of the two most unclean things, and he rubs his body and bathes and puts on his best clothes and perfumes himself, then that constriction is repelled from him and he finds in its place relaxation, happiness, and expansiveness. All of this is not for the purpose of displaying one's image before people or in order to maintain their conventions, but it is rather due solely to the decree of the rational soul.

The first state is called "ritual impurity" (*ḥadath*) and the second, "purity." The wise person, and the one who exhibits the sound properties of the species with the material capacity to submit to the properties of the specific form, knows that the two states are distinct from one another and by nature loves one state and hates the other. The ignorant man, if his animalistic side becomes somewhat weaker and he continues in a state of purity and devotion to piety, and occupies himself with recognizing both of the states; will inevitably recognize them and be able to distinguish the one from the other.

Purity is the attribute of the pneuma which most resembles the states of the Highest Council in their detachment from animalistic impurities and their rejoicing in the light which they possess. Therefore purity prepares the soul to be clothed in its perfection in accordance with the practical faculty.[1]

Ritual impurity, when it gains power over man and surrounds him on all sides, causes in him a propensity to receive the whisperings and visions of the devils by means of the sensus communis (*ḥiss mushtarik*) and in nightmares. It also causes a propensity for the appearance of darkness around him in what is close to the rational soul, and for him to imagine accursed and vile beasts. When purity gains power over a person and he becomes surrounded by it, notices it, and relies on it, it will give rise to a propensity for receiving the inspirations of the angels, seeing them, having good dreams, the manifestation of the lights, the representation of pleasant things, and other splendid and blessed effects.

2) The second virtue is humbling oneself (*ikhbāt*) before God, may He be Exalted. Its true nature is that when a person who is sound (in temperament) and detached (from other concerns) is reminded of the signs and attributes of God, may He be exalted, and applies himself assiduously to recalling these, the rational soul is alerted and the senses and the body submit to it, and become faint and disconcerted. An inclination towards the Holy arises, and his state is like that of the rabble in the court of Kings, who observe their own insignificance and the overwhelming power of those (kings) to withhold or bestow. This is the state of the pneuma nearest and most similar to the state of the Highest Council in their orientation towards their Creator while they are rapt in His majesty and engrossed in proclaiming His holiness. Therefore these states are a preparation for bringing the soul to its intellectual perfection, by which I mean the imprinting of the divine gnosis on the tablet of its mind, and its being joined to that Presence in some way, although the power of expression falls short in describing this.

---

[1] This practical faculty, along with the intellectual one, distinguishes humans from animals. Practical and intellectual perfection in acquiring virtues is discussed in Chapter 33.

3) The third quality is magnanimity (*samāḥa*). Its true nature is the soul's refusal to follow the promptings of the animalistic force, and its not having its imprints etched on it, so that the harmful effect of its pollution does not attach to it. This is because the (lower) soul, when it is preoccupied in the affair of its livelihood, desires women, seeks to amass pleasures, or wants to devour food. It strives to obtain these things until it fulfills its needs for them. Likewise, when it becomes angered or is covetous of something, inevitably it will remain for a time engrossed in this mood, and certainly will not direct its concern at what lies beyond it. Then, when this state ceases, if it (the soul) is magnanimous, it will emerge from these narrow straits as if it had never been in them at all. If it is not magnanimous, then it will become enmeshed in these moods and take on their contours as the imprints of the sealing ring are assumed by the wax.

When the magnanimous soul separates from the body and unburdens itself of the dark accumulated bonds, returning to its original state; it will not find any of those things which opposed the angelic side while in this world and it will achieve intimacy and have a most agreeable existence.

The impressions of greed are represented in the soul, as you see in the case of the person who has been robbed of a precious possession. If he is magnanimous his mind will not be disturbed, but if he has a weak soul, he will become like a madman and his losses will obsess him.

There are many names for magnanimity and its opposite depending on the pertinent conditions. If property is involved these are called "generosity" and "miserliness," and if lustful desire for sex or food, they are called "continence" or "cupidity." If they relate to the desire for luxury and remaining far from hardship, they are called "patience" or "impatience," and when they concern a motive for transgressing what is forbidden in the divine law they are called "moral conscience (*taqwā*)" or "depravity."

When magnanimity takes hold over a person, his soul remains free from worldly desires and is prepared for the higher, transcendent pleasures. Magnanimity is a cast of mind which prevents man from being controlled by anything which is contrary to the desired perfection, intellectually or practically.

4) The fourth quality is justice (*ʿadāla*). This is a habit (*malaka*) acquired in the soul which gives rise to those acts through which

the order of the city and the quarter is established with ease. It seems as if the soul is naturally disposed to these acts. The secret of this is that the angels and the souls divested of bodily attachments[2] are imprinted with what God intended in creating the world, in terms of setting right its order, and so on, and thus what pleases them becomes transformed into what is suitable for this order, and this is the nature of the pure spirit.[3] Then, when the soul separates from its body while something of this quality persists in it, it is delighted to the fullest extent and finds a means for pleasure distinct from more base pleasures. If it separates (from the body) while the opposite of this trait is within it, the state is depressing for it and it is estranged and pained. Thus, when God sends a prophet to establish the religion and "to bring people from the darkness into the light"[4] and to make the people establish justice; whoever strives to spread this light and make a path for it among the people will be shown mercy, while he who tries to deflect and extinguish it will be cursed and reviled. When justice is consolidated in a person, a cooperation is engendered between him and the Bearers of the Throne[5] and those near to the circle of angels who are the medium for the descent of generosity and blessings. This becomes an open door between him and them, and a place for the descent of their colors and influences such as enable the soul to be furnished with intuitions from the angels, and provoked to act according to these.

Once you have verified the true nature of these four qualities and understood the way that they necessitate intellectual and practical perfection, and how they prepare for following the path of the angels; and once you have realized the way that the divine laws branch out for them according to (the exigencies) of every age, then you will be given great good and you will become a person of sound judgment (a *faqīh*) in religion, one of those for whom God intends good.[6] The state composed of these (four qualities) is called the original nature (*fiṭra*). There are causes through which the original nature is fulfilled; some of them are intellectual, and

---

[2] I.e., the Highest Council.

[3] Detached from the body.

[4] An echo of Qur'ān 14:1, 14:5, 57:9, etc.

[5] The highest angels.

[6] An echo of the hadith "*man arāda allāhu bihi khairan yufqihuhu.*" Bukhārī 'Ilm, 10 Khums 7, I'tiṣām 10.

others practical, and there are veils which block man from this and ruses for rending these veils. We wish to alert you to these matters, so listen well to what is related to you, with success granted by God, may He be Exalted; and God knows better.

CHAPTER 33

The Way of Acquiring These Qualities, Perfecting The One Who is Deficient in Them, and Restoring The Missing Ones

You should know that the acquisition of these qualities is regulated in two ways, one intellectual and the other practical.

1) As for the intellectual regulation, it is required because the nature obeys the intellectual faculties and therefore we see the ceasing of passion and lust when something comes to mind which causes the mood of shame or fear to arise in the soul. Thus, when a person's knowledge becomes filled with what is suitable for the original nature, this leads to its becoming established in the soul. This is, that he will believe that he has a Lord transcendent above all human defilements, and that "not an atom's weight escapes His notice on the earth or in the heavens."[1] (He believes that) "there are no three having a secret conversation but that He is the fourth, and no five but that He is the sixth,"[2] that "He does what He wants"[3] and "decrees as He wills,"[4] that no one can deny His decree nor forbid His order, that He bestows the basis of existence and the physical and psychological blessings which are derived from it, and that He rewards him for his actions, good with good and evil with evil. This is His, may He be Exalted, saying, "My servant sinned and he knew that he had a Lord who forgives sins and (takes men to task) for sin; I have forgiven My servant."[5]

In summary, he believes in that which produces awe and the utmost reverence with a firm conviction, which does not allow there to remain or arise in his heart so much as a gnat's wing of submission to or fear of any other. He believes that human perfection is to turn to his Lord and worship Him and that the best state

---

[1] Qur'ān 34:3.
[2] As stated in Qur'ān 58:7.
[3] Qur'ān 14:27, 22:18.
[4] Qur'ān 5:1.
[5] A hadith qudsī, or Prophetic report of God's saying. Muslim Tauba 29, Siddiqi trans. IV:1439.

of human beings is to resemble and approach the angels. He also believes that these things bring him nearer to his Lord, that God is pleased by his doing them, that God has the right to these things from him, and that he must fully discharge them.

In summary, he knows incontrovertibly that (his ultimate) felicity lies in the acquisition of these qualities and that his misery lies in neglecting them and that he necessarily must have a severe punishment to strongly warn and powerfully alarm the animalistic side.

The paths of the prophets differ in this (intellectual regulation). The most important thing that God, may He be Exalted, revealed to Abraham, may peace be upon him, was reminding people of the clear signs of God, His Exalted Attributes, and His cosmic and personal blessings, until it was confirmed in a way to which nothing could be added that He is worthy of having them sacrifice all their comforts for Him, that they should prefer remembering Him over thinking of anything else, should love Him greatly, and worship Him to the furthest extent of their efforts. God added to this for Moses, may peace be upon him, reminding people of the "Days of God,"[6] that is, the explanation of God's, may He be Exalted, requital of the obedient and the disobedient in this world,[7] and His transposing of blessings for adversities, so that fearing sins and a strong inclination to acts of obedience were represented in their hearts.

For our Prophet, may the peace and blessings of God be upon him, He included along with these two things the warnings and good tidings about the events of the grave, and what follows it, and an explanation of the properties of piety and sin. Simply knowing about these things is not sufficient; rather they must be repeated and reiterated frequently and observed at all times, and kept in view until the intellectual faculties are filled with them and the limbs are guided by them.

These three things (brought by the Prophets)[8] along with two other ones, one of them the explanation of the rules of the compulsory and the forbidden, and so on, and the second of them how to

---

[6] Ayyām Allāh.

[7] Especially the punishments of previous nations who turned away from the teachings of their prophets.

[8] I.e., recalling the Days of God, giving warning and good tidings, and explaining the properties of piety and sin.

dispute with the unbelievers, are five arts which are basis of the sciences of the great Qur'ān.[9]

2) As for the practical regulation (of the acquisition of these qualities), the most important thing is choosing attitudes, acts, and things which remind the soul of the desired qualities and alert it to them, and which stir it and urge it on, either through a habitual connection between it and the quality, or due to the soul's being the expected place to find the quality by virtue of innate suitability. Just as the man who wants to arouse his soul to anger and experience it vividly imagines the abuse that someone he is angry with hurls at him and that brings him shame and so on; and the wailing woman, when she wants to resume her grieving reminds her self of the good qualities of the deceased and imagines them, and incites her thoughts to review them, quickly and slowly; and the one who desires sexual intercourse persists in whatever motivates to it. There are very many similar instances of this phenomenon which will not baffle the one who wishes to understand all aspects of the discussion. Likewise there are for each one of these qualities means by which it is acquired, and one must depend on the mystical perception (dhauq) of the people of genuine mystical experience for knowledge of these matters.

Thus, the causes of impurity are the filling of the heart with a low state[10] like the satisfaction of lust for women through his having intercourse and sexual contact; or by his harboring inside himself things which are opposed to the truth; and by the curses of the Highest Council surrounding him; and his needing to urinate or defecate, or having just urinated, defecated or broken wind, and these three things are the wastes of the stomach. (Other causes of impurity are) the body being dirty, bad breath, the accumulation of nasal mucus, growing the pubic and armpit hair, having the clothing or body soiled by dirty things which render men ritually impure; filling the senses with an image which evokes the lower state, such as filth, looking at the genitalia, watching the mating of animals, staring during sexual intercourse, slandering the angels and righteous people, and striving to harm people.

---

[9] This theme of five Qur'ānic arts is elaborated in Shāh Walī Allāh's book on interpretation of the Qur'ān, *al-Fauz al-Kabīr fi-Uṣūl al-Tafsīr*, Arabic original with Urdu translation by Maulānā Rashīd Aḥmad Anṣārī (Lahore: Maktaba Burḥān. 1963). English translation by G.N. Jalbani, *The Principles of Qur'ān Commentary*, (Islamabad: National Hijra Council, 1985). See also Chapter 7 of the present work.

[10] That is, the excess of the requirements of the animal side.

The causes of purity lie in the elimination of these things, the acquisition of their opposites, and the habitual carrying out of those things which establish his being extremely clean, such as full bathing (*ghusl*) and ritual ablutions (*wuḍū'*), putting on one's best garments and using scent. The performance of these things will alert the soul to the attribute of purity.

The causes of humility are his holding his soul to the highest states of reverence possible for him, in standing with bowed head, prostrating (in prayer), speaking words which indicate intimate conversation with and submission before God, and asking Him for the fulfillment of one's needs. These things make the soul strongly aware of the attribute of humility and submission (before God).

The causes of magnanimity are practicing generosity, giving liberally, forgiving the one who does wrong, maintaining fortitude in the face of adversity, and so on.

The causes of justice are preserving the rightly-guided practice in all of its details, and God knows better.

CHAPTER 34

The Veils Preventing the Manifestation of the Sound Original Nature (Fiṭra)

You should know that most of the veils are of three kinds:

1) The veil of nature,
2) The veil of convention (*rasm*), and
3) The veil of misinterpretation.[1]

That is because the drives for food, drink, and marriage are built into man and his heart is made an instrument of natural states like sorrow, joy, anger, fear, and other things; so that it continues to be occupied with these, since each state is preceded by the soul turning its attention toward its causes and the intellectual faculties yielding to what is appropriate for it. This combines with the soul's absorption in this state and distraction from everything else, and with the remaining of some trace of its shadow and contamination. Thus days and nights pass while someone is in this (state), not free to acquire anything else which might perfect him. Many a man may step into this slough and not come out for his whole life. Many a man may be dominated by the decree of the physical nature so that he withdraws his attention from the surveillance of convention and reason, and is not held back by censure; and this veil is called the lower soul (*nafs*). However the man whose reason is developed and whose awareness is abundant, will snatch from his moments (*awqāt*) opportunities in which to suspend his natural, physical states, and will enlarge his soul for these (higher) states and others. He will merit the emanation of other types of knowledge not realized by the requirements of the physical nature, and will be attracted to the perfection of the species in accordance with the two faculties, the intellectual and the practical. Then, when he opens the eye of his insight, he will at first see his people; with arts of civilization,

---

[1] "*Sū' al-ma'rifa.*" That is, a misinterpretation or misconception of God and His Attributes.

fashion, things they pride themselves on, and rhetorical and technical excellence; so that he will be very impressed and will take these things up completely with a firm intention and strong enthusiasm.

This is the veil of convention which is called "the world." Among people there are ones who remain absorbed in this until the time of death. Then these graces vanish completely because they only come about through the body and physical instruments, and the soul remains stripped of them and possesses nothing. What happens to him resembles the case of the owner of a garden which was struck by a cyclone, or like "ashes blown away by a strong wind on the day of a gale."[2]

If he is extremely alert and very astute, he may ascertain by some rationally demonstrated or rhetorical proof, or by following the divine law, that he has a Lord, Almighty over all of His worshippers, Who manages their affairs and bestows all blessings on them. Therefore He has created in his heart an inclination to Him, and a love for Him so that he wants to draw near to Him, raise his requests to Him, and throw himself before Him. Some men succeed in gaining this end, while others miss it.

Most errors are of two kinds; to believe that the Necessary Being[3] has the attributes of created things, or to believe that created things possess the attributes of the Necessary Being. This first is anthropomorphism (*tashbīh*) which arises from drawing analogies about the unseen on the basis of what is perceived. The second is associating others with God (*ishrāk*) which arises from seeing extraordinary effects among human beings and thinking that they are attributable to them in the sense that they create them, and that these effects essentially belong to them.

You must examine all human individuals. Do you see any inconsistency in what I have told you? I do not think that you will find any. On the contrary, every person, even if he follows some divine law, must experience moments that are drowned in the veil of nature whether these moments are few or many.

As long as he continues to carry out conventional acts there will be moments overcome by the veil of convention, and at that time

---

[2] Qur'ān 14:18 "Their works are as ashes blown away by a strong wind on the day of a gale."

[3] "*Wājib (al-wujūd)*." The one who has Necessary Existence, as opposed to everything else, the existence of which is contingent.

he will give importance to emulating the wise men of his nation in speech, clothing, morals and life style. There will be moments when he hearkens to what he is used to listening to, and does not listen to the reports about Jabarūt and the unseen management of the world, and God knows better.

CHAPTER 35

The Way of Removing These Veils

You should know that the way to manage (the lifting of) the veil of nature consists of two methods: one of them to which somebody can be commanded, incited and urged; and the other which is imposed on a person from above, and for which he is held accountable whether he wishes or refuses. The first consists of exercises and practices that weaken the animalistic, like fasting and staying awake at night. Among people there are some who are excessive and choose to alter God's creation by acts such as castration and letting atrophy[1] the important limbs such as the hand and the foot, and these are the most ignorant among people. The best things are moderate ones, and thus fasting and night vigils are like the antidote for poison which should be used in proportion to need.

The second way is undertaking the repudiation of those who, following their physical natures, go against the rightly-guided practice, by explaining the way to become free of every domination by nature, and setting a practice for this. The people should not be treated with too much severity, although verbal reprimand will not suffice in all cases and in some matters painful beatings and debilitating fines are necessary. The most deserving of these (punishments) are those extreme behaviors whose harm affects others adversely, such as adultery and murder.

The way to effect (the pulling aside) of the veil of convention is through two methods. One of them is that the recollection of God, may He be Exalted, should be included with every useful act, sometimes by memorizing the phrases which He commanded and sometimes by complying with limits and restrictions that are only observed for the sake of God. The second is to make acts of worship widespread conventions and to emphasize their preservation, whether they wish to do them or refuse, to condemn abandoning them, and

[1] Literally, "dehydrating."

to prevent a person from obtaining the things he deserves such as status, and so on, as a punishment for omitting these things. Through these two ways of regulating them the disasters of convention will be averted, and the conventions will become a support for the worship of God, and become a means of summoning to the truth.

Misinterpretation of both categories (anthropomorphism and associationism) arises from two causes. One of them is that a person is not able to truly know his Lord, due to His great elevation above human qualities and His transcendence of the features of temporally originated and tangible things. The way of regulating this is that people should be addressed only according to what their minds can comprehend.

The basis of this is that there is nothing, whether existent or non-existent, and which is conceived of as either occupying space or as an abstraction, but that the knowledge of man can grasp it, either through the direct presentation of its form (*ḥuḍūr ṣūratihi*) or through analogizing or inference, even (in the case of) absolute non-being (*al-ʿadm al-muṭlaq*) and the absolutely unknown (*al-majhūl al-muṭlaq*).

Thus non-existence is known from the aspect of knowing existence and observing that it (the non-existent) cannot be qualified by existence. He knows the meaning which is derived from the grammatical form based on the pattern of the passive participle (*mafʿūl*)[2] (i.e., for the verb "*j-h-l*" this form is *majhūl* "the unknown") and he knows the meaning of "absolute" (*muṭlaq*). Thus he combines these things and adds one to the other, so that a composite image is arranged that discloses the non-compound thing whose depiction was intended and which has neither external nor mental existence. In this way, perhaps, he may consider a theoretical concept and apply what he considers to be a genus and a differentia, so he combines the two of them and obtains a composite concept in order to disclose what he wishes to conceptualize.

Thus, for example, people should be told that God, may He be Exalted, is "a Being existent in a way unlike our existence," and that "He is Alive in a way unlike our life."

---

[2] The argument here is based on logic and the Arabic language. In this case the "absolute unknown" (*al-majhūl al-muṭlaq*) is known through comprehending what is "unknowing" (*j-h-l*) of which the "unknown thing" (*majhūl*) is the passive participle, and then combining this with the concept of "absolute" (*muṭlaq*).

In summary, (in describing God) attributes that are the source of praise in (our) perception should be used. Three senses are taken into account in whatever we (normally) perceive. Firstly: a thing that possesses these attributes and from which their effects have proceeded. Secondly: a thing that does not possess these attributes and it is not in its nature to have them. Thirdly: a thing that does not have them although it is within its nature to have them. Examples are: "living," "inanimate," and "dead." These (praiseworthy) attributes may be confirmed (about God) through the confirmation of their effects and this comparison obliges us to admit that He is not like us.

The second reason for misinterpretation (with regard to God) is the representation of the tangible form along with its embellishment, and the pleasures along with their beauty, and the filling of the intellectual faculties with sensory images, so that the heart yields to these, and attention to God is not pure. The way of rectifying this consists of religious exercises and actions which prepare man for lofty emanations even if these should occur in the next life. He should withdraw to perform devotions and eliminate distraction as far as possible, as the Prophet of God, may the peace and blessings of God be upon him, tore down the figured curtain and removed the shirt with designs on it, and God knows better.[3]

[3] 'Ā'isha told the Prophet that she had used a figured curtain for screening a room and he tore it down. *Mishkāt*, p. 941. The hadith mentioning that the Prophet removed a brocaded shirt may be found in *Mishkāt*, p. 921. Bukhārī Khan translation IV:297–99.

BOOK V

THE FIFTH INVESTIGATION: PIETY AND SIN

CHAPTER 36

Preface: Concerning the True Nature of Piety and Sin

We have already elaborated the rationale behind requital and its essential nature. Then we described the supports of civilization to which humans are naturally predisposed, and these remain in effect among people and are inseparable from them. Then we described (the ultimate) felicity and the means for acquiring it. It is now, therefore, the appropriate time to undertake the investigation into the meaning of piety and sin.

Piety is every act which a person does as a requirement of his submission to the Highest Council, his being effaced in accepting inspiration from God, and his surrendering himself to the divine purpose. It consists of every act which is rewarded by good in this world or the next, every act which benefits the supports of civilization upon which the human order is based, and every act which promotes the state of submission and removes the veils.

Sin is every act which man does as a result of his obedience to the devil and which results in his surrendering to his purpose, every act which is rewarded by evil in this world and the next, every act which damages the supports of civilization, and every act which reinforces the attitude opposed to submission and which strengthens the veils. The supports of civilization were discovered by skillful people, and then the other people followed them due to the testimony of their hearts, and all the people of the earth concurred in them, or at least those worth taking into account. Similarly, with regard to piety, there are established practices (*sunan*) that God, may He be Exalted, has inspired in the hearts of those who are supported by the angelic light who are dominated by the ethic of the original nature (*fiṭra*), in the same way in which He has inspired bees with what is suitable for their livelihood. Thus they customarily practice them, adopt and teach them, and urge to them, so that people emulate them. The people of all the religious communities in all regions of the earth agree on them, despite the distances between their lands and the differences in their religions, due to the decree of a consistency of the original nature and a requirement of the species. This difference in the forms of these practices does not preclude (this basic congruence) once there is agreement on their principles, neither does the obstruction of a defective group. If insightful people were to reflect on this group they would have no doubt that their matter had defied the form of the species and could not permit the expression for its properties. These ones with respect to other people are like a supernumerary member of the body whose excision would be more desirable it then its remaining.

There are important causes behind the extensive spread of these practices, and established ways of managing them which were determined by those who were aided by revelation, may the blessings of God be upon them. Thus they have established for themselves a great indebtedness on the part of humanity. We want to make you aware of the basic principles of these practices which have been agreed upon by the majority of the inhabitants of all of the sound regions who are members of great nations; including mystical philosophers (*muta'allihūn*), kings, and wise men, who possess sagacious opinions, and whether they are Arabs, non-Arabs, Jews, Magians, or Hindus. We will explain the nature of their originating from the submission of the animalistic force to the angelic one, and also certain of their benefits, on the basis of what we ourselves have experienced more than once; and sound reason leads to this, and God knows better.

CHAPTER 37

The Unity of God (Tauḥīd)

The most basic principle of piety and the mainstay of its various aspects is the doctrine of the unity of God. This is because humility before the Lord of the worlds, which is the greatest of the virtues in acquiring ultimate felicity, depends on this. This is the basis of the intellectual regulation of (acquiring virtue) which is the more useful of the two regulations.[1] Through this man is able to address himself completely to learning from the Unseen, and to prepare his soul to become connected to it[2] through becoming sanctified. The Prophet, may the peace and blessings of God be upon him, has informed us of its importance and of its being one of the types of piety that is comparable to the human heart, which when it is sound, everything is sound, and when it is corrupted, all is corrupt.[3] He also indicated its importance when he used the phrase about a man who died without associating anything else with God that "he entered Paradise," or that "God prevented him from going to Hell", or that "he was not veiled from Paradise,"[4] and other expressions of this sort. He reported from his Blessed and Exalted Lord, "Whoever meets Me (on the Day of Judgment) with enough sins to fill the earth, but not associating another thing with God, I shall meet him with one (an earth) like it full of forgiveness."[5]

You should know that there are four degrees of Tauḥīd.

1) The first is the restriction of necessary existence to Him, may He be Exalted, for none other than He is Necessary.[6]

---

[1] See Chapter 33 on the two measures (intellectual and practical) for acquiring the qualities which lead to happiness.

[2] Attached to the Unseen, through connections to the angels.

[3] Bukhārī Īmān 39, Also cited in the introductory section.

[4] Such as the hadith in Bukhārī Bad' al-khalq 6, Khan trans. IV:296, Muslim Īmān Ch. 41, trans. I:54–55.

[5] *Mishkāt*, p. 496. Muslim Dhikr 22, Tirmidhī, Ibn Mājah, Dārimī, Ibn Ḥanbal.

[6] The proof of the existence of God established by the philosopher Ibn Sīnā (d. 1037) was that God's existence is necessary, in contrast to the contingent existences of everything else. Henceforth God was termed the Necessary Being (*wājib al-wujūd*).

2) The second is the restriction of the creation of the Throne, the Heavens, the earth, and the rest of the substances to Him, may He be Exalted. These two levels were not discussed in the divine scriptures and the Arab polytheists, Jews, and Christians did not disagree about them; rather the Magnificent Qur'ān asserted that these were premises accepted by them.[7]

3) The third is restricting the management of the heavens and earth and what lies between them to Him, may He be Exalted.

4) The fourth is that no one besides Him is entitled to be worshipped, and these two are intertwined and inseparable due to a natural connection between them.

Some sects of people have disagreed on these latter two degrees, the majority of them falling into three groups.

a) The first are the star-worshippers who believed that the stars were worthy of being worshipped and that worshipping them would bring benefits in this world, and that raising requests to them was legitimate. They said, "We have established that they have a great influence in daily events and on man's fortune and his misfortune, and his health and illness; and that they have intelligent, transcendent souls which cause them to move, and that they do not neglect those who worship them." Thus they made images in their names and worshipped them.

b) (The second group) are associators (polytheists) who concurred with the Muslims concerning God's management of the important things and that He judges and decides, and that He leaves no choice to others. However, they did not agree with the Muslims regarding the remaining matters. They believed that the righteous ones before them had worshipped God and drawn nearer to Him, so that due to this God had bestowed divinity on them, making them worthy of being worshipped by the rest of God's creatures. This would be like the case of a servant who serves the king of kings and does his service well, so that he is given the robe of honor of the king, and the government of one of his cities is entrusted to him so that he becomes entitled to being listened to and obeyed by the people

---

That His existence (*wujūd*) was part of His essence, entailed by it, is an additional aspect of Ibn Sīnā's proof. When Shāh Walī Allāh mentions that the necessity of God's existence is asserted in the Qur'ān it is in the sense of His primacy over all other things.

[7] For example the Qur'ānic verse, "If you were to ask them who had created the heavens and the earth, they would say the Exalted, the All-Knowing, created them."

of that city. They said that the worship of God is only accepted if it is included in worshipping these ones, for indeed God is at the extreme limit of being Exalted, and thus worshipping Him does not serve to approach Him, so that rather these ones must be worshipped in order for people to "approach closer to God in proximity."[8] They also said that these beings hear, see, intercede for their worshippers, manage their affairs, and give them help; and they carved stones in their names and made the stones a focus for directing their worship toward these beings. Then there came generations after them who did not understand the difference between the idols and the one in whose image they were made, so they thought that these were the very objects of worship. Therefore God, may He be Exalted, refuted them, sometimes warning that judgment and kingship belonged to Him exclusively, and sometimes by explaining that these were inanimate objects. "Have they feet wherewith they walk, or hands wherewith they hold or have they eyes wherewith they see, or have they ears wherewith they hear?"[9]

c) The Christians (the third group) believed that the Messiah, may peace be upon him, had a special relationship to God and was higher than other created beings, and that for this reason he should not be called a "servant" or considered equal to another person, for this would be poor manners with respect to him, and would overlook his special relationship to God. Some of them inclined to express this particularity by calling him the "son of God," in view of the fact that the father is kind to the son, educates him under his supervision, and that he (the son) is above the servants, and thus this name is more appropriate to him. Some of them were inclined to call him "God" considering that the Necessary Being had become incarnated in him and entered into him, and that for this reason he could produce effects which had never been observed among humans, such as raising the dead and creating the birds.[10] They considered that his speech was the speech of God, and that to worship him was to worship God. Then after them came generations who did not understand the reason for this appellation, and they would all but make "sonship" literally true or claim that he was (God) the Necessary Being in all respects. Therefore God,

[8] Qur'ān 39:3.

[9] Qur'ān 7:195.

[10] Qur'ān 3:49, 5:110. By God's leave Jesus created a bird from clay, breathed into it, and it came to life.

may He be Exalted, refuted them, sometimes by the fact that He does not have a consort,[11] and sometimes by His being the Creator of the heavens and earth out of nothing. "But His command, when He intends a thing, is only that He says unto it: 'Be!' And it is."[12]

These are the three groups which have made extensive claims and (concocted) many fables which are not concealed from the one who pursues the matter. The Magnificent Qur'ān discussed these two levels (of the unity of God) and has fully refuted the confusion of the unbelievers.

---

[11] As in Qur'ān 6:101. "How should He have a son, seeing that He has no consort?"

[12] Qur'ān 36:81.

CHAPTER 38

Concerning the True Nature of Associationism (Shirk)

You should know that worship is the utmost act of humbleness. The utmost humbling of oneself before another must be either through the form, for example one standing while another prostrates, or through the intent, in which one intends by this act the honoring of the master on the part of servants; like the way that subjects honor their king, or pupils respect their teacher. There is no third type besides these two.

Since it is established that the angels prostrated to salute (*sujūd al-taḥiyya*) Adam, may peace be upon him, and that the brothers of Joseph, may peace be upon him, prostrated to Joseph, and that prostration is the highest form of giving respect, it necessarily follows there is no (physical) distinction (between the acts of prostration)[1] except through their intention. However the matter up until now has not been made precise, since the word "master," for example, is used in various senses. What is meant in this context is certainly "the one worshipped," thus it was construed as being within the definition of "worship." The precise rendition (of this definition) is that humility requires the taking into consideration of weakness in the humble one and strength in the other; and baseness in the humble one and nobility in the other; and obedience and submission in the humble one and domination and giving orders in the other. A person, when left to himself, necessarily perceives that there are two standards set for strength, nobility, dominance, and other similar things which are used to express "perfection." One standard is that for himself and others who resemble him, and the other standard is for the One who is completely exalted above the blemish of temporal origination and contingency.

For the one to whom something of the properties of this Exalted Being is conveyed, the knowledge of unseen things (*al-ʿilm bi-l-ghaib*)

[1] That is, between the prostration of worshipping God (*sujūd al-ʿibāda*) and that of respect for a creature (*sujūd al-taḥiyya*).

is set at two levels. (The first is) knowledge acquired through vision and ordering premises, or through intuition or in dreams, or through receiving inspiration, from which he finds himself never to be totally cut off. (Secondly) there is innate knowledge (*ʿilm dhātī*) which is inherent in the essence of the knower and which he does not learn from anyone else, nor does he acquire it by his own effort.[2] Similarly having an effect, managing, and controlling—whichever expression you use—is set at two levels. In one sense as doing something using limbs, faculties, and getting assistance from the properties of temperament such as heat and cold, and other similar things which a person is easily or remotely capable of doing. The second sense would be that of bringing something into being through other than a bodily property, and without directly contacting anything, and this is His, saying, "But his command, when he intends a thing, is only that he says to it, 'Be!' And it is."[3]

Therefore greatness, nobility, and power are fixed at two levels.

1) One of them is like the greatness of the king in relation to his subjects due to his many assistants and greater power, or like the greatness of the hero or teacher in relation to the one whose courage is weak or to the pupil, in that one finds himself sharing in the basis of the thing.

2) The second level is that which is only found in the greatly Exalted One. Do not give up the investigation into this secret until you verify that the one who realizes that the chain of contingency must terminate by reaching a Necessary One who does not need any other cause, is forced to classify these properties which are used for praise according to two levels: one suitable for what is There (the Divine level) and the other suitable for those to whom he compares himself.

The expressions employed at the two levels are close, and thus sometimes the pronouncements of the divine laws are applied in a sense other than their intent. Often a person comes to know about an effect proceeding from some human individuals or angels or someone else, which he regards as being remote from the members of his own species. The matter then becomes confused for him, so that he affirms for that person divine nobility, holiness, and a divine dominance. People are not equal in recognizing the exalted

---

[2] This type of knowledge is also known as *ʿilm ḥudūrī*, i.e., direct knowledge.
[3] Qur'ān 36:82.

level. Among them is one who comprehends the powers of those lights which encompass and overwhelm created beings, so that he recognizes that they are at this ennobled level. There are also among them those unable to do this, and every human is charged according to his capacity. This is the interpretation of what the truthful and veracious one, may peace be upon him, said about the salvation (by God) of the man who had not done good deeds and who ordered his family to burn him and scatter his ashes, fearing that God would resurrect him and have power over him. For that man was sure that God was qualified by complete power, but that this power was only over possible things and not over the impossible. He thought that the reconstitution of his ashes, half of which were scattered on the land and half in the sea would be impossible. This was not deemed to be a fault on his part so that he was held responsible according to the extent of the knowledge which he had, and was not considered to be an unbeliever.[4]

Anthropomorphism and associating (divinity) with the stars or righteous persons who can produce miracles such as disclosing facts and granting the requests of prayers, were passed down among them. Every prophet who is sent to his people must inevitably make them understand the true nature of associationism, and must distinguish each of the two levels, and restrict the level of the holy to the Necessary Being, even if the words for each are similar. This is like the case in which the Prophet of God said to the physician, "You are a person who can give kind treatment (to the ill) but God is the physician (who cures),"[5] and as he said, "The master is Allah,"[6] indicating that the words ("physician" and "master") in these hadiths have special connotations. Then when the disciples among the Companions of a Prophet and the bearers of his religion died out, "there arose after them a generation which neglected the prayers and followed their carnal desires,"[7] so that they misinterpreted the words which had been employed ambiguously. For example they misinterpreted "belovedness" and "intercession" which God, may He be exalted, had confirmed for special persons in all His laws

---

[4] Muslim Tauba 25, Siddiqi trans. IV:1438, Bukhārī Anbiyā' 54, Tirmidhī, Nasā'ī, Ibn Mājah, Ibn Ḥanbal.

[5] Hadith from *Musnad* Ibn Ḥanbal IV:163. Previously cited in Chapter 4

[6] Dārimī Muqaddima p. 7. Variant "*f-allāhu al-sayyid,*" rather than, "*al-sayyid huwa Allāh.*"

[7] Qur'ān 19:59.

without exception. They took the performance of miracles and receiving illuminations to mean that the highest levels of knowledge and domination had been conveyed to the one on whose part they witnessed these things. The truth is that all of these things derive from powers, whether human (*nāsūtī*) or spiritual, which prepare in some sense for the descent of the divine management in some aspect. This has nothing to do with bringing into being or other matters which are particular to the Necessary Being.

The ones struck by this malady are of various sorts.

a) Among them are those who totally forget the majesty and power of God so that they begin to worship only those others whom they associate with God and they raise their needs only to them, not turning to God at all, although they know by demonstrative proof that the chain of being terminates with God.

b) Among them are those who believe that Allah is the Master and Manager but that He has bestowed on certain of His servants the mantle of nobility and deification, and He has given them the right to act independently in certain special matters, and He accepts their intercession for His servants, in the way that the king of kings sends a king to every region, and entrusts them with managing this kingdom except for in the highest matters; so that they hesitate to call them "servants" of God and to equate them with others. Then they moved beyond this to call them the "sons of God," and the "beloveds of God," and they called themselves the bondsmen of these ones, such as the Servant of the Messiah and the Servant of Al-'Uzzā.[8] This is the malady of most of the Jews, Christians and polytheists, and certain of the extremists among the hypocrites within the religion of Muḥammad, may the peace and blessings of God be upon him, in our present time.

Since the basis of divine legislation is establishing an anticipated source in the place of the principle;[9] tangible things which are anticipated sources of associationism are considered to be infidelity (*kufr*). Examples are prostrating to idols, sacrificing to them, making oaths in their names, and things like these. The first disclosure of this knowledge to me was a vision I had of a group of

---

[8] A pre-Islamic goddess.

[9] The word *maẓinna*, pl. *maẓānn* has been translated here as "anticipated source" and elsewhere as "expected location" or "symbol," according to the context. It literally means "the place where something is likely or expected to be found."

people prostrating before a small poisonous fly which kept moving its tail and limbs continuously. Then it was uttered in my heart, "Do you find in them the evil of polytheism and has error encompassed them as you find among idol worshippers?" I said, "I do not find it in them for I find that they have made the fly a direction of worship (Qibla) and they do not confuse the degree of humbleness with the other (of worship)."[10]

It was said to me, "You have been guided to the secret," and on that day my heart was filled with this knowledge and I became possessed of insight into the matter, and came to know the true nature of the belief in God's unity and associationism, and those things which the divine law has set as anticipated sources for these two, and I realized the connection of worship to the divine management and God knows better.

[10] The issue of whether one could prostrate before any being other than God was, and is controversial. Shāh Walī Allāh's conclusion that a prostration indicated humbleness is not unbelief (*kufr*) is more liberal than some. For example, Shāh Ismā'īl Shahīd, Walī Allāh's grandson, declared all prostrations to other than God to be *kufr* in *Taqwiyat al-Īmān*, translated from the Urdu by Mīr Shahāmat 'Alī, *Journal of the Royal Asiatic Society*, 13 (1852), 310–372.

CHAPTER 39

The Categories of Associationism (Shirk)

The true nature of associationism (*shirk*) is that a person should think that the amazing effects emanating from a revered personage only emerge due to his being characterized by an attribute of perfection which is unknown among humans, but which rather is restricted to the Necessary Being, may His magnificence be exalted. Rather he believes that this is not found in anyone else unless God bestows the mantle of divinity upon another, or annihilates this other in His essence, and makes him subsist through His essence,[1] or something of this sort which the believer in these varieties of superstitions holds. An example is what was reported in the hadith, "The polytheists used to utter the following formula of '*talbiyya*': 'We are at Your service (*labbaik*), we are at Your service, You have no partner except an associate over whom You possess sovereignty while he does not.' "[2] Thus they humbled themselves to the furthest extent before this other, and they behaved toward this person in the way people behave with respect to God, may He be Exalted.

This concept has various embodiments and forms, and the divine law only discusses embodiments and forms of it which people practice with the intention of associationism, so that they become anticipated sources of *shirk* and customarily inseparable from it. This is similar to the practice of the divine law in establishing the causes that entail good or evil actions as being tantamount to those acts themselves. We want to alert you to those things which God, may He be Exalted, has made anticipated sources (*maẓinnāt*) of associationism in the divine law brought by Muḥammad, may there be peace and blessings upon the one who brought it, so that he forbade them.

---

[1] Annihilation (*fanā'*) and subsistence (*baqā'*) are terms used in Sufism to indicate respectively absorption of the personal ego in the Divine and the return of the transformed person after the experience of transformation.

[2] *Talbiyya* is the formula repeated by the pilgrims on their way to make the Hajj. *Mishkāt*, p. 543. Muslim Ḥajj 22.

1) Among them is their former custom of prostrating before idols and stars, so that prostration before other than God was forbidden. God said, may He be exalted, "Prostrate not to the sun nor to the moon; but prostrate to Allah Who created them."[3]

Association (of others with God) in prostration necessarily entailed associationism with regard to the divine management, as we have alluded to. The matter is not as certain Theologians think, i.e., that being exclusively worshipped is one of the commands of God, may He be Exalted, which may vary with the variation in religions and which can not be pursued through demonstrated proof. How could it be, for if it were thus, God, may He be Exalted, would not have required them (the associators) to consider Him unique in creation and management. As He, the Greatest Speaker said, "Say: Praise be to Allah, and peace be on His servants whom He has chosen! Is Allah better (or those others that they associate with Him)" and so on for five verses.[4] Rather the truth is that they had acknowledged His exclusive claim to creation and divine management in important matters and they agreed that worship is attendant on these two things, according to what we have indicated in the discussion of the meaning of belief in the unity of God.[5] Therefore God held them responsible for what He enjoined on them,[6] and "God has the convincing argument."[7]

2) Among them (the types of *shirk*) is that they used to request assistance with their needs such as in curing the sick and meeting the needs of the poor, from other than God. They would make

---

[3] Qur'ān 41:37.

[4] Qur'ān 27:59 ff. "Is He not the best Who created the heavens and the earth and sends down for you water from the sky from which We cause joyous orchards to spring forth, whose trees you have never had to cause to grow. Is there any God beside Allah? No, but they are people who ascribe equals. Is He not (best) Who made earth a fixed abode, and placed rivers in the folds thereof, and placed firm hills therein, and has set a barrier between the two seas? Is there any God beside Allah? No, but most of them know not! Is He not (best) Who answers the wronged one when he cries unto Him and removes the evil and has made you viceroys of the earth? Is there any God beside Allah? Little do they reflect! Is not He (best) Who guides you in the darkness of the land and sea. Who sends the winds as heralds of His mercy? Is there any God beside Allah? High exalted be Allah from all that they ascribe as partner (to Him). Is He not best Who produces creation, then reproduces it, and who provides for you from the heaven and the earth? Is there any God beside Allah? Say: Bring your truth if you are truthful!"

[5] See Chapter 37.

[6] To affirm that God alone should be worshipped.

[7] Qur'ān 6:149.

vows to them, expecting the accomplishment of their purposes through these vows, and they would recite their names, hoping for their blessing. Therefore, God, may He be Exalted, made incumbent upon them that they say during their prayers: "Thee alone do we worship, Thee alone do we ask for help."[8] And He said, may He be Exalted, "Call upon no one except God."[9] The meaning of "calling upon" (*duʿā*) is not "worship" (*ʿibāda*) as certain of the interpreters of the Qur'ān said, it rather means "seeking help," according to the saying of God, may He be Exalted, "No, but you call (*tadʿūna*) upon Him and He removes the thing because of which you call upon Him."[10]

3) Among them are that they used to call some of the ones whom they associated with Him, "daughters of God" and "sons of God." They were then forbidden this most strongly, and we have explained the secret behind this previously.

4) Among them is that they used to think that "their rabbis and monks were lords besides God,"[11] may He be Exalted, in the sense that they used to believe that whatever things they declared permissible were permitted, and that there was nothing wrong with them in themselves; and they believed that whatever they declared forbidden was prohibited and that they would be blamed for doing it. Then when His, may He be Exalted, saying was revealed: "They have taken their rabbis and monks,"[12] ʿAdī ibn Ḥātim asked the Prophet of God about it. He answered, "They used to say that things were permitted for them, so that they regarded them as lawful; and they used to forbid them some things so that they considered them forbidden."[13]

The secret behind this is that making something permitted or forbidden is an expression for a creative process (*takwīn*) that is operative at the level of Malakūt whereby one will or will not be held accountable for a certain thing. Thus this creative process (*takwīn*) is the reason for a person being or not being held accountable for a thing, and this is part of the Attributes of God, may He be Exalted.[14]

---

[8] Qur'ān 1:4.
[9] Qur'ān 72:18.
[10] Qur'ān 6:41.
[11] Qur'ān 9:31.
[12] Qur'ān 9:31.
[13] Tirmidhī IV: 342 5093 Tafsīr Sūra 9.
[14] Shāh Walī Allāh elaborates in Chapter 76, on the existence of these (*mithālī*) forms of divine laws at higher planes.

As for the attribution of permitting or prohibiting to the Prophet, it is in the sense of his speech being a decisive sign (*imāra*) of God's permitting or forbidding. As for the attribution of them (permitting and forbidding) to the legal interpreters (*mujtahidūn*) of his community, this is in the sense of their transmitting this element of the divine legislation based on the revealed text of the lawgiver, or their inferring the meaning of his words.

You should know that when God, may He be Exalted, sends a Prophet and confirms his message by miracles, and through him permits some things which had been forbidden to them, some people find in themselves a resistance to this,[15] so that there remains in their hearts an inclination to forbid it due to the prohibition which had existed in their community. This (vacillation) may be occur in two ways.

a) If it is due to a hesitation in confirming this divine law, then the person who hesitates is a disbeliever in the Prophet.

b) If it is due to a belief that the first ban took place in such a way as to render abrogation impossible because God, may He be Exalted, had bestowed on some human being a robe of divinity, or because this person was annihilated in God and subsisted through Him, so that his forbidding or disliking an act required that it would produce a loss in a person's wealth and his family; then this one is a polytheist (*mushrik*) who affirms that there is a sacred wrath, displeasure, forbidding and permitting on the part of someone other than God.

5) Among them is that people used to seek to gain favor from idols and the stars by sacrificing to them, either through invoking their names during the sacrifice, or by sacrificing to idols designating them, which they were forbidden to do.

6) Among them are that in order to draw closer to those whom they associated with God, they would set free a Sā'iba and a Baḥīra.[16] Thus God, may He be Exalted, said, "Allah has not made (any rule in the nature of) a Baḥīra or a Sā'iba."[17]

[15] For example, Qur'ān 6:137-140 refers to customs of the pagan Arabs which they were reluctant to give up.

[16] Types of animals which the pagan Arabs set free in honor of the idols. "The Baḥīra is the filly of the Sā'iba, the Sā'iba is the she camel which gives birth to ten fillies without an intervening colt. She is set free and never ridden, her hair is not shorn and only a guest is allowed to drink her milk." Ibn Isḥāq (trans. A. Guillaume) *The Life of Muhammad* (Karachi: Oxford U. P., 1980), 40. See also Bukhārī Tafsīr 5:103 Khan trans. VI:115.

[17] Qur'ān 5:103.

7) Among them was their belief that the names of certain people were blessed and exalted, and that to tell a lie while swearing by their name would incur a loss of wealth or an injury to family, so they refrained from doing this. They therefore used to make their opponents swear oaths by the names of those alleged "partners," then this was forbidden to them. The Prophet said, "Whoever makes an oath by other than God, commits *shirk*."[18] Some of the scholars have interpreted this as being rigorous and a threat, but I do not hold this position. What is meant, in my opinion, is the oath made in the name of other than God that one will do or will abstain from doing something in the future,[19] and an oath in which no exception is made by saying, "If God wills,"[20] according to the belief which we have mentioned.[21]

8) Among them is the performance of a pilgrimage in honor of anyone other than God, may He be Exalted. This occurs when people go to visit spots sacred to those whom they worship other than God, and believe that by alighting in those places they will draw closer to these ones. This the divine law forbade. The Prophet said, "Do not saddle your camel except (to travel) to three mosques."[22]

9) Also among them was their practice of naming their sons "'Abd al-'Uzzā," "'Abd Shams," and so forth.[23] God said,

> He it is who created you from a single soul, and made from it its spouse that he might take rest in her, (then when he covered her, she bore a light burden, and she passed [unnoticed] with it, but when it became heavy they cried unto Allah, their Lord, saying: If thou givest unto us a righteous son we shall be of the thankful. But when He gave unto them a righteous son, they ascribed unto Him partners in respect of that which He had given them. But God is high exalted above all that they associate with him).[24]

[18] Tirmidhī III: 46 #1574 Nudhūr 8, Ibn Ḥanbal II: 34, 69, 87, 165.

[19] "*al-yamīn al-mun'aqida*".

[20] "*Yamīn ghamūs*" literally means an oath which plunges its swearer into sin and then into the fires of Hell. It can mean a false oath made with the intention of cheating people.

[21] In this case, in both of these types of oaths the person would assume that power over what would happen in the future could depend on the will of one other than God.

[22] Mecca, Medina, and Jerusalem. Mishkāt, p. 140. Bukhārī Masjid Mecca 1, 6, Ṣaum 67, Ṣaid 26, Muslim, Abū Dāwūd, Tirmidhī, Nasā'ī, Dārimī, Ibn Ḥanbal.

[23] Such names refer to being the slave of the pre-Islamic God al-'Uzza or the sun (*al-shams*).

[24] Qur'ān 7:189-190. The portion in brackets was not included in the original text.

It is reported in the hadith that Eve called her son ʿAbd al-Ḥārith[25] and that this was due to an inspiration from the devil. It is confirmed in innumerable hadith reports that the Prophet changed the names of his companions ʿAbd al-ʿUzzā and ʿAbd Shams and suchlike to ʿAbd Allāh and ʿAbd al-Raḥmān and other similar names.[26]

These are embodiments and forms of associationism, which the law giver forbade due to their being forms of it, and God knows better.

---

[25] Tirmidhī IV:332 #5073 Tafsīr Sūra 7.
[26] Ibn Ḥanbal IV: 178.

CHAPTER 40

Belief in the Divine Attributes

You should know that one of the greatest forms of piety is believing in the Attributes of God, may He be Exalted, and the belief that He is qualified by them, for this opens a door between the worshipper and Him, may He be exalted, and prepares him for the disclosure of God's majesty and glory.

You should know that God, may He be Exalted, is above being compared to something which can be rationally comprehended or sensually perceived, or that attributes could adhere to Him as accidents do in their substrata,[1] and above being apprehended by ordinary minds, and having conventional language applied to him. Still, He must be made known to people in order that they will fulfill the perfection which is possible for them.

Thus it is necessary that the Attributes be applied in the sense of their utmost limit and not in the sense of their basic meaning. Hence the meaning of "Mercy" (*al-raḥma*) is bestowing blessings, not sentimentality of the heart and gentleness. It is necessary to use the words which metaphorically refer to the King's domination of his city for indicating His domination over all existent things, since there is no more apt expression for this concept. It is necessary that allegories (to human attributes) be employed on the condition that they are not to be taken literally, but rather that they indicate meanings associated with them in common usage. Thus, for example, "extending the hand" means generosity. Comparisons must also be used with the condition that those addressed must not be made to construe these as applying to things tainted by the animalistic, and this varies according to the differences in those addressed. Therefore, it will be said that "He sees" and "He hears," but it is not said that "He tastes" and "He touches." It is necessary to call

---

[1] "*Al-a'rāḍ fī maḥālhā.*" In terms of the Aristotelian thought of the time, the accidents were temporal manifestations which existed through attaching themselves to material substrata.

all of the proliferation of meanings which coincide in some aspect by one name, such as the "Sustainer" and the "Fashioner." It is necessary that everything not befitting Him should be negated, especially those things which wrong-doers have said concerning Him, for example, "He does not beget nor was He begotten."[2]

Without exception all of the heavenly religions agree that the Divine Attributes should be explained in this way, and that these expressions should be applied according to their proper meaning, and that nothing more should be discussed about them beyond this usage. In this way the upright generations (close to the time of the Prophet) passed away, then a group of Muslims became engrossed in discussing the Attributes, and in investigating their meanings without a revealed textual basis and without conclusive, demonstrative proof.

The Prophet, may the peace and blessings of God be upon him, said "Speculate about the creation, not about (the nature of) the Creator,"[3] and he interpreted God's, may He be Exalted, saying, "Your Lord, He is the utmost limit,"[4] by "there should be no speculation about the Lord."[5]

The Divine Attributes are not created, temporally originated things,[6] and speculating about them lies in asking "How is God qualified by them?" and therefore this is speculation about the Creator.

Al-Tirmidhī said concerning the hadith "The hand of God is full,"[7] that the religious leaders had said concerning this hadith that they believed it as it was given without interpreting or making conjectures about it. Likewise, more than one Imām, among them Sufyān al-Thaurī,[8] Mālik ibn Anas, Ibn ʿUyayna,[9] and Ibn al-Mubārak,[10] said that these things have been transmitted (in the hadith) and they believed them, and there was no scope for asking "how."[11] On

---

[2] Qur'ān 112:3.

[3] A hadith not found in the standard collections, but a close variant is found in Abū Nuʿaym *Hulliyat* and al-Isfahānī in *al-Targhīb wa-Tarhīb*.

[4] Qur'ān 53:42.

[5] This hadith is quoted by al-Suyūṭī, *al-Durr al-Manthūr* VI (Tehran: al-maktaba al-islāmiyya, 1957), 130, in explanation of Qur'ān 53:42, although it is not found in the standard collections.

[6] Thus our author rejects the Muʿtazilite position in this regard.

[7] Tirmidhī Tafsīr Sūra V. Bukhārī Tauḥīd, Muslim, Ibn Mājah, Ibn Ḥanbal.

[8] Al-Thaurī was a traditionalist and ascetic of the second Islamic century.

[9] Sufyān ibn ʿUyayna, born in Kufa and raised in Makkah, a traditionalist d. ca 813/14.

[10] An early ascetic and collector of traditions, d. 797.

[11] This entire report is found in *Sunan al-Tirmidhī* IV: 217 #5036 Tafsīr Sūra V.

another occasion he (Tirmidhī) said, "The application of these attributes, as they are, is not anthropomorphism. Rather it is anthropomorphism to say, 'Hearing is (literally) hearing (like our hearing) and seeing is (literally) seeing.'" And Ḥāfīẓ ibn Ḥajar said,[12]

> No definitive statement has been reported from the Prophet, may the peace and blessings of God be upon him, nor from any of the Companions, by any reliable chain of transmission, that the interpretation of any of these things is necessary; namely, the ambiguous (or allegorical) things; nor that mentioning them is forbidden.[13]
>
> And it is impossible that God should have ordered His Prophet to propagate what was divulged to him by his Lord, and revealed to him, "This day I have perfected your religion for you,"[14] and that he (the Prophet) would have left this subject (untouched) and not distinguished those things which may be attributed to Him, the Exalted, from those which may not be; at the same time as he urged the propagation of (the religion) by his saying, "Let the one who is present inform the absent one,"[15] so that they reported his sayings, acts, circumstances, and what was done in his presence, which all proves that they agreed to believe it as God, may He be Exalted, wished. His transcendence above literal comparisons to created things is required by His saying, "There is nothing like unto Him."[16] Therefore, whoever affirmed what opposes this after them (the Prophet and the Companions) has diverged from their practice.[17]

I hold that there is no difference among hearing, seeing, having power, laughing, speaking, and sitting; for regular speakers of the language understand these in a fashion inappropriate to the Holy. Is there any inconsistency in laughing except that it requires a mouth, and likewise so does speech? Is there any incompatibility in falling and descending other than that they require a hand and a leg? And

---

[12] Ibn Ḥajar al-ʿAsqalānī (d. 1449). Scholar and author of *Fatḥ al-Bārī*, a commentary on *Ṣaḥīḥ al-Bukhārī*.

[13] In Islamic theology, the Muʿtazilites said that the anthropomorphisms were allegories (*mutashābihāt*) and could be rationally interpreted in exegeting. The term "*mutashābihāt*"—ambiguous matters or allegories appears in the Qur'ān 3:7, where it is contrasted to "*muḥkamāt*"—decisive revelations. The People of the Hadith, on the other hand, shunned interpretation and said we must accept the seemingly anthropomorphic statements "without asking how."

[14] Qur'ān 5:3.

[15] Ibn Ḥanbal I: 83.

[16] Qur'ān 42:11.

[17] End of quote from Ibn Ḥajar.

likewise seeing and hearing require eyes and ears, and God knows better.[18]

These speculators behaved contemptuously toward the People of the Hadith[19] calling them corporealists and anthropomorphists and saying that they sought refuge in the formula of "without asking how" (*bi lā kaif*).[20] It has become eminently clear to me that this contempt of theirs is unfounded and that they err in their sayings both from the viewpoint of tradition and of reason and that they err in slandering the leaders of the true religion.

This can be elaborated as follows. There are two positions:

1) One concerns the fashion in which God, Blessed and Exalted, can be described by these attributes and whether they are additional to His essence or identical with it,[21] and the true nature of (His) hearing, seeing, speaking and so on; for what is manifestly understood from these words is incommensurate with the Holy.

The truth about this position is that the Prophet, may the peace and blessings of God be upon him, did not discuss this at all, indeed he prohibited his community from talking about it and discussing it. Thus it is not anyone's prerogative to attempt to do what he prohibited.

2) The second position concerns the things by which we may either describe Him or we may not describe Him, may He be exalted, according to the divine law. The truth is that it is not for us to hold opinions about His Attributes and Names, in the sense that even if we were aware of the principles on which the divine law based its explanation of His attributes, may He be exalted, as we stated at the beginning of this chapter; still many people, if they were allowed to discuss the Attributes, would go astray and would mislead others. Although describing Him by many of the Divine Attributes was basically legitimate, some of the unbelievers misapplied these word and this spread among them; therefore the provisions

---

[18] That is, why can seeing and hearing be applied if laughing cannot? The reference is to a theological argument as to whether certain attributes could be applied to God while others could not.

[19] The followers of Aḥmad ibn Ḥanbal who opposed the Mu'tazilites and took anthropomorphic references to God literally.

[20] This formula "without asking how" was an approach to anthropomorphic references in the Qur'ān and sunna taken by al-Ash'arī and others. The critical accusation that they "sought refuge in *bi lā kaif*" came from Mu'tazilites such as al-Zamakhsharī.

[21] The Mu'tazilites held that the attributes were additional to His essence since eternity is the only eternal quality of God.

of the divine law prohibited their use in order to prevent this harm. Many of the Attributes can be misconstrued by applying their outward meanings in senses contrary to what was intended, therefore this (abuse) must be guarded against. For these wise reasons the divine law did not leave them open for us to pass opinion upon, and did not permit discussing them using personal opinion (*ra'y*).

In summary, we are permitted to use "laughing," "happiness," "smiling," "anger," and "satisfaction"; but we cannot use "weeping," "fearing," and so on, although the manner of performing the two types of actions are close. The issue, as we have investigated it, is supported both by reason and by tradition. It cannot be nullified from any aspect, and the extended refutation of their arguments and opinions is more appropriate elsewhere.

It is our right to interpret them (the anthropomorphic attributes) in senses more easy to comprehend and more suitable than what they said, for clarity's sake, since holding these (specific) meanings is not stipulated, and the one investigating by means of intellectual proof is not forced to use them, and they are not preferable to others, nor do they have any exclusive merit. There is no ruling to the effect that God meant what we hold, nor is there a consensus that one should believe in them and acknowledge them; such a thing would be absurd.

Thus, we say, for example, that since you have before you three types: living, dead, and inanimate; the living is the closest likeness to God, due to His being "knowing" and His "influencing" the creation; and therefore, He must be called "Living." Since knowledge on our part is disclosure (of things) and to God all things are disclosed since they are all subsumed in His Essence, and since they (the things) took on existence in a detailed manner, it is necessary to call Him, "Omniscient." Since vision and hearing are complete disclosures of things seen and heard, and this occurs for God in a more perfect way, it is necessary that He should be called "Seeing" and "Hearing." Then, since by our saying "someone willed," we mean that he suddenly decided to act or not to do an act, and the Merciful performs many of His actions by creating a condition (sharṭ) or a capacity in the world so that through this something is required which was not necessary before, and there occurs in some of the lofty places[22] a consensus which had not existed (before)

[22] The Highest Council and Holy Enclave.

through His permission and His decree; it is necessary that He be called, "Willing." In addition, since His essential, eternal, unitary will, which is interpreted as the decree of (His) essence, attached to the entire universe all at once, following which the temporally originated events unfold day after day, it is correct to attribute to every temporally originated occurrence an event occurring individually, and to say "He (God) willed such and such a thing." Since by our saying "someone is able" we only mean that it is possible for him to act, and that he is not prevented from it by an external cause. As for the preference of one of the two things of which he is capable on the part of the one give power, this does not preclude his being designated by name of "the one who is able,"[23] and the Merciful is Able to do all things, but He does certain acts and not their opposites due to His providence[24] and His essential decree—thus it is necessary that He be called "Able." Since by our saying, "Someone spoke to someone else," we mean only that he communicated his intended meanings conjoined with words indicating them, and since the Merciful One sometimes communicates knowledge to His servant, and communicates along with it words established in his imaginative faculty which refer to it so as to make the teaching clearer, God must necessarily be called "Speaking."

God, may He be Exalted said, "It is not for any human that Allah should speak to him except through revelation (*waḥi*) or from behind a veil, or that He sends out a messenger, who reveals with His permission what He wills. Indeed, He is Exalted, Wise."[25]

"Revelation" (*waḥi*) is inspiration into the spirit (*rūḥ*) by means of a dream, or the creation of necessary knowledge[26] while he is concentrating on the Unseen World. "From behind a veil" is that he hears composed speech (*kalām manẓūm*) as though he had heard it from outside, while he doesn't see his interlocutor, or that "He sends out a messenger," and then the angel appears before him. Sometimes while he is concentrating on the Unseen and subjugating the senses, he will hear a sound like the tinkling of bells, just as

---

[23] This definition contrasts God's power in choosing an act to that of humans for whom power could only be the power to do an act but not its opposite according to Ash'arite theology.

[24] This opposes the position of the philosophers who held that God has no choice.

[25] Qur'ān 42:51.

[26] "*'Ilm ḍarūrī.*" Essential or necessary knowledge which arises without discursive reasoning.

visions of red and black colors may occur when falling into unconsciousness.[27]

There is in the Holy Enclave an order which they desire to establish among mankind, so that if people are in conformity with it (the order) they will be joined to the Highest Council and are taken out of the darkness into the light of God and His bounty, and they are given joy in their souls and other humans and the angels are inspired to treat them well. If they oppose it (the order), they remain separate from the Highest Council and are struck by their (the Council's) hatred and punished in a way that has been mentioned above. Therefore it must necessarily be said that God is "pleased" and "gives a reward of thanks," or is "angered" and "bestows curses;" all (of this), derives from the running of the universe according to the requirement of the beneficial purpose (*al-maṣlaḥa*). Sometimes the creation of the thing which is prayed for turns out to be part of the order of the universe, and this is called "the answering of prayer." Since "vision," according to our usage, is the disclosure of what is seen in the most perfect way; and since people, when they are transported to certain states which they had been promised at the Resurrection, will attain the theophany[28] which stands in the center of the World of Images, so that all of them will see Him with their eyes;[29] it is necessary to say that, "you will see Him as you see the full moon,"[30] and God knows better.

---

[27] Some of these effects are described in hadiths recounting the Prophet's experiences when receiving revelations. For example *Mishkāt*, p. 1254. Bukhārī Bad' al-khalq 6, Bad' al-wahī 1, Muslim, Tirmidhī, Nasā'ī, Muwaṭṭa', Ibn Ḥanbal.

[28] This concept of theophany (*tajallī*) refers to mystical illumination or revelation in Islamic mystical thought. For Shāh Walī Allāh there is a Great Theophany (*al-tajallī al-aʿẓam*) which is the source of the knowledge of God and approaching Him, *al-Tafhīmāt al-Ilāhiyya* II:264) and other *tajalliyyāt* which take place at certain times. This theophany in the *ʿālam al-mithāl* is mentioned in Chapter 42 of the present work.

[29] Thus he supports the theological position of a literal vision of God at the Resurrection.

[30] As promised in the hadith. *Mishkāt*, p. 1206. Bukhārī Mawāqit 16, Tauḥīd 24, Muslim, Ibn Mājah, Ibn Ḥanbal.

CHAPTER 41

Belief in Predestination

Among the highest types of piety is belief in predestination, due to the fact that through it humans contemplate the unitary management (*al-tadbīr al-wāḥid*) which integrates the universe. The one who believes in it correctly sets his sites on the divine level, seeing the world and whatever is in it as its shadow. He sees that choice on the part of human beings arises from the decree of God, like the image which is reflected in a mirror. This will prepare him most completely for the disclosure of the unitary management (*tadbīr waḥdānī*) at that higher level, even if this takes place only at the Resurrection. The Prophet, may the peace and blessings of God be upon him, informed us of its great importance among types of piety when he said, "Whoever does not believe in predestination, its good and its evil, I am free of him."[1] And he said, "No person believes until he believes in predestination, its good and its evil, and until he knows that what strikes him would never have missed him, and what misses him would never have struck him."[2]

You should know that the essential primordial knowledge of God, may He be exalted, encompassed every event that has existed or will take place. It is impossible that His knowledge should lag behind something or that something should come into effect which He did not know, for this would be ignorance, not knowledge. This is the issue of omniscience, not the issue of pre-destination, and no Islamic sect has disagreed with it. Rather, the predestination (*al-qadar*) which is indicated in abundantly transmitted and well-known hadith reports, which the righteous predecessors believed in, and which only the true investigators were given success (by God) in understanding, and concerning which the question has been raised as to whether it contradicts God's imposition of religious duties (on humans), and wherein then is the need of acting?

---

[1] *Kanz al-'Ummāl* I: 69. *Musnad* Abī Ya'lā.
[2] *Mishkāt*, p. 30. Tirmidhī III: 307 #2231 Qadar 10.

This is the compelling predestination (*al-qadar al-mulzim*). It makes events necessary before their existence, so that through this divine determination (*ījāb*) comes into being, which cannot be prevented by fleeing nor avoided through a ruse.

This predestination occurred five times.

1) The first of them is that God decided in pre-existence to create the world in the best way possible,[3] taking into account those things conducive to benefit and effecting the good in relation to their time of coming into existence. The knowledge of God determined on the appointment of a single form with which no other would combine. Thus temporal events are an ordered sequence which has a combined existence, never holding true for two multiplicities. Hence the will to bring the world into being (*ījād*) on the part of "the One from who nothing is concealed," is in itself the specifying of the form of its existence to the full extent of His command.

2) The second is that He measured the destinies, and it is reported that He recorded the destinies of all created things, and the meaning is the same; fifty thousand years before He created the heavens and the earth.[4] That is, He created the things according to a primordial providence in the imaginative faculty of the Throne[5] and He created all the forms there, and this is what is spoken of as the "*dhikr*"[6] in the divine laws. Thus, for example, the form of Muḥammad, may the peace and blessings of God be upon him, was actualized There, and His sending him to creation at a certain time, and his warning them, and the repudiation of Abū Lahab[7] and his (Abū Lahab's) soul being encompassed by error in this world, and the fire burning him in the afterlife. These forms are a cause for the coming into existence of events as they were (depicted) at that plane, like the effect of the image imprinted in our

---

[3] For a discussion of this issue in Islamic theology see Eric J. Ormsby. *Theodicy in Islamic Thought* (Princeton: Princeton University Press, 1984).

[4] Muslim, Qadar, 16. Siddiqi trans. IV:1397.

[5] The Throne in Shāh Walī Allāh's cosmology is an expression for the source of the forms which come to inhere in matter. Thus, the imagination of the Throne is like the world of images, wherein forms exist primordially. See also Mullā Ṣadrā. *The Wisdom of the Throne*, translated by James W. Morris (Princeton: Princeton University Press, 1981).

[6] *Dhikr* refers to "the Preserved Tablet" (*al-lauḥ al-maḥfūẓ*) which is under the Throne of God and on which everything is inscribed which will happen, i.e., on which the forms are imprinted.

[7] Abū Lahab was an uncle of the Prophet and the only member of his own clan to oppose him. His punishment by Hell fire is mentioned in Qur'ān 111:3.

minds of the foot slipping on the trunk of a tree placed upon two walls, and it would not have slipped if that (piece of wood) had been placed on the ground.

3) The third is that when God created Adam, may peace be upon him, to be the father of humankind, so that the human species would originate from him, He brought into being in the World of Images the forms of his descendants and represented their (ultimate) felicity and their misery by light or darkness.[8] He gave them the capacity to be charged with religious duties and He created in them knowledge of good and humbleness before Him. This is the basis of the covenant infused in their original nature,[9] so that they will be held accountable according to it, even though they may have forgotten this event, since the souls created on the earth are nothing but the shadow of the forms which existed (in *mithāl*) on that day. What is inherent in the souls is what was infused on that day.

4) The fourth of them is when the spirit is breathed into the foetus. When a seed is put in the ground at a particular time and specific conditions surround it, the person acquainted with the property of that species of palm tree, and the property of that earth and that water and climate, can forecast that its growth will be sound and that certain things about it will turn out in a certain way. Likewise the governing angels get information on that day, and there are disclosed to them the matters of a person's life span, his livelihood, and whether he will perform the actions of one whose angelic side overcomes his bestial side, or the contrary, and what will be the characteristics of the person's happiness and his misery.[10]

5) The fifth of them occurs a short before the occurrence of an event, so that the command descends from the Holy Enclave to the earth and a imaginal (*mithālī*)[11] thing is transferred, thus His decrees are spread over the earth.

---

[8] The imaginal (*mithālī*) forms which unite with them at Resurrection.

[9] The primordial covenant, referred to in Qur'ān 7:172, when all of the souls acknowledged Allāh as their Lord.

[10] This is parallel to a hadith which says, "When the embryo is four months old God sends an angel with four decrees which he writes down, viz. its actions, life spans, sustenance, and whether it is happy or miserable . . ." *Mishkāt*, p. 23–24. Bukhārī Bad' al-Khalq 6, Anbiyā' 1, Tauḥīd 82, etc., Muslim, Abū Dāwūd, Tirmidhī, Ibn Mājah, Ibn Ḥanbal. Shāh Walī Allāh puts this into a context of natural determination, like that of a seed planted at a certain place or time.

[11] The concept of the imaginal or mithālī world is discussed by Shāh Walī Allāh in Chapter 2 of the present work.

I have witnessed this on various occasions. Once when some people were quarreling about a matter and had become full of resentment against each other. I sought help from God and then I saw an illuminated imaginal point which had descended from the Holy Enclave to the earth and had begun to spread gradually. As it spread the rancor left them, and we had not departed the assembly before they had become friendly towards one another and each of them had regained his previous affection. This for me was a wonderful sign of God.

Another of these events occurred when one of my children once was ill and my mind was preoccupied with him. While I was performing the Noon prayer I witnessed his death descend and he died that night.

The prophetic traditions have clearly shown that events are created by God, may He be Exalted, in some way before they take place on earth. Hence they descend to this world and become manifest in it as they were created for the first time (in the World of Images), as the usual practice of God, may He be Exalted. Moreover, sometimes the (previously) established thing may be effaced and the non-existent thing may become established with respect to this existence. God, may He be Exalted, said, "Allah effaces what He wills, and establishes (what He wills) and with Him is the source of the Book."[12]

For example God, may He be Exalted, creates some affliction and sends it down on the one to be afflicted, and then petitionary prayer ascends and repels it. He may have created death, then some righteous action ascends and repels it. The way to understand this is that the created thing which was descending was a customary cause, like eating and drinking in relation to causing a person to remain alive, and ingesting poison and being struck with a sword in relationship to causing a person's death.[13] Many hadiths have indicated the confirmation of the existence of a world[14] in which accidents[15] are embodied, concepts conveyed, and a thing is created before its appearance on the earth. Examples are kinship's

[12] Qur'ān 13:39.

[13] The distinction between the customary and the absolute is discussed in Chapter 51 of the present work.

[14] The World of Images.

[15] In the technical sense of accidents which are not self-sufficient and inhere in other things, as opposed to substances.

being attached to the Throne, the descending of civil-wars like the falling of rain drops, and the creation of the Nile and the Euphrates at the root of the lote-tree, then their descent to earth. (Other examples are) the descent of iron and livestock; the descent of the Qur'ān to the lowest heaven all at once, and the presence of Heaven and Hell between the Prophet and the wall of the mosque so that he was able to reach the grapes and so that the heat of Hell could be felt. Others are the struggle between affliction and prayer, the creation of the seed of Adam, the creation of the intellect and its approaching or running away, the two (chapters of the Qur'ān)[16] coming forward illuminated as if they were two flocks of birds, the weighing of the deeds, and the bordering of Paradise with hateful things and of Hell with attractive ones,[17] and things like this which are not concealed from the person with the least knowledge of the prophetic traditions.

You should know that predestination (*qadar*) does not militate against the laws of causality, because it (predestination) immediately establishes an all-embracing coherency in the course of things. This is the Prophet's, may the peace and blessings of God be upon him, statement concerning whether amulets, medicine, and precautionary measures, could turn back something predestined by God. He said, "They are (also) part of God's predestination."[18] And 'Umar, may God be pleased with him, said in the story of Sargh, "Is it not so that if you graze her (a she camel) in a fertile place, you graze her by God's predestination?"[19] and so on. People have free

---

[16] Sūras II and III, according to a hadith.

[17] Many of these examples from the hadith are cited in Chapter 2 of the present work on "The World of Images."

[18] In a well-known hadith the Prophet was asked by a companion, "Do you think that the amulets and medicines we use and the precautionary measures we take, gainsay the Divine predestination?" *Mishkāt*, p. 28. Ibn Mājah Ṭibb 1, Tirmidhī, Ibn Ḥanbal.

[19] Sargh refers to a town in the Tabūk valley. The entire hadith is reported by Mālik from 'Abd Allāh ibn 'Abbāṣ concerning the story of the plague in Syria. When 'Umar came to Sargh, and heard about the plague in Syria, he ordered a turning back. Then Abū 'Ubaidah ibn Al-Jarrah said to him, "Are you fleeing from the predestination of God?" And the end 'Umar said to him, "Yes, we flee from the predestination of God to the predestination of God. Have you considered that if you had a she-camel, and you descended into a valley with two slopes, one of them fertile and the other barren, wouldn't you graze her in the fertile by the predestination of Allah. If you grazed her in the barren, it would also be by the predestination of Allah." Bukhārī Ṭibb 30, Muslim, Mālik, *al-Muwaṭṭa'*, trans. Muhammad Rahimuddin (Lahore: Ashraf, 1980), 372–373.

choice (*ikhtiyār*) in their acts, but indeed in this free choice they have no free choice due to its being caused by a form coming into the mind of what is desired and its benefit, and the arising of a motivation, and a resolution (to do the act), due to something which the person has no knowledge of, so how can there be free choice in these things? This is the Prophet's saying, "The hearts of men are between two of God's fingers which He turns as He wishes."[20] And God knows better.

[20] *Mishkāt*, p. 25. Muslim Qadar 17, Tirmidhī, Ibn Mājah, Ibn Ḥanbal.

CHAPTER 42

Belief that Worship is the Right of God, May He be Exalted, From His Worshippers Because He Bestows Blessings on Them and Requites Them at His Will

You should know that one of the greatest types of piety is that a person believes wholeheartedly, so that for him nothing which contradicts this belief is possible, that worship is God's right from His worshippers, and that they are held answerable for worshipping God, may He be Exalted, just like the rest of the rights that those having rights over them may demand. The Prophet, may the peace and blessings of God be upon him, said to Muʿādh,[1] "O Muʿādh, do you know what is God's right over his servants, and what is the servants' right with respect to God?" Muʿādh answered, "God and His Prophet know best." The Prophet said, "God's right from his servants is that they worship Him and do not associate anything with Him; and the right of the servants from God, may He be Exalted, is that the one who does not associate anything with God will not be punished."[2]

This is because the person who does not believe this with firm conviction, and holds it possible that he is futile, neglected and not required to worship, and that he is not held accountable for it on the part of a Willing, Choosing Lord—is an atheist (*dahrī*). His worship does not take hold even if he performs it with his limbs in place of his heart. It does not open a door between him and his Lord, and it is just a habit like the rest of his habits.

The basis of this is that it was confirmed by the realizations of the prophets and their heirs, may peace and blessings be upon them, that there is a plane (*mauṭin*) among the planes of Jabarūt at which there is a will and purpose in the sense of the consensus (*ijmāʿ*) upon an act, together with the equal correctness of doing the act

---

[1] Muʿādh ibn Jabal.

[2] *Mishkāt*, p. 10. Bukhārī Libās 101, Jihād 46, etc. Muslim, Tirmidhī, Ibn Mājah, Ibn Ḥanbal.

or not doing the act, with respect to that plane. Since the heavenly purpose[3] "does not let anything remain nor spare it"[4] unless God has made its existence necessary or has made its non-existence necessary, there is no existence of the state awaiting actualization at that plane nor any countenancing of the people called philosophers who claim that "the will" is in this sense.[5] They have kept one thing in mind while overlooking many others, and they are veiled from witnessing this plane, and defeated by the evidence of the external world and their own minds.

As for their veil, it is that they are not guided to a plane between the Greatest Theophany (*al-tajallī al-aʿẓam*)[6] and the Highest Council which is like the ray inhering in a jewel, and "God has the most sublime similitude."[7] At this plane there is represented a consensus about a thing which the knowledge of the Highest Council and their attitudes necessitated, after doing the act or not doing it were found to be equal.

As for the proof against them (the philosophers),[8] it is that each one of us knows spontaneously that he extends his hand and picks up a pen, for example, and that he in this is a willer and intender. In relation to him, doing the act or not doing the act are equally (possible) with respect to this intent and to those faculties embodied in his soul, even though with respect to the heavenly purpose (*maṣlaḥa fauqāniyya*) everything either necessarily must be done or must not be done. Likewise is the situation of everything which is required by a particular capacity; thus forms descend from the Maker of Forms on matter capable of receiving them, such as the answering subsequent to petitionary prayer in which there is an access for something new to come into being. Therefore you might say that this is ignorance about the necessity of the thing with respect to the heavenly purpose, for how can this (freedom to do the act or not do it) exist at one of the planes of God? I say, God

---

[3] "*Maṣlaḥa fauqāniyya*."

[4] Qur'ān 74:28.

[5] They construe God's will as having no alternatives. In the case of this phenomenon this is true but otherwise this is wrong.

[6] The Greatest Theophany (*al-tajallī al-aʿẓam*) is the self-manifestation of God's names and attributes which is the source of the knowledge of God in a transcendent and holy form.

[7] Qur'ān 16:60.

[8] That is, determination occurs at the highest level, but choice exists in terms of doing or not doing an act.

forbid! Rather this is knowledge of and doing justice to the nature of this plane. It would have amounted to ignorance had it been said that the thing was not necessary at all. The divine laws have refuted this ignorance since they have confirmed the belief in predestination and that whatever has befallen you could not have missed you, and what missed you could not have befallen you.[9] As for the case of saying that both doing the action and not doing it are possible with respect to this plane, this is undoubtedly true knowledge. For example, if you see the male animal performing masculine actions and the female performing feminine ones, then if you judge that these actions emerge under compulsion like the movement of a stone when it is rolling along, you judge falsely If you judge that these acts emerge without a cause which requires them, such that neither the masculine disposition requires this type of thing, nor does the feminine one require that one, then you likewise judge falsely. If you judge that the will embodied in their souls follows a heavenly necessity, and is dependent on it, and that the will cannot erupt in a sudden independent outburst which does not have behind it some goal, then you likewise judge falsely. Rather, the certain truth is somewhere between the two, to wit, that free choice results from causes and cannot fail to be attendant on its causes, while the act which is willed is required by causes, and it is not possible that it should not be. However, it is the nature of this free will that it delights in keeping itself in mind, and in not considering what is beyond that. Thus, if you were to convey the truth about this plane and say, "I feel that doing or not doing an act are equivalent, and I chose to act, so that free will was a cause for the act;" then you are correct and you have spoken well. The divine laws provided information about this will which is embodied at this plane.[10]

In summary, it is established that there is a will whose relationships change, that there is requital in this world and the next, and that the Manager of the world manages the world through the imposition of a divine law that people must follow, in order to benefit by it. Hence, the case is similar to that of the master using the services of his servants and requiring this from them, and being

---

[9] Parallel to the previously cited hadith, Tirmidhī Qadar 10, "whatever struck him could not have missed him."

[10] That is, it provided the information that reward and punishment based on choosing actions are true at the level of this plane.

pleased with the one who serves and angry with the one who does not render service. The divine laws have revealed what we mentioned about the Attributes of God and other things, using a mode of expression unsurpassed in eloquence, nor is there language more explanatory of truth than these (divine laws) irrespective of whether they use lexical meanings or conventional metaphors. The divine laws have enabled people to have this knowledge (of worship being God's right) which had been obscure for them, through three positions which have become accepted by them, functioning as well-known, obvious things for them.

1) One of them is that the He, may He be Exalted, is a Bestower of blessings, and thanking the Bestower of blessing is obligatory, and worship is thanking Him for His blessings.

2) The second is that He severely punishes those who turn away from Him and forsake His worship in this world.

3) The third is that He requites in the next world those who obey and those who disobey.

Then three fields of knowledge are expanded from there:

a) remembering the benefits of Allah,[11]
b) reminding of the days of God,[12]
c) and keeping in mind the next life;

so that the Qur'ān was revealed explaining these three fields of knowledge.

Divine providence considered it important to explain these types of knowledge because man has had created at the base of his original nature an inclination to his Creator, may His majesty be glorified, and this inclination is a subtle matter which is only apparent in his (man's) natural disposition and his anticipated situation (*maẓinna*). Man's natural disposition and his anticipated situation, according to that which has been established by true intuitive perception, is to believe that worship is the right of God, may He be Exalted, from His worshippers because He is a Bestower of blessings on them, and a Requiter of their acts. Thus whoever denies the human will or the confirmation of His right over His worshippers, or denies the requital (for actions) is an atheist (*dahrī*) bereft of his sound constitution, because he has corrupted for himself the

[11] As in Qur'ān 55:13 ff.
[12] "*Ayyām Allāh*", is the requital of nations by God.

anticipated source of the natural inclination (to God) placed in his constitution, and that he is his deputy and vicegerent and takes his place.

If you want to know the truth about this inclination, then know that in the spirit (*rūḥ*) of man there is an illuminated subtle spiritual center (*laṭīfa*) inclined by its nature toward God, Great and Exalted, just as iron is attracted to a magnet. This is a matter understood through spiritual intuition, for whoever devotes himself to examining his own subtle spiritual centers, and recognizes every center (*laṭīfa*) as it is, will inevitably perceive this illuminated spiritual center and understand its inclination to God, may He be Exalted. This inclination is called among the people of intuitive perception "essential love" (*muḥabba dhātiyya*), and it is like the rest of those states of mind which cannot be captured by logical proofs such as the hunger of this hungry person and the thirst of this thirsty one. Thus, if the person is in a stupor due to the determinants of his lower centers[13] he is like someone who uses an anesthetizing substance in his body so that he feels neither heat nor cold. Then if his lower centers cease competing with one another either through necessary death, requiring the dispersal of many of the parts of his pneuma (*nasama*) and the loss of many of its properties and faculties; or due to voluntary death,[14] and his persisting in remarkable techniques of psychological and physical exercise, he becomes like a man whose anesthetic has worn off so that he perceives the things he was completely unaware of before. When a person dies without having addressed himself to God, may He be Exalted, if this lack was due to simple ignorance and plain deficiency, then he is wretched with reference to the perfection of the species. Some of what is There (above) may be disclosed to him, but the disclosure will not be complete due to the deficiency of his preparedness, so he remains bewildered and perplexed. If this occurs together with the establishment of an attitude contradictory (to turning towards God) in his cognitive or practical faculties he will experience mutual conflict pulling back and forth. Thus the rational soul will be drawn in the direction of Jabarūt.

---

[13] The lower soul (*nafs*), heart (*qalb*) and mind (*'aql*) are spiritual centers close to corresponding parts of the physical body, which may be agitated by the flow of vapors distilled from the four elements ingested in food and comprising the material of the body.

[14] Through following Sufi training, i.e., death to the false self.

The pneuma, due to what it has acquired from the contradictory attitude, will be drawn towards the lower side, so that he will find himself in desolation arising from the substance of the (rational) soul having become spread out through the substance (of the pneuma). Sometimes this will necessitate the representation of events which embody desolation, in the same way that the choleric person sees fire and flame in his dream. This is the basic principle orienting the science of the gnosis of the soul.[15] The angry glance of the Highest Council is also cast on such a person, causing intuitions to enter the hearts of the angels and other agents with wills of their own, to punish him and pain him. This is the basic principle guiding the recognition of the causes of the ideas and motives which arise in the souls of human beings.

In sum, the inclination to the direction of Jabarūt and the necessity of acting so as to break the bonds of the mutual competition of the lower subtle spiritual centers, as well as his being held accountable for not doing this act are at the level of the determinants of the specific form, its faculties, and its effects. These were emanated into every individual of the species from the Maker of the forms and the Emanator of existence, in conformity to what is conducive to the universal good (*maṣlaḥa kulliyya*); not only due to human agreement, their committing themselves, and their basing their conventions on this. All of these actions, in reality, are the right (*ḥaqq*) of that illuminated spiritual center[16] which is drawn toward God, and they abundantly furnish its (the soul's) requirements and straighten its crookedness. Since this meaning is subtle and this spiritual center is only perceived by a small group, it is necessary to attribute "the right" to that towards which this spiritual center inclines (God), and to that which is its goal, and to that to which it turns; as if that were a specification of a certain faculty of the soul because of which the soul inclines, and as if

---

[15] In which the mystics observe a person's psychological and emotional states while his soul is undergoing spiritual training.

[16] The "perplexing stone" (*ḥajar-i-baht*). One of the higher or hidden spiritual centers mentioned by name in his *Alṭāf al-Quds* and other mystical works. There is a connection between the Greatest Theophany mentioned earlier in this chapter, and this spiritual center, which is said to be the point opposite it in a particular person on which its rays fall. See, for example, Walī Allāh, *Lamaḥāt*, Lamḥa 58. On this specific topic see M.K. Hermansen, "Shah Walī Allāh's Theory of the Subtle Spiritual Centers (Laṭā'if): A Sufi Theory of Personhood and Self-Transformation." *Journal of Near Eastern Studies* (January 1988): 1–25.

that were the abridgment of our saying that "the right" (*ḥaqq*) of this spiritual center is with respect to its being inclined toward God. Thus the divine laws were revealed, disclosing this secret with a simple mode of expression that people could understand through their innate knowledge. The practice of God is to bestow (this expression) through revealing subtle meanings in forms related to them at the imaginal (*mithālī*) level of existence, as one of us may receive an abstract meaning during a dream in the form of the thing customarily connected with it, or something like it and similar to it. Thus it has been said that "worship is the right of God, may He be Exalted, from His servants," and according to this an analogy should be drawn concerning the "right" of the Qur'ān, the "right" of the Prophet, and the "right" of the master, the parents, and the relatives. Thus all of these are the "right" of a person's soul over itself to fulfill its perfections, and the soul should not commit injustice against itself.

Rather (in these instances) the "right" has been attributed to whoever is involved with the person in this association, or to the one from whom these rights are claimed. Therefore, do not be one of those who stops at the externals,[17] but rather be one who makes inquiries into the matter as it really is.

[17] And thinks that these rights belong to other than the self and not the self.

CHAPTER 43

Honoring the Emblems of Allāh, May He be Exalted

God, may He be exalted, has said, "Whoever honors the emblems (*sha'ā'ir*)[1] of Allāh, it is surely due to devotion of the hearts."[2] You should know that the foundation of the divine laws is reverence for the emblems of God, may He be exalted, and drawing nearer to Him by means of them. This is due to what we have indicated to the effect that the method which God has appointed for people is the copying of those things in the transcendent realm by things which are easily accomplished by the animalistic (side). I mean by "emblems" those external, tangible things appointed for the purpose of worshipping God and made specific to Him so that revering these things becomes like revering God and neglecting them is like giving insufficient regard to God. This becomes embedded in the core of their hearts and can't be removed without cutting them to pieces. The emblems only become emblems through a natural process, which is that their souls become content with a habit and a trait and it becomes a famous and well-known thing joined with primary axioms, and about which there is no doubt. Thus at this point the mercy of God manifests in the form of things which their natures and general knowledge deem to be obligatory. Hence they accept these emblems and the veil is lifted disclosing their true nature, and the mission reaches the nearest and furthest equally, and at this point revering them becomes prescribed for them.

The situation becomes tantamount to swearing on oath in the name of God, which instills in a person's soul that it would be remiss with respect to God to violate this, so that he is held responsible for what he had inwardly promised. Likewise these practices become well known among them as things which their branches

---

[1] I have translated this word (*sha'ā'ir*) by emblems due to the author's usage and its semantic associations derived from the root (*sh-'-r*). It can also mean rites, and in some places I have preferred that connotation.

[2] Qur'ān 22:32.

of knowledge comply with. The fact that their branches of knowledge comply with these practices necessitates that the mercy of God will not manifest to them except through what they are following, since the divine management is based firstly on the easiest thing and then on what is even more easy. It is also required that they hold themselves to the highest level of reverence that they are capable of having for these emblems, because their perfection lies in reverence which is not tarnished by neglect.

God, may He be exalted, did not impose a thing on His worshippers in order to benefit Himself—He is greatly Exalted above that—but rather due to a benefit which devolves upon them. They are only perfected through this highest degree of reverence, and thus are admonished according to their state and commanded not "to be remiss regarding God."[3] The actual intention of the providence behind the divine legislation is not the condition of the individual but rather the condition of the collectivity, as if this stands for all persons, and "the conclusive argument is from God."[4]

The main emblems (sha'ā'ir) of God are four: The Qur'ān, the Ka'ba, the Prophet, and the ritual prayer.

1) As for the Qur'ān, people used to have disseminated amongst them the proclamations of the kings to their subjects, and their honoring of the kings was connected with their honoring their proclamations. The books revealed to the Prophets and the writings of others were (also) circulated, and the people's embracing their teachings is accompanied by reverence for these books and reciting them.

It is clearly impossible in the context of the passage of time for them to follow and obey this knowledge without having a book which they recite and transmit. Therefore in this situation people needed the mercy of God to become manifest in the form of a book revealed from the Lord of the Worlds, so that they could revere it. Among (the ways of revering it) is listening closely when it is recited and refraining from talking, making haste to carry out its commands such as the prostration of recitation[5] and saying "*Subḥān Allāh*" wherever this is commanded,[6] and among them is

---

[3] Qur'ān 39:56.

[4] Qur'ān 6:149.

[5] When reciting or listening to the Qur'ān, Muslims are to perform a prostration in certain places where prostration to God is mentioned in the text.

[6] At certain places where the name of God (Allāh) is mentioned during the recitation of the Qur'ān, Muslims are to say "Glory be to God" (*Subḥān Allāh*).

not touching the Qur'ān except when one is in a state of ritual purity.

2) As for the Ka'ba (being an emblem of God), the people at the time of Abraham, may peace be upon him, used to be preoccupied with constructing temples and places of worship in the name of the spiritual power of the sun and other celestial bodies. For them, turning their attention to a transcendent and intangible being without a temple constructed in his name so that he was represented and embodied by this as a means of their drawing closer to him, was something impossible and inconceivable. The people of that time needed the mercy of God to become manifest to them in the form of a structure which they could circumambulate, and thereby draw nearer to God, so that they were summoned to the House (of God) and commanded to revere it. Then generation after generation arose in the knowledge that revering it led to revering God and that being remiss toward it led to being remiss towards God's rights, and due to this the pilgrimage to it is obligatory and they were commanded to honor it. Among (aspects of reverence are) that only the ritually pure should circumambulate it, that they should use it as a direction (Qibla) to face in prayer, and that facing it or turning one's back to it while relieving oneself is considered repugnant.

3) As for the Prophet, he is only called a messenger (*mursal*) as a metaphorical allusion to the messengers of the kings sent to their subjects to inform them of the king's commands and prohibitions. Their obeying him was only consequent to their respect due arising from their honoring the One who sent him to them. Among (ways of) respecting the Prophet is the necessity of obeying him, asking for blessings upon him, and not speaking to him in a loud voice.[7]

4) As for the ritual prayer, what is intended in it is an analogy to the state of a king's servants when they stand before him, their intimate discourse with him, and their humbling themselves before him. Therefore giving praise is required before making petitionary prayer and the person should hold himself to the forms of behavior which must be complied with when speaking before kings,

---

[7] As commanded in Qur'ān 33:56, "O you who believe send peace and blessings upon him," and Qur'ān 49:2. "O you who believe, don't raise your voices above the voice of the Prophet, nor shout when speaking to him as you shout to one another."

such as clasping the hands together and not looking this way and that, and this is his saying, may the peace and blessings of God be upon him, "When one of you offers prayers, then God is facing him,"[8] and God knows better.

[8] Bukhārī Ṣalāt, 33, Adhān 94, Adab 75.

CHAPTER 44

The Inner Meanings of the Ablution and Full Bath

You should know that a human being may be snatched away from the turbidities of nature to the lights of the Holy Enclave so that these lights come to dominate him and he becomes for a time in some sense free from the determinants of nature. Thus he is taken into their group, and with respect to the stripping away of the lower soul the person becomes as if he were one of them. Then he is returned to his former condition, but longs for what corresponds to the previous state, to recapture it after it has been lost, and to make it a snare for recapturing what has passed away from the soul. He finds a state exhibiting this characteristic which is the happiness and delight obtained through forsaking filth and using things which purify, so he firmly adopts these measures. There follows him (in rank) a person who heard the truthful informer stating that this condition is the perfection of the human being, and that through it he pleases his Creator and that there are innumerable benefits in it, so that he confirmed it with the testimony of his heart and did what he was commanded, and found that what the Prophet had stated about it was true. The doors of mercy were opened to him and he was tinged with the coloring of the angels. There follows him a man who knows nothing of this, but the prophets guided him and induced him to attitudes which in his afterlife put him on a level whereby he might join the way of the angels; and these are the people who are pulled by chains to Paradise.[1]

The influence of ritual impurity is sensed immediately in the lower soul. Most of the people should be told about impurity due to the precise nature of its anticipated sources, its prevalence, and the fact that neglecting to teach about it would constitute a great harm to people.

---

[1] Perhaps an echo of hadiths which mention that some people will enter Paradise in chains. These were the early prisoners of war who accepted Islam as captives. *Mishkāt*, p. 840. Bukhārī Jihād 144.

The investigation of ritual impurity can be reduced to two main types.

1) The first of these types is the lower soul being distracted by three superfluous things which a person finds in his stomach: gas, urine, and feces. There is no person who does not know on his own that when he finds gas in his stomach or needs to urinate or defecate, his lower soul becomes noxious so that it is bound to the earth, and becomes bewildered and dejected, and there is a barrier between it and its being delighted. Then when he eliminates the gases and relieves himself, and uses something so as to alert his soul to purity such as the full bath or ablution, he experiences delight and happiness and feels as if he has recovered the state which had been lost.

2) The second is the distraction of the lower soul by the desire for sexual intercourse and its immersion in lust; for this completely turns the aspect of the lower soul toward the animalistic nature. This occurs to such an extent that even domesticated animals, when they are trained and habituated to the desired behaviors; and predators, when they are made tractable by hunger or lack of sleep and have been taught how to hunt for their master; and birds, if required to imitate the speech of humans—and in summary every animal which has exerted its effort to the utmost in eliminating its inclination to its physical nature and in acquiring that which its physical nature does not demand—when that animal gratifies the lust of its sexual urge and plunges into this pleasure for days, inevitably it forgets what it learned and returns to foolishness, ignorance, and misguidedness. Whoever reflects on this inevitably learns that the gratification of this lust works to soil the lower soul in a way stronger than any overeating or foolhardy undertaking and the rest of those things which cause the soul to incline to the animalistic nature. To test this for himself, a person should refer to what the doctors say concerning the regimen for abstemious monks if it is desired to return them to the animalistic side.

Purity, whose effect is perceived at first glance, and concerning which the majority of persons can properly be addressed due to the abundant means to achieve it in the populous regions, i.e., through water and its regulation, which is the most efficacious purifying agent for human souls and something well-known and accepted among them in its being like the natural path—will be investigated by concentrating on two types: lesser and greater purity.

As for the greater ablution, it consists of washing and rubbing the entire body, since water is a purifying element which eliminates impurities so that due to this the natural dispositions are made sound. Thus it is the proper means to alert the soul to the trait of purity. Many a person drinks wine and intoxicating beverages, so that intoxication overwhelms his physical nature. Therefore he commits some excess like unjustified homicide or squandering wealth on extremely costly things. Then all at once his soul comes to its senses, and reflects, and the defilement is thrown off from it. Many a person is weak and neither able to rouse himself, nor to undertake anything; then there occurs an event which strongly arouses the soul such as displays of anger, passion, or rivalry, so that he can undertake a very difficult task, and even cause grave bloodshed.

In summary the soul may experience an induced transition and awakening from one trait to another trait which is the basis of psychological cures. This induced state of awakening is only obtained through what firmly embeds in the core of their natures and the root of their souls that this is extreme purity, and this can only be water.

The lesser ablution is limited to washing the extremities (face, hands, and feet), and this is due to the fact that in the moderate regions these are customarily the places which are naturally exposed and extend outside of the clothing. An allusion was made to this when the Prophet, may the peace and blessings of God be upon him, forbade enfolding the smooth rocks[2] and there is no hardship in washing them while this is not so with the rest of the body members. In addition, the habit of civilized people is to clean them every day, and when entering before kings and their representatives, and when intending to perform clean actions. The understanding behind this is that filthy things quickly affect the exterior and these outer body parts are visible and show when persons encounter one another. In addition experience attests that washing the extremities and sprinkling water on the face and head strongly arouse the soul from drowsiness and heavy sluggishness. A person should refer back to his experience and knowledge and to

[2] He forbade that a man clothe himself in his robe and not leave any sides of it raised and block any openings for his hands and feet, likening this to smooth rocks which have no gaps or cracks. Bukhārī Ṣalāt 10, Mawāqīt 40, Ṣaum, 66, Libās 20, 21.

what the doctors order in treating someone who swoons or who suffers from the effects of excessive diarrhea or bloodletting.

Purity is one of the divisions of the Second Irtifāq on which the perfection of man depends, and it has become part of people's natural disposition. It effects drawing near to the angels, distancing from the devils, and it defends against the punishment of the grave, and this is the Prophet's saying, "Purify yourself after urination for generally the punishment of the grave is due to this."[3] In (purification) there is a great access to the soul's accepting the color of righteousness. It is indicated in God's saying, may He be Exalted, "And He loves those who keep pure."[4] When purity becomes established in the soul and takes hold of it, there is confirmed in it a portion of the light of the angels and a portion of the animalistic darkness is overcome. This is the meaning of the writing down of good deeds and the forgiveness of sins.

If purity is made a convention, it avails against the disasters of conventions. If the one who possesses it adheres to the attitudes which people adopt when entering the presence of royalty, and to the accompanying declaration of intention and litanies which are recited (at the time of purification rituals), it avails against the misconstrual (of God's nature).[5]

When a person comprehends that this purity is his perfection he trains his bodily parts according to his comprehension without there being a sensual motive. Moreover, purity becomes a regimen for making the physical side obedient to the intellect, and God knows better.

---

[3] Ibn Mājah Ṭahāra 26, Ibn Ḥanbal in several variants.

[4] Qur'ān 2:222.

[5] The third of the veils mentioned in Chapter 34.

CHAPTER 45

The Inner Dimensions of Prayer

You should know that a person may be suddenly snatched away to the Holy Enclave so that he becomes completely attached to the glory of God, may He be Exalted. There are revealed to him from There the holy theophanies so that these come to dominate the lower soul. In this event he witnesses things which the tongue falls short of describing. Then he is returned to his former state and he becomes restless, so that he cultivates himself with a state which is the closest of the lower states to the absorption of the soul in the gnosis of its Maker. He makes it (that state) a snare for capturing what the soul has lost, and this state is reverence, humility, and intimate conversations during actions and sayings established for this (purpose). There follows him a man who heard the truthful news-bearer inviting him to this state, and he desires it, and wholeheartedly attests to it and acts upon it, and finds what he was promised about it to be true and he progresses to what he had hoped. Next there follows a person whom the prophets induced to have recourse to the prayers, while he doesn't comprehend this, like the father who constrains his children to be instructed in useful trades although they are averse to it. A man may ask his Lord for protection from a calamity or to manifest a blessing, and most probably at that time he is absorbed in acts and words of reverence, so as to render effective his concentrated resolve which is the spirit of the request, and this is what was made a practice in the prayer asking for rain.[1]

Prayer is based on three things: 1) That the heart is humbled at the glimpsing of the Majesty of God and His Exaltedness and 2) that the tongue expresses God's Exaltedness and this humility with the clearest expression. 3) That the limbs are disciplined according to this humility.

---

[1] This prayer is described in the hadith collections, for example the Khan translation of Bukhārī II: 63–82.

(A poet testifies:)

> Three parts of my body reaped benefit from Your bounties,
> My hand, my tongue and my subliminal self.

Among the actions of reverence are standing before Him, confiding in Him with intimate prayer, and turning to face Him. Stronger than this is that he is aware of his lowliness and the high rank of his Lord, so that he bows his head. Humans and animals, one and all, are naturally disposed so that raising the neck is a sign of haughtiness and pride and bowing it is the sign of submission and humility, and this is God's saying, may He be Exalted, "Their necks bowed before it in humility."[2]

Stronger yet than this is that he place his face, which is the noblest of his bodily members and the place where his senses meet, in the dust before Him. These three acts of reverence have become disseminated among the various groups of human beings who constantly perform them in their prayers and before their kings and princes. The most excellent form of prayer combines the three positions progressing from the least to the greatest, in order to progressively to develop the sense of submission and humility. In this progression there is a benefit not found in individual actions of reverence, nor in the descending order from the greatest to the least.

Thus prayer was made the foundation of the actions which make one draw near (to God), rather than reflection on the majesty of God or continual remembrance (*dhikr*) of Him, because true and sound reflection is only attained by a group with high, developed souls and how few they are. Besides these things, if people became engrossed (in reflection) they would become apathetic, and waste their resources, rather than acquiring another benefit. The remembrance of God, if not presented and supported by some act of reverence performed by the bodily limbs which a person carries out regularly, results in agitation without benefit in the case of most people.

As for prayer, it is a curative potion (*maʿjūn*) composed of thought directed toward the majesty of God as a second purpose, and the ensuing attentiveness arising from each person. There is no restriction on the one with the capacity for going more deeply into the ocean of witnessing; rather this will serve to alert him most completely. Among the things most clearly conducive to this are

---

[2] Qur'ān 26:4.

dedicating his actions to God, turning his face to God, and restricting any of his requests for help to God. Among the acts of reverence such as prostration and bowing, each one supports and completes the other and apprises of it, thus (prayer) benefits both the common and the extraordinary person, and is a powerfully effective antidote, for it affects each one in a manner befitting the basic nature of his capacity. "Prayer is the path of ascent for the believer"[3] and a preparation for the theophanies of the other life, and this is his saying, may the peace and blessings of God be upon him, "Indeed you will see your Lord (in the next life), if you are able not to be kept from the prayers before the sunrise and the sunset, so perform them."[4] It is a great cause of the love of God and His mercy, and this is his saying, may the peace and blessings of God be upon him, "Help me to achieve this for you through much prostration"[5] and His account, may He be Exalted, of the utterance of the people of the Fire, "We were not among those who prayed."[6]

When prayer becomes consolidated on the part of the worshipper he is effaced in the light of God and his sins are forgiven. "Good deeds annul bad deeds."[7]

Nothing benefits the problem of misinterpretation (of God's unity) more than prayer, especially if its actions and words are performed from the heart and with a righteous intention. If it is made a convention it will have a clear beneficial effect against the disasters of (poor) conventions, and become a mark distinguishing the Muslim from the infidel, and this is his saying, may the peace and blessings of God be upon him, "The treaty which is between us and them is prayer, so whoever abandons it has become an infidel."[8] There is nothing like prayer for training the lower soul to make the physical nature yield to reason and follow its judgment, and God knows better.

---

[3] A Sufi Hadith not found in the standard collections or in the collections of weak and fabricated hadith. Maybūdī, *Kashf al-Asrār* II:676.

[4] Previously cited in the introduction. Bukhārī Mawāqīt 16, Tauḥīd 24, Muslim, Ibn Mājah, Ibn Ḥanbal. Found in the Siddiqi trans of *Ṣaḥīḥ Muslim*, I:307.

[5] When Rābi'a ibn Kalb (63 A. H.) asked the Prophet if he could be his Companion in Paradise he replied, "Then help me to achieve this for you through (performing) much prostration (in prayer). Muslim Ṣalāt 224, 225. Trans. II:256.

[6] Qur'ān 74:43.

[7] Qur'ān 11:114.

[8] *Mishkāt*, p. 115. Tirmidhī IV: 126 #2756 Īmān 9, Nasā'ī, Ibn Mājah, Ibn Ḥanbal.

CHAPTER 46

The Inner Dimensions of the Zakāt

You should know that when a need presents itself to the poor person, and he entreats God about it either verbally or through his condition, his entreaty knocks at the door of the Divine Generosity. Sometimes the best interest will be fulfilled by inspiring the heart of a pure person to furnish the remedy for his want. Thus, when the inspiration descends and he is provoked (to respond), he is given success, God is pleased with Him, and blessings flow to him from above and below, and from his right side and his left, and God's mercy is upon him.

One day a poor person asked me concerning a need in which he was hard-pressed and I sensed in my heart an inspiration commanding me to give, and informing me of an ample reward in this world and the next, so I gave and witnessed that what my Lord had promised me was true. The knocking at the door of (Divine) Generosity, the arousal of inspiration and its selection of my heart on that day, and the manifestation of the reward, all of that happened before my very eyes.

Sometimes expenditure of money is an anticipated source of the divine mercy, such as when there arises a motive among the Highest Council for the elevation of a community so that everyone who undertakes to carry out their command is blessed, and a person's undertaking an expenditure at that time is equivalent to (participating) in the difficult military expedition.[1] Another example would be a period of drought when a community is among the most needy of God's creatures, and what is intended is their revival.

In summary, the truthful news-bearer made from this conjectured instance a general rule, by saying that whoever gives charity to such and such a type of poor person—or in such and such a situation—God will accept this action from him. Then someone

[1] In the Arabic "*ghazwat al-'usra*", which is the name given to an expedition carried out by the Prophet towards Tabūk in hot and difficult conditions.

hears this and submits to its ruling wholeheartedly, and finds what was promised to be true.

Sometimes the lower self comprehends that the love of wealth and being miserly with it harms a person and turns him from his course, so that he suffers from this very greatly. He is not able to prevent this except by practicing giving away whatever he loves best. In his case giving is the most beneficial thing, for if he does not give, the love (of wealth) and miserliness will remain as they are, and in the Afterlife will take on the shape of a hairless snake[2] or his wealth will take on the shape of some thing which harms him and this is the hadith, "A soft sandy plain will be spread out for them,"[3] and His saying, may He be Exalted, "and those who hoard gold and silver."[4]

Sometimes the time of person's destruction may have come, and his ruin has been decreed in the World of Images, then he proceeds to spend great wealth, while he and other good persons around him entreat God. Thus he cancels out his destruction on his own through the destruction of his wealth, and this is his saying, may the peace and blessings of God be upon him, "Nothing turns back destiny except petitionary prayer, and nothing extends the life-span except piety."[5]

Sometimes it occurs that a person does a wicked act due to being overcome by his physical nature, and then becomes aware of its repugnance and repents. Then physical nature again overcomes him and he returns to it. The wisest thing to do in curing his soul is to persist in spending great wealth in amends for whatever (wrong) he does so that this (loss) will be on his mind, and deter him from (the wicked acts) which he intends.

Sometimes good character and the preservation of the extended family system consist of distributing food, giving greetings to all and sundry, and various types of charitable acts, and this is commanded and considered to be a freewill offering (*ṣadaqa*). *Zakāt*

---

[2] *Mishkāt*, p. 376. Bukhārī Zakāt 3, Tafsīr 3, 14, Nasā'ī, Ibn Mājah, Muwaṭṭa', Ibn Ḥanbal.

[3] This is a hadith that on the Day of Judgment those herds of camels, sheep, and goats on which a person did not pay *zakāt* will trample him on a flat plain. *Mishkāt*, p. 371–372. Muslim Zakāt 24,26,27,28, Abū Dāwūd, Nasā'ī, Dārimī, Ibn Ḥanbal.

[4] Qur'ān 9:34–. ". . . and spend it not in the way of Allāh, give them tidings of a painful doom. On that day it will be heated in the fire of Hell and their foreheads and their flanks and their backs will be branded therewith."

[5] Ibn Ḥanbal V: 277, 280, 282.

increases the blessing, extinguishes (the divine) anger by attracting an effulgence of mercy (from God), repels the punishment of the Afterlife which results from miserliness, and favorably disposes the invocation of the Highest Council who improve things on the earth toward this person, and God knows better.

CHAPTER 47

The Inner Dimensions of the Fast

You should know that sometimes a person comprehends, due to God's inspiring him with this knowledge, that his physical animalistic form is keeping him from what would constitute his fulfillment in terms of it's obeying the angelic side. Therefore he comes to detest the animalistic side and seeks to shatter its force. He won't find anything that helps him to do this as much as hunger, thirst, forswearing sex, and practicing vigilance over his tongue, heart, and limbs. Persistence in this is a cure for the sickness of his soul. There follows him one who learned this from the truthful newsbearer (the Prophet) with the witness of his heart, then next the one whom the prophets guided out of solicitude for him, while he did not realize it, so that he gets the benefit of this in the next life through the shattering of the (animalistic) form.

Sometimes a person becomes aware that having the physical nature obey the intellect is his perfection, while his physical nature is rebellious, obeying sometimes and not obeying at other times. Thus he needs training, so he takes up some severe practice such as fasting and imposes it on his physical side and perseveres in fidelity to his pledge, more and more, until he obtains the desired object.

Sometimes he commits a sin, so that he perseveres in fasting for many days oppressing himself in response to the sin in order to prevent himself from committing more like it.

Sometimes his lower soul craves women, but he is unable to marry and he fears committing fornication, so he defeats his lust through fasting and this is his saying, may the peace and blessings of God be upon him, "Fasting suppresses sexual desire."[1]

Fasting is a great good deed which strengthens the angelic (side) and weakens the animalistic. There is nothing like it for polishing

---

[1] *Mishkāt*, p. 658. *Bukhārī*, Ṣaum, 10, Nikāḥ, 2,3. Ibn Mājah, Nasā'ī, Ibn Mājah, Dārimī, Ibn Ḥanbal. "Those who cannot marry should fast, since fasting suppresses sexual desire." Literally, fasting is a castration for him.

the face of the spirit and subjugating the physical nature. Therefore, God, may He be exalted, said, "Fasting is for My sake and I reward it."[2] His sins are forgiven to the extent that the animalistic form is diminished and through this is achieved a great similarity to the angels, so that they love him. The attachment of love helps effect the weakening of the animalistic side and this is his saying, may the peace and blessings of God be upon him, "The bad breath of the one who fasts is sweeter to God than the fragrance of musk."[3] If it is made a common convention it avails against the damages of (poor) conventions. If a community persists in fasting their devils will be chained and the gates of the gardens of Paradise are opened to them and the gates of the fires of Hell are locked against them.[4]

When a person tries to subjugate the lower soul and eliminate its bad qualities, his act will take on a sanctified form in the World of Images. Among the purest of the gnostics is the one who concentrates on this form, for he is furnished with knowledge from the Unseen World and achieves union with the Divine Essence because of transcendence and sanctification. This is the meaning of his saying, may the peace and blessings of God be upon him, "Fasting is done for My sake and I reward it."[5]

Sometimes a man comes to understand the harm of his being preoccupied with his worldly affairs and his filling his senses with external stimulation. This is helped by devoting himself to worship[6] in a mosque designed for prayers, so that it is not possible for him to persist in that (preoccupation). Even if he cannot completely fulfill this retreat, he should not give it up completely. Thus he snatches opportunities from his situation and retreats in a mosque for as long as is decreed for him. Next there is the person who accepts this wholeheartedly from the truthful news-bringer (the Prophet), and (then comes) the ordinary person who is overcome, as was previously mentioned.

---

[2] A hadith qudsī, reporting from God. *Mishkāt*, p. 417. Bukhārī Ṣaum 2, Tauḥīd 35, 50, Libās 78; Muslim, Nasā'ī, Ibn Mājah, Muwaṭṭa', Ibn Ḥanbal.

[3] *Mishkāt*, p. 417. Bukhārī Ṣaum 2, 9, Tauḥīd 35, 50, Libās 78, Muslim, Tirmidhī, Nasā'ī, Ibn Mājah, Muwaṭṭa', Ibn Ḥanbal.

[4] Based on a hadith. See *Mishkāt*, p. 417. Bukhārī Ṣaum 5, Bad' al-khalq 11; Muslim, Nasā'ī, Ibn Mājah, Dārimī, Muwaṭṭa', Ibn Ḥanbal.

[5] A hadith qudsī. *Mishkāt*, p. 417. Bukhārī Ṣaum 2, Tauḥid 35, 50, Libās 78, Muslim, Nasā'ī, Ibn Mājah, Muwaṭṭa', Ibn Ḥanbal.

[6] This refers to the practice of "*i'tikāf*" or performing a retreat in a mosque.

Sometimes a person may fast but he is not able to purify his tongue except through retreating to the mosque. Sometimes he may make a request on the Night of Power,[7] and seek a connection with the angels, and this contact is not possible except through doing a retreat in the mosque. You will be given the meaning of the Night of Power, and God knows better.

[7] The Night of Power (*lailat al-qadar*). A night during the last ten days of Ramaḍān. The Prophet used to carry out a retreat in the mosque at this time. It is believed that the angels descend on the Night of Power, and that petitionary prayers are more readily answered

CHAPTER 48

The Inner Dimensions of the Pilgrimage

You should know that the true nature of the Hajj is the assembling of a great group of righteous people at a time which commemorates the blessed ones among the prophets, truthful ones (*ṣiddīqīn*), martyrs, and righteous ones, and in a place "in which are clear signs" (of God) which groups of the leaders of the religion used to visit, in veneration of the emblems of God; entreating, desiring, and full of hope for good and the forgiveness of sins from God. For if concentrated resolve is focused in this manner the descent of mercy and forgiveness will not lag far behind. And this is his saying, may the peace and blessings of God be upon him, "The devil has not experienced a day in which he is more insignificant, rebuffed, ignominious and infuriated than on the day of ʿArafāt."[1]

The basis of the pilgrimage is present in every community, for inevitably they have a place in which they ask for blessing since they have seen the manifestation of the signs of God in that place, and they have sacrifices and ritual forms handed down from their ancestors in which they persevere because they are a reminder of those who drew near to God and what they used to do.

The (place) most worthy of having pilgrimage made to it is the house of God (the Kaʿba), "in which there are clear signs (of God)." Abraham, may the blessings of God be upon him, whose goodness is attested to by most of the nations, built it at the command of God and at His revelation, after the earth (there) had been desolate and trackless. Pilgrimages undertaken to other sites either involve associationism or some baseless innovation.

From the aspect of the purity of the soul, coming to visit a place which the righteous ones did not cease to glorify and visit, and which they filled with the remembrance of God; attracts the attachment (to a person) of the concentrated resolve of the lower angels and inclines sympathetically to him the universal invocation of the

[1] *Mishkāt*, p. 557. *Iḥyāʾ ʿUlūm al-Dīn* in the chapter of the virtues of the ḥajj.

Highest Council for the people of good. Thus, if he stays there their hues come to predominate over his soul, and I have seen this with my own eyes.

Under the topic of remembering God, may He be Exalted, comes viewing the emblems of God and honoring them. When they are seen, God is remembered, just as seeing something brings to mind whatever is connected to it. This is especially prevalent when holding oneself to attitudes of reverence, restrictions, and rules and limits which alert the soul very strongly. Sometimes a person is very drawn to his Lord, so that he needs something by which to satisfy this longing, and this can only be done through the pilgrimage.

Just as the ruling authority requires a test after every period of time by which to distinguish the sincere advisor from the deluder and the obedient from the disobedient, and to raise its repute, promote its words, and to let its people know about each other, likewise the religion needs a pilgrimage to distinguish the sincere one from the hypocrite, to show how people have entered the religion of God in throngs, and so that they can see one another and each can benefit in what he didn't have before, since objectives are obtained by mingling and seeing one another.

Since the Hajj was made a well-known convention it assists (in removing) the harms of (false) conventions and there is nothing like it for reminding them of the condition of the founding leaders of the religion and inciting them to persevere in it.

Since the Hajj is a distant journey and a difficult undertaking which is only accomplished through strong individual effort, and whose performance is dedicated solely to God; it causes sins to be forgiven and eradicates what preceded it as is the case with (the acceptance) of the faith.[2]

---

[2] That is, when a person accepts Islam previous sins are forgiven and likewise performing the Hajj eradicates previous sins.

CHAPTER 49

The Inner Meanings of Various Types of Pious Actions

Dhikr is one of the pious actions,[1] because there is no veil between it and God, may He be exalted, nor is there anything like it for curing misinterpretation (of the nature of God). This is his saying, may the peace and blessings of God be upon him, "Should I not apprise you of the best of your actions?"[2] (There is nothing like this) in acquiring the presence (of God) and repelling harshness, especially for the one whose animalistic side is weak by innate disposition or acquired weakness, and by the one whose imaginative faculty is naturally reticent to mix the abstract with the determinants of tangible things.

Another pious action is petitionary prayer (*duʿā*), for it opens a great door to the presence of God (*muḥādara*), and effects complete submission to and reliance on the Lord of the Worlds in all of the situations a person encounters. This is his saying, may the peace and blessings of God be upon him, "Petitionary prayer is the core of worship."[3] It is the embodiment of the lower soul's turning to the Source with the attribute of seeking which is the secret in attracting the thing which is prayed for.

Among pious actions is the recitation of the Qur'ān and listening to religious exhortations. Whoever listens to these, and lets them have ascendancy over him will become colored with the states of fear, hope, and perplexity[4] at the glory of God, and be submerged in the blessing of God, and states other than these. This clearly serves to calm the physical nature, and to prepare the lower soul for the emanation of the coloration of what is above it. Therefore this is a most beneficial thing in the afterlife and this referred to in the saying of the angel to the person in the grave, "You didn't

---

[1] The recitation of pious phrases and the Divine Names.

[2] I.e., *dhikr*. Tirmidhī V: 127 #3437 Duʿā 6.

[3] *Mishkāt*, p. 472. Tirmidhī V:125 #3431 Duʿā 1.

[4] These are three states referred to in Sufi literature of the refinement of the soul and the spiritual progress of the person.

know[5] because you didn't recite the Qur'ān."[6] Through the Qur'ān the lower soul is purified from base attitudes, and this is his saying, may the peace and blessings of God be upon him, "Every thing has a polish and the polish for the heart is the recitation of the Qur'ān."[7]

Among these pious actions is maintaining the bond of kinship and neighborliness, good social relations with the people of the town and coreligionists, and ransoming captives in order to free them, for this prepares for the descent of mercy and contentment. Through this the order of the second and third levels of the supports of civilization is perfected, and through this a person attracts to himself the invocation of the angels.

Among them is *Jihād.* The prime example is that God curses a corrupt person who harms the collectivity, whose extirpation is more suitable for the universal welfare than his remaining. Then inspiration manifests itself in the heart of a righteous man to kill him, so that there is called forth from his heart an anger without a physical cause, and he is effaced from his own will and subsists through the will of God, and is absorbed in the mercy of God and His light, and humanity and countries benefit from that. The next case is that God decrees the cessation of the rule of tyrannical states where they do not believe in God and conduct themselves evilly. He orders one of God's prophets to make war against them so that the motivation to wage the *Jihād* is inspired into the hearts of his people and thus they become "a people brought out for mankind" and the divine mercy comes to include them. Another case is that a group becomes aware through the comprehensive outlook of the goodness of saving the oppressed ones from the predatory ones and undertaking the punishment of the disobedient ones and forbidding evil, so that this becomes a cause for the peace and contentment of the people and thus God rewards them for their action.

Among the pious actions are ways of drawing near to God which happen to a person involuntarily such as afflictions and illnesses. They are accounted as being under the topic of piety in a number of ways.

1) Among them is that the divine mercy, when it is directed to

---

[5] The answer to the first question—"Who is your Lord?"

[6] Bukhārī Janā'iz 67, 86. Abū Dāwūd, Nasā'ī, Ibn Ḥanbal.

[7] *Mishkāt*, p. 482. Baihaqqī transmitted it. Not in the six collections.

a person by reason of the righteousness of his actions while other causes decree difficulty for him, these (difficulties) serve to perfect his soul, so that his sins are atoned for, and good deeds are recorded for him. Just as if when you block a channel for water, the water will flow over the top and below and the overflow is attributed to the blockage, and the secret of this is preserving the relative good.

2) Among the ways (of drawing nearer to God) is that the believer, when calamities become very severe for him, finds, "the world becoming straitened for him where once it was expansive,"[8] so that the veils of physical nature and convention are rent and his heart becomes detached from everything except God. As for the disbeliever, he does not cease recalling what has been lost and being immersed in the life of this world so that he becomes even more wicked than he was before he was struck by that affliction.

3) Among them is that the faculty bearing hardened evils is the coarse filthy animalistic side, so that when a person becomes ill and weakened, and more (matter) is dissolved than he takes in, much of the faculty bearing these (evils) fades, and (the sins) which were borne diminish proportionately. Thus we see that the lust and anger of the sick person ceases, and his morals change and he forgets much of what he used to be involved in as if he is no longer the person who he used to be.

4) Among them is that the believer whose angelic side has become somewhat disengaged from his animalistic side is usually punished for his sins during this worldly life. This is the hadith, "The believers' share of punishment is the hardship of this world,"[9] and God knows better.

---

[8] Qur'ān 9:118.

[9] A Hadith not found in the standard collections.

CHAPTER 50

The Ranks of Sin

You should know that just as there are actions which are embodiments, symbols, and practices which are conducive to the animalistic side's yielding to the angelic, likewise there are actions, symbols, and ways of acquiring the state which is absolutely contradictory to this yielding, namely, sins.

They are at various degrees:

1) The first degree of sins totally block a person's means to the desired perfection and the majority of these are of two types.

a) One of them are those going back to the Source in that he does not know that he has a Lord, or he conceives of God as described by the attributes of created things, or he believes that a created thing may have some of the attributes of God. The second (of these first type of sins) are anthropomorphism (*tashbīh*), and the third of them is associationism (*ishrāk*). The soul will never be sanctified until it sets its sights on the heavenly transcendence and the comprehensive divine management encompassing the world. If this is lacking the soul will remain self-absorbed or preoccupied with what is similar to itself in being so limited in every one of its concerns that the veil of rejection cannot be punctured by so much as a needle prick, and this is the greatest of calamities.

b) Type two of (sins at the first level) are that a person believes that there will be no creation of the soul beyond this bodily one and that there is no other perfection beyond this which a person must seek. If the soul harbors this idea it will not raise its sights to perfection at all.

Affirming a perfection other than that of the physical body is only realizable by the majority of people through conceiving of a condition which is the opposite of the present one from every aspect. If this is not done then the rational and the tangible fulfillments would seem to be opposed to each other so that a person would incline to the tangible and the rational would be neglected. Therefore a touchstone (*maẓinna*) for this is assigned which is faith that he will meet God, and faith in the Last Day, and this is God's saying, may He be exalted, "Those who don't believe in the Hereafter, their hearts refuse to know, and they are proud."[1]

In summary, when a man dies at this level of sin, his animalistic side is effaced, and hatred will completely envelop him from above such that he finds no way ever to be free of it.

2) The second degree of sins is that he is arrogant with his animalistic arrogance toward what God, may He be exalted, established in order to make people attain their perfection, and what the Highest Council intend with the furthest extent of their resolve to propagate and elevate through the messengers and the divine laws. Thus he denies these and is hostile to them. Then when he dies all of their resolve is disposed to hate him and to pain him, and "his sin surrounds him"[2] so that he finds no means to escape, but this state of his failing to attain his perfection or of his only attaining something insignificant does not detach from him. In all of the divine laws this level expels the person from the religion of his prophet.

3) The third level is abandoning that which will earn him salvation and doing that which would convene, at the Plane of Dhikr,[3] curses on the one who does it due to its being an anticipated source, in most cases, of a great corruption on the earth and an attitude opposed to the refinement of the lower soul. (Under this level come):

a) A person's not carrying out anything significant of the divine laws that results in obedience (to God) or that will prepare him for obedience. This varies with the differences in souls, except that the ones immersed in weakly bestial attitudes are the people most in need of having the laws augmented, and the communities whose animalistic sides are strongest and most contaminated are the ones most in need of the augmentation of the strictest laws.

b) Predatory acts which provoke a severe curse such as murder,

c) lustful actions,

d) and harmful sources of income such as gambling and usury.

In each one of these things which have been mentioned is a great breach in the soul from the aspect of undertaking what opposes the necessary practice as we mentioned, and a cursing from

---

[1] Qur'ān 16:22.
[2] Qur'ān 2:81.
[3] The World of Images and the angels.

the Highest Council which surrounds him, so that the combination of the two things results in punishment. This (third) level comprises the greatest of the major sins. Their prohibition is convened in the Holy Enclave, together with the cursing of those who commit them. The Prophets constantly told of what was convened There, and most of this is agreed upon in the divine laws.

4) The fourth degree is disobedience to the divine laws and the codes which vary due to the variation of the communities and eras. This occurs because when God, may He be exalted, sent a prophet to a people, "to bring them out of the darkness into the light," to correct their deviation, and to govern them with the best polity, this mission included making compulsory those things indispensable for reforming and governing them. Thus for every goal there is a common or enduring symbol (*maẓinna*) according to which they must be held accountable, and about which they must be addressed, and setting times (for each act) has rules which require it. Many things motivate to a harm or benefit and thus people should be commanded according to what they will be motivated to do.

Some things are definitely commanded or forbidden, and some things are commanded or forbidden indeterminately. A minority of these (variable rules) are what the external revelation (the Qur'ān) revealed, and a majority were only confirmed through the independent reasoning (*ijtihād*) of the Prophet, may the peace and blessings of God be upon him.

5) The fifth rank (of sins) are what the law-giver did not specifically rule on, nor was their ruling convened among the Highest Council. Rather, a worshipper turned to God with the complete intent of his concentrated resolve and thus it struck him that a thing was supposed to be forbidden or commanded due to analogical reasoning, or derivation, or something like that. This is like the effect which certain medicines seem to have according to common people due to their incomplete experience, or due to reliance on the ruling about the cause made by the proficient doctor; while they don't know the reason for the effectiveness, nor has the doctor stipulated it. Someone like this is not relieved of responsibility unless he acts with caution. If he does not do so a veil will arise between him and his Lord in what he supposes, and thus he will be held accountable for his opinion.

The element which is pleasing to God concerning this (fifth) level is that it should be omitted and not heeded despite the fact that there exist people who want to impose it on themselves, so that the Generous One increases for them what they have imposed on themselves. Concerning this is His, may He be exalted, saying, "I am as My servant thinks Me to be"[4] and His saying, may He be exalted, in the Great Qur'ān, "But monasticism they invented—We ordained it not for them—only seeking Allāh's pleasure,"[5] and the Prophet's, may the peace and blessings of God be upon him, saying, "Don't be too severe with yourselves or God will be severe with you,"[6] and his saying, "Sin is something which has a strong impact on your heart."[7]

The same goes for someone's disobeying the ruling of a legal scholar (*mujtahid*) if that person himself is a muqallid who is performing *taqlīd*[8] of a scholar who held this (opinion), and God knows better.

[4] In a hadith qudsī found in Tirmidhī IV: 23 #2496 Zuhd 39.
[5] Qur'ān 57:27.
[6] A hadith cited by al-Suyūṭī from al-Ṭabirī in interpretation of the Qur'ānic verse 2:67 *al-Durr al-Manthūr* I (Tehran: al-maktaba al-islāmiyya al-ja'fariyya, 1957), 77.
[7] Muslim Birr 14, Tirmidhī, Dīrimī, Ibn Ḥanbal.
[8] *Taqlīd* refers to following the legal rulings of a *mujtahid* in one of the Islamic legal schools.

CHAPTER 51

The Harmful Influences of Sins

You should know that major and minor sins can be designated from one another in two respects. One of them is according to the wisdom behind piety and sin and the second according to the divine laws and codes particularized for one time and not another.

As for the major sin, according to the wisdom behind piety and sin it is a sin that strongly requires punishment in the grave and at the Resurrection and that greatly harms the sound civilized order and is radically opposed to the original nature.

The minor sin entails some of the aforementioned, or often leads to it or requires something of it from one aspect but not from another, such as the man who spends in the path of God while his family goes hungry. Thus he repels the vile quality of miserliness but destroys the domestic order.

As for their (designation) according to the particular divine laws, a major sin is something that the divine law specifically forbade or for which the law-giver promised Hell, or legislated a legal punishment (*ḥadd*), or for which he designated its perpetrator as an unbeliever outside of the community as an expression of its repugnance, and to emphasize its disgusting nature. Sometimes a thing may be a minor sin according to the wisdom behind piety and sin, and a major sin on the basis of the sharīʿa. An instance of this derived from the pre-Islamic period is that sometimes their religion practiced a sin (of lesser gravity) until it became spread as a convention among them that could not be easily uprooted. Therefore the divine law forbade it, but those people insisted on doing it and were stubborn, so that according to the divine law condemnation and threat were proportionately produced, so that perpetrating this act became tantamount to severe hostility to the Islamic religion and no one ventured to commit something like this except a mutinous person who was recalcitrant and unashamed before God or the people. Due to this it became recorded as major sin. In summary, we will postpone discussing the major sins on the basis of the divine law until the second section of this book because that is its appropriate location. Here we will give information about the harmful effects of the major sins on the basis of the science of piety and sin as we did to some extent about the types of pious actions.[1]

People disagree about what is a major sin. When a disobedient person dies having committed a major sin, and does not repent, is it possible for God to forgive him or not? Each faction put forth a proof from the Qur'ān and Sunna. The solution of the disagreement according to me is that God's actions are of two types. Some follow the customary course of nature while others are breaches in custom. The propositions which people discuss are determined by two (different) modalities—one the customary mode and secondly, the absolute.[2] The condition for a contradiction is the conflation of the two types, like what the logicians decided in the case of modal propositions (*qaḍāyā muwajjaha*); sometimes the mode may not be indicated and thus (in this case) the contexts must be pursued. Thus our saying "Whoever ingests poison, dies," means, "according to usual custom"[3] and our saying, "Not everyone who ingests poison dies," refers to the breach of custom, thus these are not contradictory.

Just as God, may He be exalted, has actions in the world which defy the customary course of things, and actions which follow the usual course of things, likewise in the next life there are actions which defy custom, as well as customary ones. As for the usual course of things, it is that the disobedient person, if he dies unrepentant will be punished for a long period, and sometimes there may be a breach of custom and this is the situation of the rights of human beings. As for the perpetrator of a major sin being eternally punished,[4] this is not true and it is not God's wisdom to treat the major sinner and the unbeliever equally, and God knows better.

---

[1] In Chapter 49.

[2] The "mode" (*jiha*) of a proposition in the terminology of the Islamic logicians refers to whether it formulates a necessary, possible, or absolute case in terms of its predicate. The mode refers to the word signifying the quality of the existence of the predicate in respect to the subject, i.e., "Zaid necessarily walks" versus "Zaid possibly walks" and "Zaid is walking."

[3] I.e., it is in the customary mode.

[4] As was held by some theologians among the Kharijites and the Mu'tazila.

CHAPTER 52

The Acts of Disobedience of a Person which Have an Impact on Himself

You should know that the angelic faculty of the person is enclosed by the animalistic faculty on all sides, and in this it resembles a bird in a cage, whose felicity lies in escaping from that cage and reaching its original environment of luxuriant meadows, eating nourishing seeds and delicious fruits from there, and joining the company of its own kind, so that it would be wholly delighted by them. The greatest misfortune of man is to become an atheist (*dahrī*), and the true nature of atheism is to be opposed to the intrinsic knowledge (of God) which was created in him. We have explained that the person has an inclination to the Creator, may His Majesty be Exalted, at the root of his natural disposition, and an inclination to glorify Him with the greatest reverence possible. This is what is indicated in His, Blessed and Exalted, speech, "When your Lord brought forth from the children of Adam,"[1] and his saying, may the peace and blessings of God be upon him, "Every child is created according to the original nature."[2] The ultimate reverence is only possible through a belief that his Creator acts with the intent, choice, requital, and imposition (of religious obligations) on people, and legislates for them. The one who denies that he has a Lord to whom the chain of creation goes back or believes that God is devoid of attributes and is not operative in the world, or that He is forced to act without willing or that He will not requite people for the good and evil that they do, or thinks that his Lord is like the rest of creation or that He shares some of His attributes with people, or believes that He did not assign them a divine law by means of a prophet—he is the atheist. Reverence for his Lord is

[1] Qur'ān 7:172. "... from their reins, their seed, and made them testify of themselves (saying): Am I not your Lord?" This is the verse concerning the Primordial Covenant (*mithāq*) between God and man.

[2] Mentioned in Chapter 17. *Mishkāt*, p. 26. Bukhārī Janā'iz 2, 80. Tafsīr 30,1; Qadar 3, Muslim, Ibn Ḥanbal.

not constituted in his soul, and there is nothing at all in his knowledge which influences the Holy Sphere. He is like a bird imprisoned in a cage of iron which contains no means of egress nor even a hole the size of a pin. When he dies the veil (of the physical nature) becomes translucent and the angelic light shines through to some extent, and the inclination of his original nature is stirred, while impediments hinder him in his knowledge of his Lord and his attaining the sphere of the Holy, so that a great desolation arises in his lower soul. Its (the soul's) Creator and the Highest Council regard it while it is in this bad state, so that the glance of wrath and contempt is fixed on it and inspirations of harshness and punishment filter down in the souls of the angels. Thus he is punished both at the level of the World of Images and externally.

Or (he may be) an unbeliever, disdaining the mode through which God, may He be exalted, effects transitions as He said, "Every day He is in a mode."[3] I mean by a mode (sha'n) that the world has eras and developments depending on the divine wisdom. Thus when an era begins, God, may He be exalted, reveals His command for it in every heaven, and the Highest Council regulates things as befits this (wisdom) and a divine law and a beneficial purpose is prescribed for people There.

Then He inspired the Highest Angels to agree to put this (new) framework into effect in the world so that their consensus became a cause for inspirations in the hearts of men. Thus this mode followed the Pre-Existent level that was not contaminated by temporal origination. This also explains certain perfections of the Necessary One, may His majesty be Exalted, such as the first level.[4] Anyone who opposes this mode, hates it, and turns others away from it is followed by a severe curse from the Highest Angels which envelops his soul so that his actions fail and his heart becomes hardened, and he is not able to acquire benefit from pious actions, and this is what is referred to in His, may He be exalted, saying, "Those who keep hidden the proofs and guidance which We revealed after We made it clear in the scripture, these ones are cursed by God and those who have the power to curse,"[5] and His saying, "God

---

[3] Qur'ān 55:29.

[4] That is, the relationship between God's Pre-Existence beyond time at the first level and His creative activity through changing modes.

[5] Qur'ān 2:159.

has sealed their hearts and hearing."[6] This person is like a bird in a cage which has avenues of escape but which is enveloped from above by a great stupor.

Less serious than this (atheism and disbelief) is to believe in the unity of God and glorify Him as should be done, but to abandon compliance with what was commanded in the wisdom underlying piety and sin. The example of this is like that of a man who recognized courage and its benefit, but could not be described by it because really achieving courage is not the same as having its intellectualized form. This is a better state than that of someone who does not even know the meaning of courage. He is like the bird in a latticed cage which can see vegetables and fruits, and used to be out among them for a time. Then it became unexpectedly confined so that consequently it longs for what is outside and flaps its wings and pokes its beak in the holes, but does not find a way to escape from it. These are the major sins according to the wisdom underlying piety and sin.

Less serious than these is to carry out these things (commanded by God) but not according to their necessary conditions. Such a person is like a bird in a broken cage from which it is difficult to escape. Escape cannot be envisioned without scraping its skin and pulling out its feathers. Thus it is able to escape from its cage, but only by diligent effort, and it is not able to fully disport itself among its kind nor to eat the fruits of the fields as it should, due to having been scratched and having its feathers plucked out. These are the ones who "mixed a righteous act with another evil one"[7] and their obstacles are the minor sins according to the wisdom behind piety and sin. The Prophet, may the peace and blessings of God be upon him, indicated these three things in the hadith about the *Ṣirāṭ* bridge when he said, "Some fell in the fire, and some fell down and were saved, and some were lacerated and were saved,"[8] and God knows better.

---

[6] Qur'ān 2:7.
[7] Qur'ān 9:102.
[8] Muslim Īmān 329. *Mishkāt*, p. 1185.

CHAPTER 53

The Sins which are Between a Person and Other People

You should know that the animal species are at many levels. Among them are those generated like the worms from the earth, the condition of which is to be inspired by the Maker of Forms how to feed but not how to manage households.

Among them are those which have reproductive capacity and the male and female cooperate in raising the young. It was their right according to God's wisdom to be inspired with the additional knowledge of how to manage households. Thus He inspired the birds with how to feed and fly, and He also informed them how to mate, build a nest, and feed the chicks. Among the species, human beings have a civilized nature and can't survive without cooperating with others of their species, as they don't feed on wild grass and raw fruit, nor are they kept warm by fur and so on, which we previously explained. It is human nature to be inspired with how to manage cities along with managing households and the proper way to obtain a livelihood. In contrast to the rest of the species which receive instinctive inspiration according to need, man does so only in a small portion of the sciences living such as in sucking the breast during nursing, coughing when hoarse, opening the eyelids when desiring to see, and so on. This is because his imagination is skillful and eager, enabling him to acquire knowledge of how to manage households and cities, and to use convention and imitation of those supported by the angelic light in what was inspired to them, and through experience and observation of the regulation of the Unseen, and reflection through intense study, analogical deductions and demonstrative proof. Similarly, their receiving the universal command whose emanation from the Creator of Forms is necessary, despite the variations arising because of their varying capacities, is like their receiving visions in dreams which emanate to them from the realm of higher knowledge. These are represented to them by suitable shapes, and thus the variation of the forms is due to a quality in those who receive the emanation, not in the Emanator.

Among the branches of knowledge emanated to all human individuals, whether Arabs or non-Arabs, city-dwellers or Bedouins—even if their way of apprehending them differs—is the prohibition of qualities which would destroy the order of their cities. These are of three types:

1) Lustful acts, 2) rapacious acts, 3) acts springing from the improper observance of transactions.

The basic principle in this is that human beings are identical in lust, jealousy, and greed. The male among them resembles the male animal in desiring females and not allowing multiple mates for one female. However, the male animal will fight until the strongest and the most single-minded predominates, and those below this are defeated, or else they don't have a sense of rivalry because they don't see each other mating.

The human being is clever, thinking a thought as if he sees and hears it. He has been inspired because of this that hostilities will ruin their cities since they cannot maintain a civilization except through mutual cooperation from people, and the male participates more in these civilizing affairs than the female. Hence it was inspired to institute each man's being specified to his wife, and to forswear competing for the woman set aside for another. This is the basis of the prohibition of fornication. Beyond this the form of this distinguishing of wives became a matter assigned to convention and to the divine laws. The male human also resembles the male animal in so far as the soundness of their original nature requires that they only feel sexual desire for females and not males, just as animals do not give this type of attention except to females. However males who are overcome by corrupt lust are like someone who finds eating mud and lumps of charcoal delicious, i.e., they become deviated from the sound original nature. Such a person satisfies his lust with men and he becomes a sodomite who takes pleasure in something which is not part of the sound nature. This results in an alteration of their temperaments and an illness in their souls. Along with this it causes disregard for procreation in so far as they satisfy the need that God, may He be exalted, destined for them in order to multiply their descendants, in some other manner. Thus they alter the order according to which God, may He be exalted, created them, so that the repugnance of this became absorbed by their souls. Therefore profligate people do it, and do not acknowledge it, and if it were attributed to them they would die of shame, unless theirs is really a powerful deviance, so that they flaunt their vice openly and are not ashamed. Their punishment will not be slow in coming, as occurred in the time of the Prophet Lot, may peace be upon him, and this is the basis for the prohibition of sodomy.

The livelihood of human beings, their domestic management, and the governance of their cities are only carried out through reason and discriminating judgment. Addiction to alcoholic beverages causes a great disturbance of their order, and engenders belligerent acts and grudges. Despite this fact, some people whose vile craving has overcome their reason engage in this vice, and corrupt the supports of civilization for themselves. If convention did not forbid them from such acts people would be destroyed and this is the basis of the prohibition of addiction to alcohol. As for the prohibition (of drinking) small or large quantities of it, this will be explained under the investigation of the divine laws.

The male human resembles the male animal in becoming angry with the one who keeps him from his goal, and who causes him pain in his mind or body. The male animal, however, only heads for a tangible or imagined goal, while the human seeks the imagined and the rational goal, and he is more greedy than the animal. Animals fight one another until one is put to flight, then they forget the rancor except for ones like male camels, cattle, and horses. Man hates and does not forget. Thus if the door to fighting one another were opened their city would be harmed and their livelihood disordered. Thus they were inspired with the prohibition of killing and striking one another except for important causes such as retaliatory punishments, and so forth.

Sometimes hatred was stirred up in the breasts of some of them as it had been stirred up in the breasts of the first ones, while they feared retaliation, so they turned to surreptitiously administering poison in food or killing by sorcery. This is at the same level as murder, in fact it is worse. For murder is overt and can be eliminated while this cannot. They also stooped to false accusation and carrying it to those in power, in order to murder.

The means of livelihood which God, may He be exalted, provided for his worshippers, are gathering from permitted terrain, herding, agriculture, manufacturing, trade, and running the city and the nation, and every way of earning besides these has no role in their civilization. Some of them stooped to harmful means of gain such as theft and extortion, and all of these are destructive of the city, so they were inspired to the effect that these were forbidden, and all human beings agreed on this. If rebellious ones practiced these things, due to the contumacy of their lowers souls, the just kings made efforts to eradicate and extirpate them. Some of them became conscious of the effort of the kings to eradicate them, thus they stooped to false claims, perjury, false witness, giving short weights and measures, gambling, and interest compounded multiply. The ruling on these harmful activities is like that of the harmful professions, indeed assessing a extortionate land tax is like highway robbery, only worse.

In summary, due to these reasons the prohibition of such things entered into the souls of human beings and those among them of strongest intellects, most sound opinions, and those most knowledgeable of the comprehensive interest undertook to forbid these, generation after generation, until this became a disseminated custom, and entered their primary axioms like the rest of the famous widespread matters. Once this happened there went back to the Highest Council a coloration from them according to the actions to which they had stooped, that inspired the angels with this prohibition and (the knowledge) that these were the most harmful things. Then whenever a human being would do any of these acts, the angels would become as if pained by it, as is the case if one of us puts his foot on a hot coal so that it is communicated to the perceptual faculties at that moment and due to this he is pained. Therefore due to the angels' being pained, radiating waves come to encircle that disobedient person and there enters the hearts of those predisposed among the angels and others that they should harm him if possible. His predestined course, which is called in the divine law the "inspiration of the angels" as to what will be his sustenance, his term, his life span, and whether his future life will be difficult or happy[1]—and in astrology the decrees of the nativity—may grant him a dispensation (for a time) so that when he dies and this destiny ceases to be effective, his Creator will finish with him as He said, "I will finish with you, O you two dependents,"[2] and then He will punish him most fully, and God knows better.

---

[1] An echo of the hadith in Bukhārī Anbiyā' 1, Ḥaiḍ 17, Qadar 1; Muslim, Ibn Ḥanbal, *"fa mā al-rizq fa mā al-ajal"*.

[2] Qur'ān 55:31. According to the commentators the two dependents are mankind and the jinn.

BOOK VI

THE SIXTH INVESTIGATION: THE REGULATION OF RELIGION

CHAPTER 54

The Need for the Guides to the Paths and the Founders of Religions

God, may He be Exalted, said, "Indeed, you are a warner and every people has a guide."[1]

You should know with regard to the practices which make the animalistic submit to the angelic, and the sins which oppose this, that although sound reason indicates them and understands the benefits of the former and the harm in the latter, yet people are negligent about them because veils come to dominate them. Therefore their intuitive awareness (*wijdān*) becomes corrupted, as in the case of the person of choleric temperament, so they do not conceive of the desired state or its benefits, nor of the feared state and its harms. Therefore they need one who has knowledge of the rightly guided practice to govern them, to command this, to spur them on to it, and to renounce what opposes it.

Among them are persons with harmful outlooks who in fact desire only what is contrary to the desired way so that they are misguided and misguide others. The nation can only be made sound through suppressing them and nullifying their reputation. Among them are persons who in general have rightly-guided judgment but who only realize a deficient portion of guidance so that they observe some matters but miss many others; or they think in themselves that they are perfect and have no need of another to perfect them, and therefore they need someone to alert them to their ignorance. In summary, people inevitably need a person with full knowledge who is preserved from errors.

The city, despite the sense of livelihood (*al-ʿaql al-maʿāshī*) found in many people through their independently realizing the system which is beneficial for it, is forced (to have) a person who correctly recognizes what is conducive to its benefit and undertakes its regulation. Therefore, what do you think of the case of a very

[1] Qur'ān 13:7.

large community, who combine capacities which vary greatly in a manner which can only be deeply grasped by the fittest ones among the people of pure constitution or those who have the most mature ability to abstract? Only those ones will be guided to this way who are at the highest degree of the ranks of souls, and how few they are! In similar manner, since the crafts of the blacksmith and the carpenter and their like do not arise from the mass of people except through practices inherited from their ancestors and masters who guide them to these ways and encourage people in them, then what do you think of those noble pursuits to whom only those granted success by God, may He be Exalted, are guided and to which only the most sincere are inclined?

Therefore it is necessary that this knowledgeable man publicly prove that he is cognizant of the rightly-guided practice (*al-sunna al-rāshida*) and that he is totally preserved from error and misguidance in what he says, that he is free from realizing only a portion of rectification (*iṣlāḥ*) while leaving aside some other portion which is also necessary.

This is reducible to two aspects.

a) He reports from a man before him who had the final word since all concur on his perfection and his infallibility; and that the report has been preserved among them. Thus it is possible for him to hold them accountable on the basis of their belief in him, and to advance it as an argument with them, in order to silence any of their objections.

b) Alternatively he himself may be the one with the final word and upon whom they all agree.

In sum, the people must have an infallible person upon whom the consensus of all agrees, whether he is present among them or his report is preserved among them. His knowledge of the state of obedience, the arising of these practices from it and the aspects of their benefits; and his knowledge of sins and their harms cannot be obtained through rational proof, nor by the kind of thinking which operates in matters of livelihood, nor by sense perception. Rather, these are matters whose truth is only disclosed through intuitive perception (*wijdān*). Just as hunger, thirst, and the effect of warming or cooling medicine is only comprehended through a sense of intuitive awareness, likewise is the knowledge of whether a thing is suitable for the spirit or is at variance with it: there is no way to this except sound spiritual experience.

His being himself secured from error is only through God's creating in him immediate knowledge (*'ilm ḍarūrī*)[2] that all that he comprehends and discerns is truly identical with reality, in the way that a person perceives as he sees. Thus, when he sees a thing, it is not possible according to him that his eye is defective or that his perception conflicts with the real. This is like the knowledge of the referents of words. The Arab, for example, does not doubt that the word "water" (*mā'*) refers to that element, while the word "earth" (*'arḍ*) likewise (refers to earth), despite the fact that no rational proof was established for him about this, and that between them there is no reasoned connection, for despite this God created in him "immediate" knowledge.

However in most cases this only obtains if the person has a natural disposition in his soul through which he constantly receives knowledge by automatic intuition of the correct practices. The fact that these intuitions successively follow one another, and the experience of his intuition being correct is repeated among people, confirming among them by many proofs, both rationally demonstrated and rhetorical, that what he summons them to is the truth, that his path is righteous, and that falsehood is remote from it. They see in him the influences of proximity (to God) such as miracles and the answering of prayers, so that they have no doubt of his high position in the heavenly management, that his soul is one of the holy souls connected with the angels, and that one like him is never disposed to lie about God nor to carry out any act of disobedience.

Then after this, events take place which join them into a great union and which make him more beloved among them than their possessions and their children, and than cold water is to the thirsting one. Due to this a community cannot take on the desired state without him. Therefore those who are occupied with similar acts of worship constantly trace back their practice to the one they believe had these qualities, whether they are right or wrong, and God knows better.

[2] *'Ilm ḍarūrī* is intuitive or essential knowledge which originates without thought, reflection, or proof.

CHAPTER 55

The True Nature of Prophecy and Its Characteristics

You should know that the highest ranks of people are "those who are instructed" (*mufahhamūn*),[1] and they are people of integration,[2] whose angelic side is of the very highest. They may be sent to establish the desired order by a heavenly motive (*dāʿiyya ḥaqqāniyya*), and divine knowledge and states from the Highest Council showers down upon them. Characteristic of "the one who is instructed" is that he is of balanced temperament and harmonious disposition and that his disposition is not excessively agitated by partial opinions (*ārāʾ juzʾiyya*), nor excessively intellectual so that he is not drawn away from the universal to the partial, or from the spirit to the form in any way. Nor is there to be found in him excessive stupidity from which he cannot escape to the universal, and from the form to the spirit. He is the person who is most committed to the rightly-guided practice, having good manners in his acts of worship, and justice in his dealings with people. He loves the universal management and is inclined to the common benefit, and causes pain to no one except in circumstances when the common good depends on this or when it is entails it. He persists in his inclination to the Unseen World and the influence of this inclination can be perceived in his speech and face, and in all of his nature it may be seen that he is aided by the Unseen. The least spiritual practice opens to him a proximity to God and a tranquillity (*sakīna*) which are not disclosed to others.

The people "who are instructed" are of many types and varying capacities:

---

[1] "Those who are instructed" or "those who are made to understand (by God or the angels)." *Mufahhamūn* is Shāh Walī Allāh's terminology for people of high angelic nature who free themselves from the attachments of the pneuma and become connected with the angels. *Hamaʿāt*, p. 110. The verbal form from which this term is derived is found in Qurʾān 21:79, "We made Solomon understand it" (*fahhamnā*). "*Tafhīm*" or making understood is found in the title of the author's work, *al-Tafhīmāt al-Ilāhiyya*" (The Divine Inspirations). See in particular Vol. II Tafhīm no. 122.

[2] Inner integration or reconciliation (*iṣṭilāḥ*) of the angelic and animalistic sides.

a) The one whose state is the greatest in receiving from God the sciences of the refinement of the soul through worship, is "the perfect one" (*kāmil*).

b) The one whose state is the highest in receiving cultivated virtues and the sciences of domestic management and so forth, is "the wise one" (*ḥakīm*).

c) The one who generally apprehends comprehensive policies and then succeeds in establishing justice among people and defending them from tyranny is called "a Caliph."

d) The one who has been visited by the Highest Council, and who was taught by them, spoken to by them, to whom they appeared, and from whom types of spiritual graces (*karāmāt*) have manifested is known as "the one aided by the holy spirit."[3]

e) The person among them whose tongue and heart are illuminated, so that people benefit from his company and his preaching, and from whom tranquillity and light are transferred to the disciples among his companions so that they reach, due to his mediation, the high degrees of perfection, while he is unflagging in guiding them, is called "a pure guide" (*hādi muzakkī*).

f) The one whose knowledge consists mainly of knowing the rules for the religious community and their benefits and who incites (them) to observe those rules which have been effaced is called "the leader" (*imām*).

g) The one who is inspired to inform them of the calamity destined for them in the world, or who realizes that God had cursed a people and who informs them of this; or who is divested of his lower soul at certain times so that he knows what will come to pass in the grave and on the Judgment Day and who informs them about these things is called "a warner" (*mundhir*).

When the divine wisdom requires that one "who is instructed" (*mufahham*) be sent to humanity so that he becomes a cause for the people's being brought out of the darkness into the light, God makes it a duty for His worshippers to accept this person both externally and in their hearts. Satisfaction is established among the Highest Council for those who follow and join him, and curses upon those who oppose and resist him. God informs people of this

---

[3] "Aided by the holy spirit" is a description of Jesus in the Qur'ān (2:87, 2:253, 5:10) and this description is applied to other prophets as well. This concept was discussed in Chapter 3 on "The Highest Council."

and makes them obey him, and such a person is "a prophet." The greatest of the prophets in rank is the one whose mission has an additional dimension and this is that the intention of God, may He be Exalted, for him is that he be a cause for bringing the people "out of the darkness into the light" and that his people be "the best community brought out for mankind," and thus his mission encompasses an additional mission.

An indication of the first (type of mission) lies in His saying, may He be Exalted, "He it is who has sent down to the unlettered ones a messenger of their own,"[4] and of the second (type) in His saying, may He be Exalted, "You are the best community brought out for humanity,"[5] and in the Prophet's saying, may the peace and blessings of God be upon him, "You have been sent to make things easy (for the people), and not to make things difficult."[6] Our Prophet, may the peace and blessings of God be upon him, encompassed all of the fields of the Instructed Ones (mufahhamūn), and was necessarily granted the fullest form of both types of mission, while among the prophets before him there had been those who had attained one or two fields, and so on.

You should know that the decree of the divine wisdom that prophets be sent was only to encompass the relative good considered in the divine management in sending (them). Only the One (God) most knowledgeable of the Unseen knows the true nature of this, but we know for certain that There (above) causes exist due to which the prophets were inevitably sent, and that the imposition of obedience is due to the fact that God, may He be Exalted, knows that the sound condition of a community lies in their obeying God and worshipping Him. Since their souls are not worthy of receiving knowledge from God (directly), the sound state for them is limited at that time to following a prophet. Therefore God decrees in the Holy Enclave that the prophet should be followed, and this command becomes established There.[7] This is either due to the fact that it is the time for the beginning of the rise of a empire (*daula*)

---

[4] Qur'ān 62:2.

[5] Qur'ān 3:110, repeated a few phrases earlier and at various other times in this work.

[6] Bukhārī Wuḍū' 58, Adab 80, Abū Dāwūd, Tirmidhī, Nasā'ī, Ibn Ḥanbal.

[7] This relates to the theory of *tadallī*. That is, an emanation comes from the Holy Enclave at the time for the establishment of a new revelation consonant with the conditions of that time.

and for other empires to be subdued by it, so that God, may He be Exalted, sends the one who will establish the religion of the members of that nation, just as He sent our master, Muḥammad, may the peace and blessings of God be upon him. Alternatively God, may He be Exalted, may decree the continuation of a people and their being selected from among humanity, so that He sends one who will correct their deviation and "teach them the book,"[8] as He sent our master, Moses, may peace be upon him. The course which was determined for a people in the continuation of their empire or religion may require the sending of a renewer such as David and Solomon and some of the prophets of the Children of Israel, may peace be upon them. God decreed that these prophets would be given victory over their enemies, as He said, "And indeed our word went forth previously to Our servants, the ones who were sent, that verily they would be given victory, and Our army would certainly overcome."[9] In addition to these (prophets) there is a group of people sent to complete the proof, and God knows better.

If a prophet is sent it would be incumbent upon those to whom he is dispatched to follow him, even if they had already been on a sound course. This is because being hostile to this one whose state is elevated would bring down a curse from the Highest Council and a consensus to forsake them, so that their path to drawing nearer to God would be blocked and their labor would be futile, and when they died the curse would surround their souls; although this case is hypothetical,[10] not real. You have a lesson in the Jews who were the most in need of the sending of a prophet due to their excesses in religion and their distortions of their scripture.

The confirmation of God's proof (*ḥujja*) for His servants in His sending the prophets is that most people are created in such a way that they are not able to know their rights and their duties, without an intermediary. Rather, either their capacity is weak so that it becomes strengthened by the messages of the prophets, or among them there are evils which cannot be prevented except by compulsion in spite of themselves, while they are to be held accountable in this world and the next. The grace of God necessitates that when certain higher and lower causes combine, He will inspire the purest

---

[8] Qur'ān 2:129, 3:164, 62:2.
[9] Qur'ān 37: 171–173.
[10] Of a righteous community to whom a prophet would be sent.

among the people to guide them to the truth and call them to the straight path. In this He is like a master whose slaves are sick, so that he orders one of his special attendants to make them drink medicine whether they want to or refuse. Even if he forced them to do this it would be right, but the fullness of his grace requires that he inform them first that they are ill and that the medicine is beneficial, and that he does extraordinary things to convince them that he spoke truly; and he mixes the medicine with something sweet, so that then they do what he ordered after comprehending it and desiring it. Therefore, miracles, the answering of prayers, and so on, are only things extrinsic to the basis of prophecy, connected with it in most cases.

The manifestation of the majority of miracles is due to three causes:

1) One of them is his being one of "the instructed" (*mufahhamūn*) which requires the disclosure of certain events to him and is a cause for the answering of prayers and the manifestation of blessings in what he blesses. Blessing (*baraka*) can occur through increasing the benefit of a thing, for example his (the prophet) making them (his enemies) imagine that his army was great so that they lost their courage, or by his changing the nature of food to a beneficial mixture, so that it is as if a person ate several times that amount of food, or through some increase in the thing itself in that its airy matter is transformed to that form by incorporating into it a force from the World of Images, and due to other causes of this type, which are difficult to enumerate.

2) The second (cause of miracles) is that the Highest Council are resolved to put his (the prophet's) order into force so that this requires inspirations, transformations, and approaches which were previously unfamiliar, therefore the friends (of the prophet) triumph and the enemies are forsaken, and the command of God triumphs "even though the unbelievers detest it."[11]

3) The third cause of miracles is that events occur due to their external causes, such as the punishment of the rebellious, and powerful things which take place in the atmosphere. Then God, may He be Exalted, makes them his (the prophet's) miracle in some manner, either by his having predicted them or by having them fall as a punishment on those who oppose his command, or by having them

[11] Qur'ān 9:32, 40:14, 61:8.

correspond to what he announced as being the means of punishment, or other things like this.

There are three reasons for immunity from sin (*'iṣma*). The first is that God creates the person pure and free from base desires, especially those which pertain to keeping within the limits of the divine law. (Secondly) that He reveals the goodness in what is recommended and the repulsiveness of the repugnant and their result, and thirdly that God intervenes between him and any base desires which he might harbor.

Then you should know that it is part of the conduct of the prophets, may peace be upon them, that they did not command speculating about the essence of God, may He be Exalted, and His Attributes, for most people are not capable of doing this. This is the Prophet's saying, "Reflect on the creation of God and do not speculate about God,"[12] and his interpretation of the verse, "The ultimate limit stops at your Lord,"[13] when he said, "There should be no speculation about the Lord."[14] Rather they were commanded to contemplate only the blessings of God, may He be Exalted, and the greatness of His power. Part of the prophets' conduct was that they only addressed people in accordance with the intelligence with which they were created, and according to the knowledge which they had achieved through the root of their constitution. This is because the human species, wherever it is found, has at the base of its constitution a limit of understanding greater than that of the other animals, except for the person whose material component strongly defies (the usual constitution). Humans may also achieve branches of knowledge which they can only derive through a break in usual habit, such as in the case of the holy souls who are prophets and saints; or through arduous practices which prepare their soul for realizing what they didn't know though calculations, applying the rules of philosophy (*ḥikma*), theology (*kalām*), or the sources of law, and so on, over a long period of time.

Therefore the prophets only addressed people according to the simple manner of understanding put in them at the root of their

---

[12] Not in the six collections. Abū Nu'aym, *Hillyat al-Auliyā*, Isfahānī, *Targhīb*. Cited by 'Irāqī in *Iḥyā' 'Ulūm al-Dīn* IV:464.

[13] Qur'ān 53:42.

[14] Not in the six collections but among the hadiths cited by al-Suyūṭī in *al-Durr al-Manthūr* VI:130, in explanation of Qur'ān 53:42. Also cited in Ch. 40 of present work.

constitution, and they paid no attention to matters which rarely arise and whose existence is unlikely. Therefore they did not oblige people to know their God through theophanies and visions, nor through discursive proofs and analogies, nor to know Him as being transcendent of all aspects, for this is at the level of something impossible for someone who has not been occupied with spiritual practices and who has not associated with exponents of the theoretical sciences over a long period. They did not guide them to the methods of inference (*istinbāṭ*), deductions (*istidlālāt*), the ways of applying discretion rather than accepted legal principles (*istiḥsānāt*), and the difference between the "similar" and the "equivalent" through the delicate use of propositions and the rest of those things about which the Rationalists exalt themselves above the People of the Hadith. In their conduct (the prophets) did not occupy themselves with what had no relevance for refining the soul and the governing of the community. They did not explain the events of the atmosphere like rain, eclipses, rings around the sun or moon, the wonders of plants and animals, the measurement of the paths of the sun and moon, the causes of daily events, and the stories of prophets, kings, countries, and so on, unless they used simple words with which their audience was familiar, and that their minds would accept if these were used to remind them of the blessings of God and the requital of God, unless this was by way of digression into a general discussion which is permitted in cases like this, through introducing metaphors and similes. This was basically the case when they asked the Prophet about the reasons for the waxing and waning of the moon, and God the Exalted turned from that to the explanation of the benefits of that waning and waxing, and said, "They ask you about the new moons, say they are times appointed for the people and the pilgrimages."[15] You may see many people whose taste is corrupted because of familiarity with these (technical) fields, or due to other causes, so that they apply the words of the prophets to things to which they did not refer, and God knows better.

---

[15] Qur'ān 2:189.

CHAPTER 56

An Explanation of the fact that the Basis of Religion (Dīn) *is One and the Religious Laws* (Sharā'i') *and Codes* (Manāhij) *Vary*

God, may He be Exalted, said, "He has ordained for you as a religion (*dīn*) that which He commanded unto Noah, and that which we had inspired in you (Muḥammad) and that which we commanded unto Abraham and Moses and Jesus, saying: Establish the religion (*dīn*) and be not divided therein."[1] Mujāhid[2] explained this as, "We entrusted you, O Muḥammad, and them, with one religion."[3]

God, may He be Exalted, said, "This community of yours is one community, and I am your Lord, so be pious; but they have divided their affair among themselves into sects, and each faction is happy with its tenets."[4] That is, the religion of Islam is "your religion," and "they were divided" i.e., the polytheists, Jews, and Christians.

And God, may He be Exalted, said, "For each of you We have appointed a divine law and a code (*minhāj*)."[5] Ibn 'Abbās said that this refers to "a way" (*sabīl*) and "a practice" (*sunna*).

God, may He be Exalted, said, "To each community We have given a devout ritual which they are to follow devoutly,"[6] that is, a divine law according to which they must act.

You should know that the basis of religion (*dīn*) is one, upon which the prophets, may peace by upon them, agree, and that any variations are in the divine laws and the codes (*manāhij*). This is to be elaborated as follows. The prophets, may peace be upon them, concur that God, may He be exalted, is unique in being worshipped and being asked for help, and that He transcends everything inap-

---

[1] Qur'ān 42:13.

[2] Mujāhid ibn Jabr al-Makkī (d.722/3). The best known second generation (*ṭābi'ī*) interpreter of the Qur'ān and a pupil of Ibn 'Abbās.

[3] In al-Ṭabarī *Jāmi' al-Bayān* XV (Beirut: Dār al-Ma'rifa), 10.

[4] Qur'ān 23: 52–53.

[5] Qur'ān 5:48.

[6] Qur'ān 22:67.

propriate to Him, and that heresy about His names is forbidden, and that the right of God over his servants is that they should magnify Him with a reverence untainted by neglect, and that they should submit externally and internally to Him, and that through God's religious rituals and emblems they should draw nearer to Him. (They agree that) God determines all events before He creates them and that God has angels, "who never disobey what He orders, and who do what they are commanded,"[7] and that He reveals the Book to whomever He wishes among His servants, and that obeying Him is a duty for people. (They agree that) the Day of Judgment is true, and that resurrection after death is true, and that Paradise is real and Hell is real. Likewise they agree on the types of pious actions, whether purity, prayer, the alms tax, fasting, pilgrimage, the drawing near to God through supererogatory acts of worship such as petitionary prayer, the remembrance of God (*dhikr*), and the recital of the scripture which was revealed from God. In like manner they agree on marriage, forbidding adultery, establishing justice among people, forbidding oppression, establishing punishments for the disobedient, striving against the enemies of God, and making efforts to spread the ordinance of God and His religion.

This is the basis of religion (*dīn*) and for this reason the Noble Qur'ān only discussed what God willed of the "why" and "wherefore" of these things, for they were agreed upon by those in whose languages the Qur'ān[8] was revealed. Disagreement only concerned the forms and embodiments of these matters. Thus, in the law of Moses, may peace be upon him, the direction of prayer was Jerusalem and in the law of our Prophet, may the peace and blessings of God be upon him, the Ka'ba. The law of Moses prescribed stoning (for the fornicator) while our law has stoning for the married one and flogging for the unmarried one. In Moses' law there is retaliatory punishment only,[9] while in our law there is both retaliation and blood money. Likewise are the variations they contain about the times of acts of worship, the proper manner of performing them, and their essential principles. In summary, the particular situations according to which the types of pious actions and the

---

[7] An echo of Qur'ān 66:6.

[8] That is, the Muslims and previous communities who had received God's message.

[9] For murder.

supports of civilization are set out and founded, are the divine law and the code.

You should know that the acts of obedience which God, may He be Exalted, commanded in all the religions are actions arising from the psychological attitudes which in the next life will be held for or against the souls. Actions are extensions of these psychological attitudes and illustrate them, and they are their embodiments and representations. Inevitably, these attitudes are the measure of them (the acts of worship), and their foundation, and thus the one who doesn't recognize them does not have insight into actions. Perhaps he will be content with what is not sufficient and may pray omitting the recitation of the Qur'ān and neglecting calling upon God, so that this will be of no benefit. Thus, it is necessary that the policy of the person with true knowledge be to render precise the hidden, ambiguous thing with clear signs, and to make that concrete and distinguishable by high and low, and to make it unambiguous for them, in order that it may be required of them and that they may be held accountable for it, according to a proof from God and their own capacity.

Sometimes sins are confused with something which is not a sin, such as the saying of the polytheists, "Usury is just like trade,"[10] either due to lack of knowledge or due to a worldly desire which corrupts their insight. Therefore circumstances demand signs by which that whatever is sinful may be distinguished from what is not. If the times had not been determined, some of them would consider excessive only a few prayers and a little fasting, and that much would not avail them anything, while it would not be possible to punish their stealthy laxity and deception. If the pillars and the conditions (of worship) were not specified for them, they would act haphazardly; if the punishments had not been given, the transgressors would not be deterred.

In summary, the religious duties of the majority of the people will not be fulfilled except through setting times, pillars, conditions, punishments, general rules, and so on. If you want to know the method of the divine legislation, then study the case of the skillful doctor when he strives to treat the sick. He informs them of what they don't know and orders them to do things whose fine points they do not understand. See how he takes physical symp-

---

[10] Qur'ān 2:275. ". . . . whereas God permits trade and forbids usury."

toms and then traces them to hidden things such as associating red skin and loss of blood from the gums with an excess of blood. See how he considers the strength of the sick person, his age, his city, and the season, and how he considers the potency of the medicine and all of these factors. Thus he intuits the specific quantity of the medicine which will be suitable for the condition, and then prescribes it. Sometimes he formulates a general principle due to establishing an anticipated source of the cause of the illness, setting the quantity which he calculates of medicine to effect the elimination of the troublesome material or to change its corrupt form. Thus, for example, he says, "If someone's skin is red and his gums bleed, then the judgment of medicine says that upon waking on an empty stomach, he should drink jujube or honey water, and the one who doesn't do this is on the brink of destruction." Or he says, "The one who eats the weight of a *mithqāl*[11] of a certain medicinal paste[12] will be cured of a certain sickness, and protected from a certain sickness." Then these general rules (of medicine) are transmitted from him and become acted upon, so that God sets much benefit in this. Then contemplate the case of the wise king attending to the improvement of the city and the administration of the armies: how he considers the lands and their yield, and the cultivators and their provisions, and the guardsmen and their provisions, so that he sets the *ʿushr* and *kharāj* land taxes according to this. See how he sets visible formalities and evidence for the virtues and traits of character which must be found among assistants, and selects them according to that ordinance. Look how he considers the needs which must be fulfilled and the quantity of assistants, so that he distributes them (in various tasks) so as to suffice the goal, and not to overwork them.

Then contemplate the case of the teacher of children in relation to his pupils, and the master with regard to his slaves. The former wants to instruct them and the latter wants the intended task to be carried out by the slaves' hands, and neither the children nor the slaves understand the nature of the benefit involved nor do they desire to accomplish this, and they slink off and make excuses and resort to tricks. See how these two anticipate the breach before it occurs, and they block the gaps. They address them only in such a

---

[11] A weight of about five grams.

[12] *Maʿjūn*. A confection of medicines mixed with honey or juice.

way, day and night, and night and day; that they can't find a way to be tricky and are not able to slip away, and thus the goal is achieved whether they (the children and slaves) realize it or not.

In summary, everyone who is responsible for the reforming of a very large number of people of various capacities, who have neither insight into the situation, nor desire it; is forced to determine things, set times, and appoint occasions and attitudes to be a support for requiring and punishing.

You should know that since God, may He be Exalted, wanted through the mission of the prophets to, "bring the people out of the darkness into the light," He therefore revealed to them His command and cast His light into them and inspired them with the desire to reform the world. At that time the guidance of the people could only be achieved by matters and principles, all of which God's wisdom required be included in His will to send the prophets. It was also required that the obligation to obey the prophets and submit to them include the necessity to follow the principles for reform and everything else which is not attained by reason or custom except due to this. Therefore, this is a totality and these things entail one another. There is nothing hidden from God and there are no arbitrary things in the religion of God, so that no thing is determined while similar things are not, except due to wise reasons and causes that those thoroughly versed in knowledge understand. We wish to point out a good number of these wise reasons and causes, and God knows better.

CHAPTER 57

The Causes of the Revelation of the Divine Laws Relating to One Age Rather than Another, and to One People Rather than Another

The basis of this is His, may He be Exalted, saying, "All food was lawful to the Children of Israel except that which Israel had forbidden to himself before the Torah was revealed. Say: Produce the Torah now and read it if you are truthful."[1]

The interpretation of this is that Jacob, may peace be upon him, became very ill, then he made a vow that if God restored him to health he would give up the food and drink that he loved best. When he was restored to health he forbade himself camel's meat and their milk and his sons imitated him in forbidding these, and thus centuries passed until they internalized the idea that it would be neglectful toward the prophets and in opposition to them to eat these things, therefore this prohibition was revealed in the Torah.

When the Prophet, may the peace and blessings of God be upon him, explained that he followed the religion of Abraham the Jews said, "How can he be of his religion, when he eats camel's meat and milk." Then, God, may He be Exalted, replied to them that all food had been permitted originally and that the camel had only been forbidden due to a (historical) accident affecting the Jews, so that when prophecy appeared among the descendants of Ishmael who were free from that historical accident, they didn't have to observe it.

The Prophet, may the peace and blessings of God be upon him, said concerning the prayer of *Tarāwīḥ*,[2] "What I have seen you doing has continued to be your practice, until I feared that it would be made compulsory for you, for had it been assigned to you, you would not have been able do it. Therefore, pray it, O people, in your homes."[3] Therefore the Prophet, may the peace and blessings

[1] Qur'ān 3:93.

[2] The prayer performed during the nights of Ramaḍān.

[3] *Mishkāt*, p. 270. Bukhārī Tahajjud 5, tarāwīḥ 1, adhān 80. Muslim, Abū Dāwūd, Nasā'ī. Cited in the introduction.

of God be upon him, restrained them from making it common and widespread among themselves lest it become one of the emblems of the religion, so that they would think that abandoning it would be remiss with regard to God and that it had been assigned as a duty for them.[4] (Other examples are) the Prophet's, may the peace and blessings of God be upon him, saying, "The Muslim who harms Muslims most is the one who asked about a thing so that it was forbidden because of his questioning,"[5] and his saying, "Abraham made Mecca a sacred place and asked for blessing on it as I did with Medina, and I asked for blessings on it and its provision, as Abraham prayed for Mecca."[6] and his saying to one who asked him whether the pilgrimage should be performed every year, "If I were to say yes, it would become a necessity, and if it were required you would not be able to perform it, and if you weren't able to perform it you would be punished."[7]

You should know that the divine laws of the prophets, may peace be upon them, differ due to reasons and beneficial purposes. This is because the religious rituals of God were rituals for intended purposes and the quantities in their legislation take into account the situation and customs of those on whom they were imposed.

Since the temperaments of Noah's people, may peace be upon him, were extremely severe and harsh, as was declared by God,[8] may He be exalted, they were required to be ordered to do uninterrupted fasting in order to resist the strength of their animal side. Since the temperaments of this nation[9] were weak they were forbidden to do this. Likewise God, may He be Exalted, did not make booty permissible to the earlier peoples, but permitted it to us because He considered our weakness.[10] What the prophets, may peace be upon them, intended, was the reforming of the supports of civilization which people already had, so these were not deviated from

---

[4] People had been praying this prayer in the mosques, then the Prophet permitted them to pray it at home.

[5] *Mishkāt*, p. 42. Bukhārī I'tiṣām 3, Muslim, Abū Dāwūd. Mentioned in the introduction to the present work.

[6] *Mishkāt*, p. 587. Literally, "I pray for its *madd* and *ṣā'*" (two measurements of provisions and crops). Bukhārī Buyū' 53, Muslim Ḥajj, 455 etc., Ibn Ḥanbal.

[7] Ibn Mājah Manāsik 2. *Mishkāt*, p. 535.

[8] As reported in Qur'ān 71:7.

[9] The Muslims.

[10] As in the hadith, "Booty was not lawful for anyone before us, because God saw our weakness and incapacity and made it allowable for us." *Mishkāt*, p. 848. Muslim Jihād 33, Ibn Ḥanbal.

in favor of something at variance with the familiar unless God willed it. The anticipated sources of things conducive to benefit, (*maṣāliḥ*) differ in accordance with the differences in the eras and customs and therefore abrogation is justified. An example of this would be the doctor who intends to preserve a balanced temperament in all conditions, so that his prescriptions differ with the differences among individuals and seasons. Therefore he commands the youth to do things which he doesn't command the old man, and he orders sleeping outside in the summer because he considers that the air is a likely location of moderation at the time; and in the winter he orders sleeping inside the house, because he considers the outdoors a likely location of cold at that time.

Whoever recognizes the basis of religion (*dīn*) and the causes for the variation of the codes will not concede any change or alteration, and therefore the divine laws were matched to their peoples. Blame was attributed to them when they deserved it according to their capacity, and they urgently requested these (varying codes and practices) according to their (inner) condition. This is His, may He be Exalted, saying, "But they (humankind) have broken their religion among them into sects, each sect rejoicing in its tenets."[11]

Therefore the special position of the community of our Prophet, may the peace and blessings of God be upon him, became apparent when they were entitled to have Friday designated[12] due to their being unfamiliar with and free from the acquired sciences, while the Jews had been entitled to Saturday due to their belief that it is the day that God finished the creation and that it is the best time for performing worship. This does not alter the fact that both days were appointed due to the command of God and His revelation. The divine law in this is like the duty which they were initially commanded to perform, and then excuses and excessive difficulty occurred in this, so that dispensations were legislated for them in consideration of their personal circumstances. Thus sometimes through this some blame attaches to them, due to their deserving that blame because of the way they are, as God, may He be Exalted, said, "Verily God does not change the condition of a people until they change what is in themselves."[13] The Prophet, may the peace and

---

[11] Qur'ān 23:53.
[12] As the day of the congregational prayer.
[13] Qur'ān 13:11.

blessings of God be upon him, said, "I did not see any more deficient in intelligence and religion; any who could more sweep away the heart of the sensible man than one of you women," and he explained the deficiency of their (the women's) religion saying, "Don't you see that she, when she has her monthly period, does not pray or fast?"[14]

You should know that the causes for the revelation of the codes (*manāhij*) in a particular form are many, but they are reducible to two types.

1) One is like a command of the natural order which requires their being charged with these rulings. Thus, for example, all the human individuals have a physical nature and conditions which they inherited from the species which require their being charged with rules. For example, a person born blind does not have colors and forms in the storehouse of his imaginative faculty, but only words and tactile perceptions and so on; so that if he receives knowledge from the Unseen in a dream or spiritual event or something like that, his knowledge would be embodied in a form which his imagination has stored and not in another. Or, for example, a speaker of Arabic who only knows the Arabic language, when knowledge is represented to him through verbalization, will find it only represented to him in Arabic, not in some other language. Likewise in those countries where elephants and other animals of ugly appearance are found, the visitation of the jinn and the frightenings of the devils appear to the (local) inhabitants in the form of these animals, while this is not so in other countries. In those (countries) in which certain things are extolled, and in which are found certain fine foods and types of clothing, blessings, and the joy of the angels, are represented to their inhabitants in these forms while it is not so in other countries. Another example is that when an Arabic speaker is about to do something, or intends to take a certain road, and he hears the word "the guided" or "the successful" it is an indication of the auspiciousness of what he is turning to, but not for the non-Arabic speaker, and the reports of the Prophet contain some examples of this type of thing. Similarly, in the divine laws are expressed branches of knowledge accumulated among the people and beliefs internalized by them, and customs which literally flow in their veins.

---

[14] *Mishkāt*, p. 9. Bukhārī Ḥaid 6, Ṣaum 41, Trans. I:181.

Therefore the forbidding of the meat and milk of the camel was revealed to the Children of Israel but not to the Children of Ishmael, and therefore the wholesome and the harmful among kinds of food were left up to the customs of the Arabs. Therefore our sister's daughters were forbidden to us but not to the Jews, for they reckoned them among their fathers' family and did not mingle with them, nor have a linkage, nor companionship, and thus they were like strangers (for them) unlike (the case among) the Arabs. Therefore cooking a calf in its mother's milk is forbidden according to them, but not to us. The theory of this being an alteration in the creation of God and in opposition to His management, in so far as it uses what God created for the development and growth of the calf to decompose its body, was firmly held among the Jews and commonly accepted among them, while the Arabs were the people furthest from this theory, so that even if it were reported to them they wouldn't have understood it, and they wouldn't have understood the goal to be related to the rule. What is considered in the revelation of the divine laws is not only the branches of knowledges, conditions and beliefs (which come to be) represented in their hearts, but rather, more greatly considered than these, and having precedence over them are things which are innate in their make-up to which their minds are impelled, whether they realize it or not. For example you see this in the connections which represent a thing in the form of something else, such as the forbidding of people to eat in the morning[15] (being represented) in the form of a seal on their mouths, such that the seal depicted the ban for the people whether they could envision it or not.[16]

God's right over His worshippers is essentially that they give Him the ultimate reverence and do not dare to oppose His command in any way. What is required among people is that they undertake the promotion of affection and cooperation, and that no one should harm another unless the comprehensive outlook (*al-ra'y al-kullī*) so demands it, and so on. Therefore, the one who has sexual intercourse with a woman thinking that she is a stranger has lowered a veil between himself and God and this is recorded as an offense on his part against God even if she is really his wife because

---

[15] During Ramaḍān.

[16] The example of the interpretation (by Ibn Sīrīn) of a dream symbol for fasting in Ramaḍān.

he dared to oppose the order of God and His rule. The one who has intercourse with a stranger believing that she is his wife will not fail to be excused for that, in what is between him and God. The one who makes a vow to God to fast will be held to his vow and not the one who has not made one, and the one who is strict in religion will find it made strict for him. Striking the orphan for the purpose of discipline is good, and in order to cause pain is evil, and the mistaken or the forgetful person is forgiven in many judgments. This basic principle is absorbed by the branches of knowledge and customs of the people, both hidden and manifest, of the people; then the divine laws are individually particularized for them in accordance with this.

You should know that the Arabs and Persians, as well as the entirety of the inhabitants of the moderate regions and the people of temperaments which are capable of accepting noble virtues, agree on many of the latent customs and branches of knowledge, such as sorrow for the dead person and preferring gentleness in his case, pride in noble descent and lineage, going to sleep when a quarter or a third of the night or something similar has passed; waking in the early morning and other things which we pointed out among the supports of civilized life. Therefore these customs and branches of knowledge are the things most worthy of consideration. Following them are customs and beliefs particularized for the people to whom the Prophet was sent, for these also are considered; "and God has made a measure for everything."[17]

Know that prophecy frequently follows (the heritage of) the religion, as God, may He be Exalted, said, "The religion of your father Abraham,"[18] and as He said, "and Abraham was from his (Noah's) following."[19]

The secret of this is that many generations arise with piety in religion and respect for its emblems so that its rules become so well-known and wide-spread as to become attached to the primary, self-evident truths which can scarcely be denied. Then another prophetic mission comes to correct what had deviated from these and to reform what had been corrupted due to confusing the reports about its prophet. This (prophetic mission) investigates the rules

[17] Qur'ān 65:3.
[18] Qur'ān 22:78.
[19] Qur'ān 37:83.

accepted by them and does not alter what is just and in accordance with the principles of the policy of the religion, rather it summons to them and urges (the people) to them. Those (principles) that have become faulty and distorted it alters to the necessary extent, and what deserves to be increased it increases according to their condition. Frequently, this prophet will make reference to what remained among them from the previous divine law in what he requires, and therefore in this case it is said that such and such a prophet is from the religion of a certain prophet, or from his following. Often the differences among prophetic missions are due to the differences among the communities to which that prophetic mission is revealed.

2) The second type (of causes for the revelation of the codes in particular forms) is like a contingent accident, and this is that God, may He be Exalted, although He is far beyond temporality, has a connectedness in some sense with time and temporality. The Prophet, may the peace and blessings of God be upon him, told that God decrees a great event after every hundred years,[20] and Adam and other prophets, may peace be upon them, informed in the hadith about intercession something along this line in so far as each of the Prophets said, "My Lord, may He be Blessed and Exalted, is angered today with an anger such as he has never had before, and such as He will never have again."[21] Then when the world is prepared for the emanation of the divine laws and the determination of the divine statutes (*ḥudūd*), God produces a theophany (*tajallī*) revealing religion to them and the Highest Council becomes filled with a strong concentrated resolve, due to which at that time even the least accidental cause is sufficient to knock at the door of divine generosity; and whoever knocks at the door of divine munificence, shall have it opened to him. You have an analogy in the spring season when the slightest effort at planting and sowing has an efficacy which multiplied several times will not avail in another season. The concentrated resolve of the Prophet, may the peace and blessings of God be upon him, and his honoring a thing, praying for it, longing for it and seeking it are a strong cause for the revelation of the divine decree in that matter. If his prayer can revive the pure practice,

---

[20] May refer to the hadith "God will send to this community at the beginning of every century someone to renew its religion for it." Abū Dāwūd IV:109 #4290 Malāhim 1.

[21] Bukhārī Anbiyā' 3, Tafsīr 17:5, Muslim, Tirmidhī, Ibn Ḥanbal.

overcome a large group of people, and increase food and drink to a visible extent, then what do you think of the revelation of the ruling which is a subtle spirit (*rūh laṭīf*), and is only specified through a similitudinary (*mithālī*) existence (in the World of Images). On this basis it is necessary to conclude that the occurrence of a great and imposing event at that time which startled the Prophet, may the peace and blessings of God be upon him, is a cause for the revelation of rules and the disclosure of the true state of affairs to him, such as the case of the story of the falsehood,[22] and the question which an inquirer asked him repeatedly and the Prophet discussed it with them, as in the story of the *Ẓihār*.[23] The fact that it takes a long time for the people to become obedient and habituated to yielding and that they are predisposed to disobedience, and likewise their desire for something, their seizing it firmly, and their believing that there is negligence with regard to God in abandoning it—is a reason that the firmest compulsion and the strongest forbidding should be strictly demanded of them. The example of all of this in the calling down of divine generosity is that of the righteous man with strong concentrated resolve who targets the hour of the spreading of spiritual energy and the power of happiness, and then asks God for this with the full extent of his concentrated resolve and his answer is not slow in coming. It is to these meanings which His, may He be Blessed and Exalted, saying refers:

> O you who believe. Don't ask about things which if they were known to you would trouble you, for if you ask about them while the Qur'ān is being revealed they will be made known to you.[24]

The element which is pleasing (to God) is that these sorts of causes for the revelation of the divine laws be few, since this leads to a revelation of something in which the determination of beneficial purposes particularized for that (specific) time predominate, and frequently this is burdensome for those who come later. Therefore the Prophet, may the peace and blessings of God be upon him, disliked (needless) questions (*masā'il*) and used to say, "Leave me (and don't ask me) as I have left you, for those before you were

---

[22] The accusation that the Prophet's wife 'Ā'isha had committed adultery.

[23] *Ẓihār* was a pre-Islamic form of divorce whereby men declared to their wives, "You are as my mother's back to me." It was condemned by the Prophet. See Qur'ān 33:4, 58:3.

[24] Qur'ān 5:101.

destroyed through the excessiveness of their questionings and their differing over their prophets,"[25] and he said, "The Muslim who harms other Muslims most is the one who asks about a thing, and it is forbidden because of his questioning."[26] And it was reported that, "If the Children of Israel had sacrificed any cow they wanted, it would have been sufficient (punishment) for them, but they were more hard on themselves, therefore God made it harder for them,"[27] and God knows better.

[25] *Mishkāt*, p. 535. Bukhārī I'tiṣām 2, Muslim, Tirmidhī, Nasā'ī, Ibn Mājah. Khan trans. IX:289–90.

[26] *Mishkāt*, p. 42. Bukhārī and Muslim transmitted it, previously cited in this chapter. Khan trans. IX:290.

[27] Hadith cited by al-Suyūṭī, *al-Durr al-Manthūr* I:77 in explanation of Qur'ān 2:67.

Chapter 58

The Causes for Being Held Accountable for the Codes (Manāhij)

Let us investigate the codes and the divine laws which God, may He be Exalted, has imposed on his servants, and whether reward and punishment are a consequence of them resulting from the principles of righteousness and sin; or whether they are only a consequence of what are made anticipated sources (*maẓinnāt*), embodiments and models (*qawālib*) of them. Thus, will the man who doesn't pray on one occasion while his heart is sure in submission to God, be punished for not praying? And will the one who prays the prayer following the pillars and conditions in accordance with what is assigned, while this does not arise from submission at all, nor does it penetrate the inner depths of his heart, be rewarded for his action? This discussion does not concern whether disobeying the codes is a great evil from the aspect of its being a depreciation of the rightly guided practice, an opening of the gate of sin, disloyalty relative to the community of Muslims, and a thing harmful to the district, the city, and the region; as in the case of a torrent whose course has been dammed up for the welfare of the city, then a man comes and makes a breach in the dam and saves himself, and destroys the people of his city. Rather the discussion of this concerns what reverts to his soul in its being enveloped by evils or by good things.

All of the people of the religions without exception believe that the religious codes necessarily bring about rewards and punishments in themselves, and the ones of correct opinion among them, those firmly grounded in knowledge, and the disciples among the followers of the prophets, may peace be upon them, understand along with this the nature of the relationship and the connection of these forms and models with their principles and spirits. The generality of the religious scholars and those mindful of the divine laws are satisfied with the first, while the Islamic philosophers believe that punishment and reward only occur due to the psychological characteristics and the virtues adhering to the lowest part of the spirit, and that their exemplifications and forms being mentioned in the divine laws is merely an aid which facilitates understanding and brings the subtle meanings closer to the comprehension of the people. This is an explanation of the position according to the outlook of people.

I hold that: the truth is what the ones of true opinion among the people of the nations hold. The explanation of this is as follows. The divine laws have preparatory circumstances and causes which particularize them, and which show certain of their possible applications to be more appropriate than others. God is aware that people are not able to implement religion except through these divine laws and codes, and He knows that these statutes are the most suitable for being imposed on them, and thus they were included in God's providence for the people in pre-eternity. Then when the world was prepared for the emanation of the forms of the divine laws and the bringing into being of their similitudinary individuations (*al-shukhūṣ al-mithāliyya*) He brought them into being and emanated them and their authority was confirmed There, and they (the laws and codes) became one of the principles (of reward and punishment). Then when God inspired this knowledge to the Highest Council and revealed to them that the anticipated sources (*maẓinnāt*) would stand in the position of the principles, and that they were their embodiments and representations, and that religious obligations could not be imposed on a people except through them, a certain consensus (*ijmāʿ*) obtained in the Holy Enclave that they (the laws) were like the word in relation to the reality which is its referent, and like the mental image in relation to the external reality from which it is derived, and like the painted portrait in relation to the one of whom it was drawn in its portraying him, and like the written form in relation to the words which it puts down. For, in all of these things, when the relation between the signifier and the signified is strengthened there arises between them a permanent bond and a close association with each other, and in a certain scope, one is the other. Therefore the form of this knowledge or its reality filtered down into the perceptions of all of the descendants of Adam, Arabs and non-Arabs, and they agreed on it, so that you will never see anyone but that he has secreted in his soul a portion of this. Sometime we call it (this symbolization) "an existence which is a likeness to the referent," and sometimes this existence has amazing effects which are not concealed from the one who pursues (this), and provision was made for some of this in the divine laws. For this reason alms come about from the filth of the alms givers[1] and the most repulsive work is done for wages.[2]

Therefore when the Prophet, may the peace and blessings of God be upon him, was sent, and supported by the Holy Spirit, and he was inspired in his heart to reform the people, there was opened at the core of his spirit broad access to strong concentrated resolve in the matter of the revelation of the divine laws and the emergence of the similitudinary individuations, so that he resolved upon this with utmost determination, praying with his fullest concentrated resolve for those who agreed with him and cursing the opponents. Their (the prophets') concentrated resolve penetrates the seven heavens. When they prayed for rain and there was not a puff of cloud in the sky, clouds spread like mountains even while they were still praying. They invoked and the dead were brought to life through their prayers, confirming the convening of satisfaction and displeasure in the Holy Enclave and this is the Prophet's saying, "Abraham, Your Prophet and servant prayed for Mecca and I pray for Medina."[3]

Consequently, a person, if he knows that God, may He be Exalted, commanded him to do a certain thing, and that the Highest Council endorses the Prophet, may the peace and blessings of God be upon him, in what he commands and forbids, and he knows that neglecting the latter and undertaking the former is defiance of God and neglect in this regard; and yet he undertakes the action purposefully and intentionally while he sees and has insight—this is due to nothing except great envelopment in veils and a complete breach with the angelic side, and this effects the establishment of a sin in the soul. If he performs a difficult action from which his nature shrinks, not in order to show off before the people, but rather to draw nearer to God and to preserve His satisfaction, this does not occur except through a great envelopment in righteousness and a complete breach with the animalistic side, and this effects the establishment of a good deed in the soul. As for the man who didn't pray on some occasion, it is necessary to investigate why he didn't do it, and what thing led him to this. If he forgot it, or slept

[1] According to hadiths such as Muslim Zakāt, 167, 168. This refers to the purifying effect of giving *zakāt*.

[2] Based on a hadith in which the alms are compared to impurities which are eliminated through giving.

[3] Variant of hadith cited in Ibn Mājah Manāsik 104, Muwaṭṭa' Madīna 2, Ibn Ḥanbal. *Mishkāt*, p. 587.

through it, or was ignorant of its obligatory character, or was unavoidably occupied during it, religion specifies that he has not sinned. If he knowingly abandoned it, while he remembered it and was fully able to do it; then this was inevitably due to some thing making his religion waver, and a satanic or egotistical enveloping which clouded his insight, and it will revert to his soul. As for the man who performed his prayer and fulfilled the obligation which was upon him, it is necessary to investigate this also. If he did it to show off and for his reputation, or in order to conform to the custom of his people, or frivolously, then the religion specifies that he is not obedient and he is not given credit for his act. If he did it to draw nearer to God and performed it with belief, taking into account and having faith in the divine promise, and calling to mind his intention and worshipping God sincerely, then inevitably a door opens between him and God, even if it is as small as the head of a pin. As for the one who destroyed the city and saved himself, we do not agree that he saved himself (due to his personal capability). How could this be, when There above are angels of God whose utmost concentrated resolve is praying for the one who strives to improve the world, and against the one who tries to corrupt it, and their invocation knocks on the door of divine generosity and is a cause for the descent of requital in some fashion. In fact, God, may He be Exalted, has There a providence for people which requires this, and due to the subtlety of grasping it we made the invocation of the angels a sign for it, and God knows better.

CHAPTER 59

The Secrets of the Ruling (Ḥukm) *and the Reason for Legislation* ('Illa)

You should know that human beings perform actions because of which the Lord of the worlds is pleased with them, and actions due to which He is angered, and actions which do not incur pleasure or anger. His far-reaching wisdom and His complete mercy required that He send prophets to them and that He inform them through the prophets of the connection of pleasure and anger with these actions, and that He seek from them the first type and forbid the second, and that He inform them about what is other than this. "That whoever perished might perish by a clear proof and that he who survived might survive by a clear proof."[1]

Thus the connection of pleasure and anger with the act or its being indifferent with regard to them, and a thing being sought from people or forbidden to them, while they are free to choose it whichever you wish, call this the "ruling" (*ḥukm*).

What is desired from them may be definite (*mu'akkad*),[2] requiring pleasure and reward for doing the desired action and anger and punishment for not doing it; or it may be indefinite, requiring pleasure and reward for doing what is sought but not anger and punishment for omitting it.[3]

Likewise the prohibition of thing may be definite, which necessarily brings pleasure and reward for shunning it because of the prohibition, and anger and punishment for carrying out the forbidden act; or it may be indefinite (*ghair mu'akkad*), requiring pleasure and reward for shunning it because of the prohibition, but not anger and punishment for carrying it out. Consider the words you know for requiring and forbidding, and the expressions people have for this, and you will find a duality of each division from the aspect of the

[1] Qur'ān 8:42.
[2] This refers to a type of sunna (action of the Prophet), which has a stronger normative force rather than being merely recommended.
[3] This refers to the two levels of the sunna.

coming into effect of either pleasure or anger; in contradistinction to what was said at the beginning, as a natural thing that is unavoidable. Thus there are five (degrees of) rulings: compulsory, recommended, permitted, reprehensible, and forbidden. That which is advanced in addressing the people who are commanded to perform the duties cannot be the precise circumstance of every one of their acts, due to these being infinite and due to the incapacity of people to comprehend this knowledge. Hence it is necessary that they be addressed through universal rulings controlled by a unity which gives order to a multiplicity, so that intellectually they can encompass them and recognize from them the status of their actions. Here you have a lesson in the comprehensive skills that were made an established principle in specific fields. The grammarian says that the active participle (*fāʿil*) is in the nominative case, and the one who hears this bears in mind his saying and through it recognizes the state of Zaid in our saying, "Zaid stood up," and ʿAmr in our saying, "ʿAmr sat down," and so on. This unity which orders the multiplicity is the reason for legislation (*ʿilla*) upon which the ruling operates. It has two categories.

1) Under one category is considered the state of those on whom religious duties are imposed. It is not possible that this be a permanent state which is never separated from them so that the meaning of their being addressed would be their being permanently obligated with the thing, since they are incapable of this. Such obligation is applicable only in the case of faith (*īmān*) which is a case by itself. Thus it must inevitably be considered a situation composed of an inherent property inseparable from the one obligated which justifies his being addressed, and of an accidental condition which occurs from time to time. This category applies most often to the acts of worship and the form, whether of time or a capacity within easy reach, or the anticipated source of hardship, or the will to do a thing, and this sort of element such as the saying of the divine law:

> The man who attains the time of a prayer and is sane and mature is required to pray it, and the one who witnesses the start of the month (of Ramaḍān) and is sane, mature, and capable, is required to perform the fast, and whoever owns the minimum amount of property when the year has passed, is obligated to pay *zakāt* on it. The one who is on a trip is allowed to reduce the prayers and to break the fast, and the one who wants to pray and has lost his state of ritual purity must perform the ablution.

In cases like this, sometimes the properties considered in the majority of the commands will not be articulated, while that property through which certain of them may be distinguished from others is singled out. One would be careless in calling it a "reason for legislation" (*'illa*), for this would be saying that the reason for the legislation for the prayer is attaining the time, and that the reason for the legislation of fasting is the witnessing of the new month. Sometimes the law-giver allots an influence to some of these properties and not to certain others. For example, paying *zakāt* a year or two in advance is permitted for the one who owns the minimum amount of property, and not for the one who does not own it. Thus the jurisprudent (*faqīh*) gives each thing its proper due so that some of these properties are designated as a cause (*sabab*)[4] and others as a condition (*sharṭ*).[5]

2) Then there is a category (of reasons for legislation) which takes into account the condition in which the action occurs or that accompanies it. This is either a characteristic inherent in it such as the saying of the law-giver, "Drinking wine is forbidden, and eating pork is forbidden, and the eating of all wild beasts that have tusks, and all birds that have claws is forbidden, and marrying one's mother is forbidden," or it is an accidental characteristic which may occur from time to time, such as God's, may He be exalted, saying, "Cut off the hands of the male and female thieves."[6] And His saying, may He be exalted, "Whip the adulteress and the adulterer each with one hundred strokes."[7]

Sometimes this category may combine two, or more than that, of the circumstances in which the act occurs, as in the saying of the law-giver, that stoning the adulterer who is married is obligatory, and whipping the fornicator who is unmarried. Sometimes it combines the condition of the one who is obligated, and the situation in which the act occurs such as the saying of the law-giver that gold and silk are forbidden to the men of the community and not to the women.[8]

---

[4] The cause (*sabab*) of a ruling would be that which brings it into effect such as the arrival of the time for prayer.

[5] The condition (*sharṭ*) for an action would be a thing required for it to be effective such as being in a state of purity being a requirement for the performance of the prayer.

[6] Qur'ān 5:38.

[7] Qur'ān 24:2.

[8] In numerous hadith, for example Bukhārī Libās 30.

There is no arbitrariness in the religion of Allah, and pleasure and anger are not attached to these acts except due to a reason, and this is that there are specific things to which pleasure and anger are attached in reality, and these are of two types.

a) One of them entails piety and sin, the supports of civilization and their neglect, and whatever follows this.

b) The second of them is whatever is connected with the divine laws and the codes in terms of blocking access to distortion, guarding against slippage, and so on.

These two types of specified acts have causes and attendant factors to which divine pleasure and anger are linked incidentally, and they are related to them (causes and attendant factors) in a broader sense. The parallel to this is what is said about the reason (*ʿilla*) for the cure being taking medicine, while the reason in reality is the becoming sound of the humors or their elimination. This is a thing that customarily follows taking the medicine; the medicine itself is not the cure. Likewise it is said that the reason for a fever may be sitting in the sun, it may be a tiring undertaking, and it may be eating hot food, but the real reason is the heating of the humors and it is basically one, although these are a means to it and forms of it. Sufficing with the principles and forgoing consideration of the multiplicity of the means and instances is the language of those deeply absorbed in the theoretical disciplines and not of the masses. The divine law was revealed in the language of the general public, and it is necessary that the reason for the legislation of the ruling be a property recognized by common persons and one whose truth is not concealed from them, nor should its presence be indistinguishable from its absence for them. It should be an anticipated source (*maẓinna*) of one of the principles connected with divine pleasure or anger, either due to this instance leading to the principle or its being very close to it, and so on. An example is the drinking of wine which is an expected location for evil acts to which anger is connected, due to turning away from righteousness, inclining to the world, and the corruption of the order of the city and the home; anger is usually a consequence of these, therefore the prohibition was applied to all types of alcoholic beverages. When a thing has many attendant factors and means, none of these is specified as the reason for legislation, except whatever can be distinguished from the rest of those involved, due to its being more apparent and determined, or more closely linked to the basic principle, and so on. For example, the permission to shorten the prayers or break the fast is contingent on being on a journey or being sick, but not on other situations in which hardship can be expected. This is because the professions involving hard labor such as agriculture and blacksmithing, involve hardship, yet this would be damaging to the acts of worship because the person who practices these professions perseveres in them continuously and his livelihood depends on them. As for (hardship arising from) the weather being hot and cold, this is not precisely definable since these have various levels whose enumeration and determination by signs and dis-tinguishing marks would be difficult. Therefore in the investigation only those situations where hardship was expected are considered that were prevalent and well-known according to the people of the Islamic community in the early period. Being on a journey or being ill was an unambiguous matter among them, although today there is some confusion due to the dying out of the early Arabs and people becoming hair-splitters concerning the permissible interpretations, and ultimately their sound taste which was found among the pure Arabs has become spoiled, and God knows better.

CHAPTER 60

The Beneficial Interests (Maṣāliḥ) *which Require the Determination of the Obligatory Religious Acts, the Pillars, the Proper Behaviors, and Their Like*

You should know that it is necessary in order to regulate the religious community that two limits should be set for every act of obedience: a higher and a lower. The higher is what leads to the intended result in the most complete sense, and the lower is what leads on the whole to the intended result without omitting anything worthy of consideration. This (determination of limits) occurs because there is no way that something may be demanded from them while its components, form, and the amount of it required are not explained to them, for this would contradict the substance of the law. There is no way of requiring from the masses the performance of (all of) the proper behaviors and the perfected decrees, because this would be imposing something impossible for those who are occupied with work or who are in straitened circumstances. Rather the regime of the community should be built on moderation and not on extremes. There is no way to neglect the higher and to suffice with the lower, for the former was the belief of the early Muslims and the portion of the sincere persons, so that neglecting something like this is not courteous.

In this case there is no way to avoid clarifying the lower and recording that people be required to do it, while recommending whatever exceeds it as being non-compulsory. Whatever is recorded as being required of them is divided into a specified amount of worship, such as the five daily prayers, and the fast during Ramaḍān, and into the elements of these without which they won't be considered fulfilled, such as pronouncing the *Takbīr*[1] and reciting the opening chapter of the Qur'ān during the prayers. Such obligatory elements are called "pillars," and matters extrinsic to the acts themselves, without which they won't count, are called "conditions" (*shurūṭ*), such as performing the ablution before the prayer.

---

[1] Saying "God is great."

Know that a thing may be made a pillar due to a cause which makes it resemble the natural way (*al-madhhab al-ṭabīʿī*) or it may be made a pillar due to an incidental cause.

In the first case the act of worship would not be correct and would not achieve its benefit without it; such as bowing and prostrating in the prayers and abstinence from food, drink, and sexual intercourse during the fast. Or it may make precise a concealed ambiguity as is necessary for the act of worship such as pronouncing the *Takbīr*," for it renders precise the intention and readies one for prayer; and like reciting the opening chapter of the Qur'ān, for it renders the petitionary prayer correct, and like the *Salām*[2] for it renders precise the emergence from ritual prayers through a righteous act which is consistent with dignity and respect.

In the second case something may be compulsory due to some other cause, and it is made a pillar of the prayer because it perfects it and fulfills its intended purpose, and establishes the best time for it. An example is the recitation of a chapter from the Qur'ān according to the opinion of those who make it a pillar, for the Qur'ān is one of the emblems of God that must be held in reverence, and it should not be treated carelessly. The best way to establish the timings was commanding them in the case of the most important acts of worship (i.e. the prayer), the most frequent, and those most stressed as being religious duties.

(Another reason for a thing being made a pillar due to an incidental cause) is that distinguishing between two similar things, or between what precedes something and the thing itself, may depend on a certain action so that this is made a pillar and becomes commanded; such as standing up between bowing and prostrating during the prayer, which differentiates between bending over which precedes the prostration, and the bowing that is an independent act of reverence. Similar to this are the offer, the acceptance, having witnesses, the presence of the guardian, and the consent of the woman in marriage; for the distinction between fornication and marriage is only achieved through these. It is also possible that certain pillars are derived based on the two aspects combined.

It is necessary to draw an analogy from what we said about the pillar to the case of the condition (*sharṭ*). Sometimes a thing is

---

[2] Turning the head to the right and left and reciting, "May the peace and the mercy of God be upon you," at the end of the prayer.

compulsory due to a reason, so that it is made a condition of respect for the emblems of the religion in order to be a sign for its elevation. This condition will not be fulfilled unless this act of worship is perfected by including it, such as facing the Qibla.[3] Since the Ka'ba is one of the religious emblems of God it must be respected, and it is among the greatest tokens of reverence that it should be faced in the best of people's states.[4] The facing of a particular direction in this case is one of the religious rituals of God. One is that the person praying with humility and submission is reminded the attitude of servants who are standing before their masters; and therefore facing the Qibla was made a condition of the prayer.

Sometimes the thing does not result in any benefit without a certain formality, so this is made a condition in order to make this substantial, such as declaring one's intention, for actions only take effect due to their being the embodiments of psychological attitudes. Prayer is the embodiment of submission, and there is no submission without the declaration of intention. Facing the Qibla would be such an formality according to another interpretation. Since the attention of the heart is a hidden thing, in its place was set up facing the Ka'ba which is among the emblems of God. Other such (formalities) are performing the ablution, covering the *'aura*,[5] and avoiding whatever is dirty. Since reverence is a hidden thing, the formalities which a person takes upon himself before kings and their like were established and considered to indicate reverence, and this later became internalized by people, and the Arabs and the non-Arabs agreed on its status.

When acts of worship of God are specified as being obligatory it is necessary to examine certain principles.

1) Only what is easy should be imposed, and this is the Prophet's saying, may the peace and blessings of God be upon him and his family, "Were it not that I would have been hard on my community, I would have commanded them to clean their teeth before every prayer with a *siwāk*."[6] The interpretation of this is what was

---

[3] In order to perform the ritual prayer a condition is turning to face the Ka'ba in Mecca.

[4] Of time, place, purity, and respect.

[5] Those parts of the body which must be covered to maintain proper modesty.

[6] *Siwāk* is a wooden brush used to clean the teeth. *Mishkāt*, p. 79. Bukhārī Juma' 8, Tamanna 9, Ṣaum 27, Muslim, Abū Dāwūd, Tirmidhī, Nasā'ī, Ibn Mājah, Muwaṭṭa', Ibn Ḥanbal.

given in another version, "Were it not that I would be hard on my community I would have imposed the duty of cleaning their teeth with a *siwāk* at every prayer on them, as I had made ablution a duty for them."[7]

2) When the community has come to believe about some degree of practice that abandoning it or neglecting would be a shortcoming with regard to God, and they have become satisfied about it, either due to its being transmitted by the prophets and agreed on by the pious ancestors, or due to something like that—then it is wise to decree that degree for them which they deem necessary, such as the forbidding of camel's meat and milk to the Children of Israel;[8] (in line with) this is the Prophet's saying, may the peace and blessings of God be upon him, about getting up to pray the (*tahajjud*) prayer during the nights of Ramaḍān, "I feared that it would become prescribed for you."[9]

3) Nothing should be recorded as having been imposed as a religious duty unless it is clear and well-defined, and not obscure for them, thus modesty and the rest of the virtues were not made pillars of Islam, although they are part of Islam.

Next, the lowest limit (in fulfilling) the acts of worship may vary according to variations in conditions of comfort and hardship, so that standing up is made a pillar of the prayer in the case of the one who is able to stand, while sitting replaces it in the case of another person.

As for the highest limit, it may be increased in quantity and mode. In the case of quantity, this would consist of treating the supererogatory practices as if they were obligatory practices, such as performing the sunna prayers and the supererogatory prayers,[10] performing the prayers in the middle of the night (*tahajjud*), fasting three days out of every month, giving extra charity, and so on. As for mode, it consists of formalities, litanies, and avoiding what is not suitable for a devout attitude. These were commanded in worship in order to perfect it and lead to the intended result in the most complete way; such as rubbing the creases of the skin which is

---

[7] This version in Ibn Ḥanbal I:214.

[8] In a hadith interpreting Qur'ān 6:146, the forbidding of camels' meat is mentioned. Bukhārī tafsīr. Khan trans. VI:123.

[9] *Mishkāt*, p. 270. Transmitted by Bukhārī Tahajjud 4. Khan trans. II:128.

[10] *Rawāṭib* prayers are supererogatory actions such as prayers or fasting done at certain times.

commanded in the ablution to perfect the state of cleanliness, and beginning with the right side[11] which is commanded so that the soul will be made conscious of the sublime status of worship, and it will have this in mind whenever it prepares itself to carry out important works.

You should know that if a man wants to acquire some virtue, and that his soul should be affected by it, and that he will thoroughly comprehend it; the stratagem for this is that he should hold himself to actions and attitudes appropriate for that virtue, even if in small matters which most people do not consider. For example the one who is practicing courage does not shrink from plunging into the mire, walking under the blazing sunlight, traveling in the dark night, and so on. Likewise the one who is practicing humility before God preserves the most polite manners in every situation, so that he doesn't sit to defecate except modestly with bowed head, and whenever he mentions God he holds his hands and feet together, and so on. The one practicing justice gives every thing its due and proceeds with the right hand in eating and in performing pure actions, and with the left when eliminating impurities. This is the secret of what the Prophet, may the peace and blessings of God be upon him, said in the story of the tooth cleaning, "Respect seniority!"[12] and his saying in the story of Ḥuwayṣah and Muḥayṣah, "Respect seniority!"[13] and this is a basic principle of the sub-divisions of proper manners.

You should know that what my Lord, Blessed and Exalted, made me understand concerning the inner meaning of the Prophet's saying, may the peace and blessings of God be upon him, "The devil eats with his left hand"[14] and similar statements, in attributing certain

---

[11] In the ablution and all other righteous actions based on the Sunna of the Prophet.

[12] A hadith transmitted from Abī 'Umar, may God be pleased with them (him and his son), the Prophet said, "I saw in a dream that I was cleaning my teeth with a toothpick and two men came to me, one older than the other, and I extended it to the younger of the two and it was said to me, "respect the older," so then I presented it to the older of them." *Mishkāt*, p. 80. Bukhārī Wuḍū' 78, Khan trans. I:154, Dārimī.

[13] Ḥuwaiyiṣah and Muḥaiyiṣa were the two youngest sons of Mas'ūd, and when 'Abd Allāh ibn Sahl was killed in Khaibar and his murderer was unknown, 'Abd al-Raḥmān, the brother of the slain and Mas'ūd's two sons went the Prophet and 'Abd al-Raḥmān began to speak, although he was the youngest, and the Prophet said to him, "respect the eldest." Bukhārī Khan trans. VIII:105. Adab 89, Muslim, Tirmidhī, Nasā'ī.

[14] Muslim Ashraba 105-107. Libās 70, 71, Abū Dāwūd, Tirmidhī, Dārimī. Muwaṭṭa', Ibn Ḥanbal.

acts to the devils—was that God, may He be Exalted, enables the devils to materialize in the dreams of people and in their sight while awake, with forms given by their temperaments or through states they experience at the time of the apparition. The people of sound intuitive perception have learned that it is those peoples' temperaments that cause them to adopt disgusting behaviors, things which make a person incline to frivolity and irritability, approaching ritually unclean things, hardheartedness at the mention of God and the corruption of every desired and commendable order. I mean by disgusting acts those which make people shrink from whoever performs them, so that their skin erupts in goose flesh, and their tongues burst forth with curses and imprecations. This is like a natural belief (*madhhab ṭabīʿī*) of human beings which the human specific form provides, and all segments of the nations agree on it, not due to their following the custom of one nation rather than another, or one religion rather than another religion. Examples of these disgusting actions are that a man grasps his private parts and leaps and dances, puts his finger between his buttocks, soils his beard with nasal mucus, or that he has a cut-off nose and ears and a blackened face, or that his clothes are upside down so that the top of his shirt is at the bottom, or that he rides an animal facing its tail, or that he has one shoe on and the other foot bare, and other detestable acts and attitudes of this sort which no one sees but that he curses, insults, and reviles, and I have on certain occasions seen devils doing some of these acts. By actions which incline to frivolity I mean a person's playing with his robe or pebbles, or moving his hands and feet in an abominable way.

In summary, God disclosed these acts to His Prophet, may the peace and blessings of God be upon him, and explained that they are caused by a devilish temperament, so that the devil does not appear to anyone, asleep or awake, unless he has adopted some of these. What pleases God on the part of the believer is that he stay as far away as possible from the devils and their attitudes. Therefore the Prophet, may the peace and blessings of God be upon him, explained these acts and attitudes, despised them, and commanded their avoidance. On this topic is the saying of the Prophet, "These gardens are inhabited by jinns, and devils,"[15] and his saying, "The

---

[15] By "gardens" is meant the common place for public relief of bodily needs. The Prophet recommended that people be modest in these places. *Mishkāt*, p. 76. Dārimī Ṭahāra 3, Ibn Mājah, Ibn Ḥanbal.

Devil makes sport with peoples' posteriors,"[16] and that he laughs when people yawn.[17] Then compare to this the preference for the attitudes of the angels, and this is the Prophet's saying, "Why don't you get in rows as the angels get in rows,"[18] and this is another basic principle of the divisions of proper behaviors.

You should know that among the reasons for something being made an obligation of Muslims which is sufficed if only some of them perform it,[19] is that if all the people began to do it at once it would ruin their livelihood and lead to the destruction of the supports of their civilization, and due to the fact that it is not possible to assign certain people to it and certain others to something else. Such as in the case of *Jihād*, for if all rallied to fight and abandoned agriculture, trade, and manufacturing, their livelihood would be destroyed. At the same time, it is not possible to assign certain people to *Jihād*, others to trade, others to agriculture, and others to judging and imparting knowledge, things come easily to some persons and not to others, and the one predisposed to do one of these things is not designated by names and categories so that the ruling could depend on these.

Among the reasons for something being made an obligation for Muslims which is fulfilled if some of them perform it (*farḍ kifāya*) is that the beneficial purpose intended by this is the existence of a certain order, while the corruption of the soul and the dominance of the animalistic does not result from a person's not carrying out this action. Examples of this are giving legal judgments, giving instruction in the religious sciences, and discharging the office of Caliph, for these things were legislated for the sake of the order, and are achieved by one person undertaking them. Other examples of this are visiting the sick or performing the funeral prayers, for the intended result is that the sick or the dead person should not be neglected, and this is attained as long as some people undertake these things, and God knows better.

---

[16] Therefore a person should be modest and conceal himself. *Mishkāt*, p. 75. Abū Dāwūd I:9 #35 Ṭahāra 19, Ibn Mājah, Dārimī, Ibn Ḥanbal.

[17] As cited in the hadith Bukhārī Adab 125, 127, Bad' al-khalq 11; Muslim, Tirmidhī. "*fa amā al-tathā'ub ḍaḥika minhu al-shaiṭān.*"

[18] To pray. Muslim Ṣalāt 119, Abū Dāwūd, Nasā'ī, Ibn Mājah, Ibn Ḥanbal.

[19] "*Farḍ kifāya*" is a religious obligation which only some capable persons in the Muslim community have in order to fulfill the obligation of the whole to be met.

CHAPTER 61

The Inner Meanings of the (Appointed) Times

The regulation of the religious community is not fulfilled except through fixing the times for its acts of worship. The basic principle in appointing times is intuitive recognition of the situation of those who must perform the worship and choosing what is not oppressive for them, while this suffices to achieve the intended result. In addition to this there are in it wise principles (*ḥikam*) and beneficial purposes of which those thoroughly versed in knowledge are aware, and these go back to three principles.

1) One of them is that although God, may He be Exalted, transcends temporality, still verses of the Qur'ān and hadiths have demonstrated that at certain times He draws nearer to His servants, that at certain times actions are presented before Him and that at certain times He decrees events, and other changing conditions; even though the true nature of these is only known to God, may He be Exalted.

The Prophet, may the peace and blessings of God be upon him, said, "Our Lord descends every night to the lowest heaven when the last third of the night remains,"[1] and he said, "Human actions are presented before God on Mondays and Thursdays,"[2] and he said about the middle night of Shaʿbān, "God will appear in it" and in another version, "He will descend on it to the lowest heaven,"[3] and the hadiths on this subject are many and well known.

In summary, among the axioms of the religion (*dīn*) are that there are times at which a spreading of spiritual energy takes place on the earth and a force from the World of Images flows through it. There is no better time for the acceptance of acts of worship and the answering of prayers than these times, for then, at the least effort, a great door is opened for the animalistic to obey the

---

[1] *Mishkāt*, p. 255. Bukhārī Mawāqīt 11, Adhān 104, Tahajjud 14, etc. Muslim, Tirmidhi, Nasā'ī, Dārimī, Muwaṭṭa', Ibn Ḥanbal.

[2] Tirmidhī II:124 #744 Ṣaum 44, Ibn Ḥanbal.

[3] *Mishkāt*, p. 271. Ibn Mājah Iqāma 199, Tirmidhī, Ibn Ḥanbal.

angelic. The Highest Council does not recognize the spread of this spiritual energy and the flowing of this force by means of calculating the celestial revolutions, but rather through mystical sensibility (*dhauq*) and direct intuitive perception that something is being imprinted on their hearts. Thus they know that There a divine decree is descending and that spiritual energy is spreading, and so on. This is what is expressed in the hadith "like an iron chain clattering upon smooth rocks."[4]

The prophets, upon whom be peace, have these types of knowledge imprinted on their hearts by the Highest Council due to which they also recognize the times through direct intuitive perception without calculating the celestial revolutions. Therefore they make efforts to appoint a symbol (*maẓinna*) for that hour, and then they order the people to uphold it.

Among these times are:

a) Those which follow the course of the years and this is His, may He be Exalted, saying, "We have sent it down on a blessed night; We are ever warning, therein every wise command is determined, as a command from Our presence; We are ever sending."[5] And on this night[6] the spirituality of the Qur'ān was fixed in the lowest heaven and it so happened that this night was in Ramaḍān.

b) One which revolves with the weekly cycle and this is the brief hour in which it is hoped that prayer will be answered and that acts of worship will be accepted. When people pass on to the life to come this will be the time when God theophanizes to them and draws near to them. The Prophet, may the peace and blessings of God be upon him, explained that its expected occurrence is Friday,[7] and gave as proof of this that great events occurred on this day such as the creation of Adam,[8] may peace be upon him, and that sometimes animals learn from the Lower Council knowledge of the greatness of this time and become startled and alarmed like someone who is frightened by a great noise,[9] and he said that he had witnessed this on Fridays.

---

[4] This hadith was cited previously in the chapter on the Highest Council. According to it, the sound of their wings is like an iron chain on rocks as they communicate God's decree. *Mishkāt*, p. 959. Bukhārī Tauḥīd 32, Tafsīr 15:1, 34:1, Tirmidhī, Ibn Mājah.

[5] Qur'ān 44:4–6.

[6] The Night of Power.

[7] For example Bukhārī Friday Prayer 35, Khan trans. II:28.

[8] As in Muslim Munāfiqīn 27.

[9] *Muwaṭṭa'* Juma 16, Nasā'ī.

c) Those which revolve with the revolution of the days, and the spiritual energy at these times is weaker than at the others. People gifted with mystical knowledge that they received about these from the Highest Council agree that they are four times: a little before sunrise, just after mid-day, after sunset, and from midnight to just before daybreak. At these times and just before and after them, spiritual energy spreads and blessing manifests. There is no religion on the earth which does not know that these times are the best for worship to be accepted. The Magians, however, distorted religion and began to worship the sun instead of God. Therefore the Prophet, may the peace and blessings of God be upon him, sealed the entry point of distortion and changed these times to others not far from them so that the basic object was not lost. He did not make it obligatory for them to pray at midnight because of the trouble-someness of this, and it has been confirmed that the Prophet, may the peace and blessings of God be upon him, said, "Indeed during the night there is an hour in which no Muslim could be consistent in asking God for good in worldly matters and in the next world but that God would bestow this on him,"[10] and that it occurs every night. The Prophet is reported to have said, "The best prayer is the midnight prayer, yet how few perform it."[11] He was asked which prayer was more heeded, and replied, "The one in the middle of the night."[12] He said concerning the noon-time, "This is the time when the gates of heaven open, and I like to have a good deed ascend during it for me."[13] He also said, "The angels of the day ascend to Him before the angels of the night, and the angels of the night ascend before the angels of the day."[14] God indicated these meanings in unambiguous verses of His book when he said, "So glory be to God both in your evening hour and in your morning hour. His is the praise in the heavens and the earth, and at the sun's decline and in your noontide hour."[15] Revealed statements about this subject are plentiful and well-known and I have observed this to be a great time.

---

[10] *Mishkāt*, p. 256. Muslim Musāfirīn 166, 167, Ibn Ḥanbal.

[11] *Kanz al-'Ummāl* III:301. Baihaqqī from Abū Dharr.

[12] *Mishkāt*, p. 257. Tirmidhī Īmān 8, Du'āt 78, 118, Nasā'ī, Ibn Mājah.

[13] *Kanz al'Ummāl* III:313 from Ibn 'Asākir.

[14] A similar hadith is Bukhārī Bad' al-Khalq 6, Nasā'ī, Tirmidhī, Ibn Ḥanbal II:396.

[15] Qur'ān 30:17,18.

2) The second principle is that the time of turning attention to God should be a time when a person is free from physical distractions such as excessive hunger, overeating, being overcome by sleep and fatigued, or when he needs to urinate or defecate. (He should also be free of) mental disturbances such as having the hearing filled with disquieting tumult and uproar, and having the sight with various forms and jumbled colors, and so on, with all other types of distractions. These times vary according to customs. Early morning and dusk, however, are times which seem to constitute a natural pattern (*madhhab ṭabīʿī*) for Arabs and non-Arabs, and people of the East and the West, and this should appropriately be taken as a rule among the universal laws to which any contrary thing should be considered an aberration. Man needs a polish which will eliminate rust after it has taken root in his lower soul, and this polishing occurs when he betakes himself to his bed and drifts toward sleep; for this reason the Prophet banned the holding of lengthy conversations and reciting poetry after dinner.

The regulation of the religious community is only effected through commanding watchfulness over the lower soul after every moment of time, so that waiting for the prayer, preparing for it before performing it, and the remaining of its mood and the vestiges of its light after it has been performed, are considered to be part of the prayer, thus the inclusion of most of the times will be effected even if it is not possible to include all of them. We have experienced that one who goes to sleep with the resolution of getting up during the night does not become submerged in animalistic slumber, and that the one who divides his mind between worldly occupations and observing the time of prayer or private recitation (wird) so as not to miss this, does not become absolutely devoted to the animalistic. This is the secret of the Prophet's saying, "The one who gets up during the night"[16] and God's, may He be Exalted, saying, "Men whom neither merchandise nor selling tempt from the remembrance of Allah."[17]

---

[16] . . . and says "There is not God but God, no partner to Him, His is the Kingdom and the praise, and He is able to do anything," and "Glory to God, praise be to God, there is no God but God, God is great, there is no power and no strength except through God" then who says, "O my God, forgive me," if he asks God for something, it will be answered for him, and if he performs the ablution and prays, his prayer will be answered. *Mishkāt*, p. 252. transmitted by Bukhārī Tahajjud 20, Khan trans. II:140, Tirmidhī, Ibn Mājah, Dārimī, Ibn Ḥanbal.

[17] Qur'ān 24:37.

It is proper that between every two times (of prayer) the interval should be calculated based on a quarter of the daytime, and that this should contain three hours, and this is the first limit usually applied for the quantity used by the Arabs and non-Arabs in dividing up the night and the day; concerning it is the report, "The first one to divide the day and the night into hours was Noah, may peace be upon him, and his sons inherited this."[18]

3) The third principle is that the time for performing worship should be the time which is a reminder of a blessing from God. For example God gave Moses victory over Pharaoh on the day of 'Āshūra', so the Prophet fasted on it and ordered fasting, and like the month of Ramaḍān during which the Qur'ān was revealed, and this was the beginning of the emergence of the Islamic religion, or as a reminder of the prophets of God worshipping their Lord and His accepting that from them, such as the Day of Sacrifice which recalls the story of the sacrifice of Ishmael, may peace be upon him, and his being ransomed "with a great sacrifice."[19] The performance of worship at a particular time may also be an allusion to certain of the rituals and emblems of religion, such as holding the *'Īd* prayer on the day of breaking the fast of Ramaḍān and giving alms on this day are in honor of Ramaḍān and an expression of gratitude for God's grace in enabling one to succeed in keeping the fast, or like the Day of Sacrifice on which there is an imitation[20] of (those on) the pilgrimage and a display of the blessings of God intended for them. These times of worship may also follow the practice of those early righteous ones whose goodness is well attested among the nations, since they worshipped God at this time. Examples are the five times of the daily prayers, according to the saying of Gabriel, "This is your time and the time observed by the prophets before you,"[21] and like the month of Ramaḍān according to one version of the interpretation of His, may He be Exalted, saying, "Fasting is commanded for you, even as it was commanded for those before you,"[22] and like fasting on the day of 'Āshūra' in relationship to us. It seems that the third principle is taken into account most of the time, while the two first principles are the most essential, and God knows better.

---

[18] Hadith not found in the standard collections.
[19] Qur'ān 37:107.
[20] In performing the sacrifice wherever they may be.
[21] *Mishkāt*, p. 119. Tirmidhī Mawāqīt 1.
[22] Qur'ān 2:183.

CHAPTER 62

The Inner Meanings of the Numbers and the Fixed Quantities

You should know that the divine law did not specify a certain number or a certain quantity rather than another, unless due to wise reasons and beneficial purposes, even though total reliance was on intuition based on recognizing of the situation of those on whom these who must perform the worship what would be suitable for them given their regime. These wise reasons and beneficial purposes are based on certain principles.

1) The first is that the odd number (*witr*) is a blessed number which should not be exceeded as long as it is sufficient, and this is the Prophet's saying, "God is single and loves the single, so pray the odd numbered (witr)[1] prayer, O people of the Qur'ān."[2] The secret of this is that the beginning of any multiplicity is unity, and the multiplicity closest to unity is what is odd, since every stage of the number has within it a false unity through which it achieves that stage. For example ten is a combination of units which are considered a unity, not as five and five, and so on following this analogy. This (false) unity is a model of the real unity at these stages, and it inherits its position. The odd number contains this (false) unity and another unity like it, and this is unity in the sense of not being divisible into two equal whole numbers; thus it is closer to unity than the even number. The nearness of anything existing to its origin goes back to its nearness to God, because He is the Origin of all origins, and the most perfect in unity is what is based on the quality of God.

Therefore you should know that oddness (*witr*) is at several levels. There is oddness resembling evenness and inclining to it, such as the numbers nine and five, for after you subtract one they divide into two pairs, and nine, even if it is not divisible into two equal

---

[1] *Mishkāt*, p. 264. The *witr* prayer is one of three or another odd number of *rak'ats* prayed after the *farḍ* and sunna of the *ishā'* (evening) prayer.

[2] Bukhārī Du'āt 69, Muslim, Abū Dāwūd, Tirmidhī, Nasā'ī, Ibn Mājah, Dārimī, Ibn Ḥanbal.

integers is still divisible into three equal integers. Likewise the even numbers also can be at levels of evenness resembling oddness, for example twelve—for it is three fours, and like six which is three two's. The leader of the odd numbers and the one furthest from resembling the even is the "one," and its regent among them and its caliph and heir are "three" and "seven," and (numbers) other than these are part of the people and the nation of the one. Therefore the Prophet, may the peace and blessings of God be upon him, chose the one, the three, and the seven for many of the quantities. Whenever wisdom decreed that he command more than this he chose a number obtained by raising one of them, such as raising the one to ten, one hundred, and one thousand, and also to eleven; and such as raising the three to thirty, and thirty-three, and three hundred; and the seven to seventy, and seven hundred. What is obtained by raising is something like the original number itself, and therefore the Prophet, may the peace and blessings of God be upon him, made a practice of (saying) a hundred words after every prayer, then he divided them into three times thirty-three, and he made one left over so that the sum of all of this (100) would be odd, going back to the leader or its deputy.[3] Likewise for every category of substance and accident there is a leader and a deputy; for example the point is a leader and the circle and the sphere are its deputies, and the forms nearest to it.

My father, may God sanctify his secret, told me that he saw a great vision in which were represented "life," "knowledge," "will," and the rest of the Divine Attributes, or he (may have) said "the Living," "the Knower," "the Willer," and the rest of the Divine Names—I don't know which—in the form of illuminated circles. Then he pointed out to me that the representation of the simple (not compound) thing in the form of geometric figures is only by means of those figures closest to the point; on the plane this would be the circle, and in three dimensions this would be the sphere, and thus ended his speech.

You should know that the practice of God proceeds in such a way that the descent of the unity to multiplicity only occurs through similitudinary (*mithālī*) connections, and according to these

---

[3] Usually the litany, "Glory be to God" (*Subḥān Allah*) 33 times, "Praise be to God" (*Al-ḥamdu lillāh*) 33 times, and "God is the Greatest" (*Allāhu Akbar*) 34 times, adding up to 100, which is based on the unity.

connections events are given forms, and these very connections are what the interpreters of the primordial language (the Qur'ān) took into account as far as was possible.

2) The second principle concerns the disclosure of the secret of what was previously explained concerning incitement to good and deterrence from evil and the like, with respect to the numbers.

You should know that sometimes qualities of piety and sin were shown to the Prophet, may the peace and blessings of God be upon him, and the superior aspects of the former and the defects of the latter were disclosed to him, thus he reported what God had taught him, and mentioned some number whose state he knew at that time, but his goal was not restriction (to that number). For example, he said, may the peace and blessings of God be upon him, "The acts of my community were presented to me, the good and the bad, and I found among their good acts, the dangerous object removed from the road, and I found among their bad actions spit which was in the mosque and had not been buried."[4] He also said, "The rewards of my community were presented before me, even the impurities which a man removed from the mosque; and the sins of my community were presented to me, and I saw no greater sin than a chapter of the Qur'ān which a man memorized and then forgot."[5]

According to this principle it is necessary to interpret his saying "Three (types of people) who will have two reward . . .,"[6] and his saying, "Three to whom God, may He be Exalted, will not speak . . .,"[7] and his saying, "Forty virtues the highest of which is giving someone a female goat,[8] no servant (of God) acts with one of these virtues hoping for its reward, or believing in its promised reward, but that due to this God makes him enter the garden."

[4] *Mishkāt*, p. 144. Muslim Masājid 57.

[5] *Mishkāt*, p. 145. Abū Dāwūd Ṣalāt 16, Tirmidhī.

[6] "A man from the People of the Book who believed in his prophet and then believed in Muḥammad, and the slave when he executes the rights of God and the rights of his master, and a man who has a female slave and teaches her goodness, and educates her in the best way, then he frees her and marries her, and therefore he will have two rewards." *Mishkāt*, p. 7. Transmitted in Bukhārī 'Ilm 31, Muslim, Tirmidhī, Nasā'ī, Ibn Ḥanbal. Khan trans. I:78.

[7] "And they will not become pure: the aged adulterer, a king who lies, and an administrator (*'āmil*) who is haughty." *Ṣaḥīḥ Muslim*, I:61. Bukhārī Aḥkām 48, Tauḥīd 24, Muslim Abū Dāwūd, Nasā'ī.

[8] That is, he gives someone a female goat for a certain period to make use of its milk and hair, and then it is returned. Bukhārī Hibba 35, Abū Dāwūd, Ibn Ḥanbal.

Sometimes the merits of an action or the portions of a thing were disclosed to him in a general way, so that he exercised his independent reasoning in order to find a way of making it precise, and he assigned a number to narrow it down that occurred many times or had great significance, and so on; then he made this known. According to this must be interpreted his saying, may the peace and blessings of God be upon him, "The congregational prayer is better than the individual prayer by twenty-seven degrees,"[9] for this number (27) is three times three times three, and he saw that the benefits of the congregational prayer proceed from three categories. The first is what derives from the benefit to one's soul in refining it, manifesting the angelic, and conquering the animalistic; (secondly) is what accrues to the people from spreading the rightly-guided practice among them, their competing in carrying it out and becoming refined by it, and their unanimity about it; (thirdly) is what goes back to the religion of Muṣṭafā, may the peace and blessings of God be upon him, in its continuing to flourish, and remain fresh, unadulterated by corruption and neglect. In the first are three (benefits): drawing nearer to God and the Highest Council, good being recorded for them, and forgiveness of their sins. In the second there are also three (benefits); the putting in order of their district and their city, the descent of blessings upon them in this world, and their interceding for each other in the next world. In the third also are three benefits: putting into effect the consensus of the Highest Council, grasping on to the lifeline extended by God, and the reflection of the lights of some of them back and forth on each other. In each one of these nine are three (benefits); God's being pleased with them, the prayers of the angels for them, and the withdrawing of the devils from them. In another version of the hadith it says twenty-five (times better than the individual prayer)[10] and this is because the benefits of the congregational prayer are five times five; the rectification of their souls, the sociable coming together of their congregation, the support of their religion, the rejoicing of the angels, and the driving away of the devils from them. Then in each one (of these five) are five others: God's being pleased with them, the descent of blessings in this world on them, good deeds being recorded for them, the forgiveness of their sins,

[10] For example in Bukhārī Adhān 30, 31, Ṣalāt 87. Khan trans. I:277, 351. Nasā'ī, Tirmidhī, Ibn Ḥanbal.

and the intercession of the Prophet and the angels for them. The reason for the difference in the versions of this is due to the difference in the ways of reckoning, and God knows better.

Sometimes a number is given to show the greatness and magnitude of a thing, therefore this number should be interpreted as a symbol, like what may be said about the love for so-and-so in my heart being like a mountain, or that the measure of a certain person reaches the clouds in the sky. According to this one must interpret the Prophet's saying, "The grave (of the believer) will be extended by seventy cubits,"[11] together with his saying, "As far as (the range of someone's) eyesight,"[12] or his saying, "My cistern[13] is (the size of) what is between the Kaʿba and Jerusalem,"[14] together with his saying, "My cistern is longer than the distance between Ailat[15] and Aden."[16] In examples like this he would sometimes mention one quantity and at another time some other quantity. There is no contradiction in this with respect to what relates to the intended meaning.

3) The third principle is that a thing must only be reckoned by a known, apparent, quantity used by those addressed in administering the ruling, and this quantity must be appropriate for the scope of the ruling and its underlying rationale. Thus dirhams must only be calculated by ounces, and dates weighed only by *ausāq*.[17] It is necessary that a fraction not be made something which could only be computed by experts in mathematics, such as fractions of one-seventeenth or one twenty-ninth. Therefore God, may He be Exalted, mentioned in the context of shares of inheritance only fractions which could easily be halved and doubled and whose calculation could easily be performed. These are of two divisions: one of them one-sixth, one-third, and two-thirds; the second of them one-eighth, one-fourth, and one-half. The secret of this is that an amount in excess or deficient becomes apparent at first glance, and calculating the cases becomes easy in the case of both

---

[11] When the believer answers the questioning angels Munkar and Nakīr with a steady reply, they say to him, "We knew that you would say that," then his grave is extended seventy cubits by seventy. Bukhārī Janā'iz 87, Muslim, Tirmidhī, Ibn Ḥanbal.

[12] Tirmidhī Qiyāma 26.

[13] The cistern of Kauthar outside the gates of Paradise.

[14] *Kanz al-ʿUmmāl*. VI:91. Ibn Mājah.

[15] A town near where ʿAqaba (Jordan) now stands.

[16] *Mishkāt*, p. 1179. Muslim trans. Siddiqi I:157, Ṭahāra 36, 38

[17] One "*wasq*" equals sixty *ṣāʿ*. A varying cubic measure.

the nearest and the furthest relative. Whenever circumstances require a quantity other than the amount considered previously, and the relationship between them is not an exact multiple, then it is necessary not to exceed two-thirds between one-half and one, and not to exceed one-third between the quarter and the half, because the rest of the fractions are more recondite than these. If what is desired is a calculation which gives a larger sum, it is suitable to multiply by three, and if what is desired is more than that, then the suitable thing is to multiply it by ten. Then if the thing may be little or much, the proper course is to take the lowest limit and the upper limit and then select the median. What is taken into account for the alms tax are (fractions) of one-fifth, one-tenth, half of one-tenth, and a quarter of one-tenth, because an increase in alms given depends on the abundance of surplus and the smallness of the requisite for subsistence. Since the net incomes of the multitude of people of the regions were only classified according to four levels, it is proper that the distinction between each two of these levels be apparent—as clear as it can be—and this is that one of them should be double that of the other, and a detailed explanation of this will be given to you. Then when circumstances require that, for example, what constitutes affluence be assessed, it is necessary to take into account what is considered affluence according to common understanding, and to demonstrate whatever in this are the determining factors of affluence.

All of this is in accordance with the custom of the multitude of those on whom religious duties are imposed, from the East to the West, the Arabs and the non-Arabs, and in accordance with what is like a natural belief among them if no factor precludes it. If the matter was not based on the custom of the masses their situation would become fragmented, so what is taken into consideration is the situation of the early Arabs in whose language the Qur'ān was revealed, and according to whose customs the divine law was determined. Therefore the law defined the amount of the "*kanz*"[18] as five awāq because this amount would suffice the people of a small household for a whole year in most of the inhabited regions—except in a drought year, or in a very large city or in its hinterland. A small flock of sheep was set at forty, and a large one at

[18] *Kanz* is that amount of a person's property on which *zakāt* is not required to be paid.

one hundred and twenty.[19] An abundant crop was set at five *ausāq*;[20] because the smallest household consists of a man, wife, and a third, either a servant or a son, with them, and the most a person consumes in a day is a *madd* or a *raṭal* and in addition to this some additional condiments or food will be required, and this amount will suffice them for a whole year. A great amount of water is set as being two large jars full[21] because this is the minimum amount of water in ordinary springs, and the maximum amount that water containers customarily are able to hold among the Arabs, and the cases of the rest of the measured quantities are analogous to these, and God knows better.

[19] One sheep is given for a flock of forty to one hundred and twenty, and two for a flock between one-hundred and twenty to two hundred.

[20] Each *wasq* is sixty *ṣā'* or five camel loads. *Zakāt* is due on anything above this.

[21] A "*qulla*" is a large jar holding 950 Baghdadī *raṭals*, and this (2 *qullas*) is the minimal amount of standing water which is used by animals in the wilderness, out of which water may be taken (as pure) for ablution. That is, this quantity remains pure even if some polluting substance (*najasa*) enters it unless one or more of its qualities change noticeably. Abū Dāwūd I:17 #65 Ṭahāra 33, Tirmidhī, Nasā'ī, Ibn Mājah, Dārimī, Ibn Ḥanbal.

CHAPTER 63

The Inner Meanings of Making Up (Qaḍā') *(for Missed Religious Obligations) or Receiving Dispensations* (Rukhṣa) *(to Diminish Them)*

You should know that part of the regulation is that if something is commanded or forbidden, and the ones addressed do not know the goal of this with sure knowledge, then it is necessary that it should be made for them something which is effective through its own property, and something whose effectiveness is believed, although the cause behind the effect is not perceived. An example is curative litanies,[1] the cause of whose efficacy cannot be discerned. Therefore the Prophet, may the peace and blessings of God be upon him, refrained from explicitly explaining the inner dimensions of the commands and the prohibitions in most cases. He rather intimated something about them only to those in his community who were thoroughly versed in knowledge. Therefore the concern of the bearers of the religion such as the rightly-guided caliphs and the leaders of the faith with establishing the forms of the religion was greater than their concern for establishing the inner dimensions (of these forms). Thus it was reported of 'Umar, may God be pleased with him, that he said "I calculate the taxes of Bahrain while I am praying, and I think of equipping an army while I am praying." Therefore the practice of the Muftis[2] both in the past and currently has not been to adduce this as an argument (*ḍalīl*) about a case when giving a legal opinion (*fatwā*).

It was also necessary to definitively record adopting what was commanded, to lay the strongest blame for disregarding it, and to make them desire it, and be fond of it with a true desire and fondness, so that the motivation to do right would envelop them externally and internally.

---

[1] "*Ruqā.*" That is, those litanies recited for curative purposes. Ibn Mājah Ṭibb 1, Tirmidhī, Ibn Ḥanbal.

[2] Those who provide legal opinions on questions of proper religious conduct.

Since this is the case, it became necessary that in the case where someone was prevented from doing what was commanded by a compelling obstacle, that a substitute should be legislated in its place. This is because in this case the one on whom this was imposed would be in either one of two situations.

a) He would have to be required to carry out the action despite the hardship and trouble involved, and this opposes the object of the divine law. God, may He be Exalted, said, "God desires for you ease, he desires not hardship for you."[3]

b) Secondly, the matter might be totally thrown aside and thus the soul would become accustomed to leaving it aside, and would give itself over to neglecting it. Rather than this the soul should be given the training of a wild riding animal from which tameness and motivation must be ensnared. The person who is occupied with disciplining himself or teaching children or who trains riding animals and so forth, knows through repeated experience how to obtain tameness, and due to this the work becomes easier. He knows how tameness is lost through abandonment and neglect, so that the soul finds the task oppressive and it weighs heavily on it, and thus if he wants the task repeated he will need to inculcate the tameness once again.

It is indispensable, in that case, that a way of making up (*qaḍā'*) be legislated in cases where the time of the action has passed. Excuses or special dispensations (*rukhaṣ*) for the act should also be legislated to make the action feasible for such a person and facilitate it for him. This is based on speculation that takes into account the situation of those on whom the obligation is imposed, the goal of the action, and those of its portions which are indispensable for obtaining this goal. In addition to this there are principles known to those well-versed in knowledge.

1) One of them is that the pillar (*rukn*) and the condition (*sharṭ*) have two aspects. One aspect is the essential one, which is intrinsic to the nature of the thing, or inseparable from it, without which it cannot be counted with respect to the basic element that is intended by it; such as the invocation of God, and the act of bowing indicating respect, and alerting to the two attributes of purity and humility. Among the characteristics of this division is that it cannot be abandoned whether inconvenient or pleasant, since nothing

[3] Qur'ān 2:185.

of the action comes into effect when it is abandoned.

The second aspect is the supplementary (*takmīlī*) which is only legislated due to its being compulsory in another sense, which requires setting a certain time which is the best for this act of obedience, or because it is a proper means for carrying out the basis of the intended goal completely and abundantly. The character of this division is that special dispensations may be given in times of adversity and discomfort. According to this principle it is necessary to give a special dispensation in facing the Qibla[4] to the one seeking in the darkness and so on, and a special permission to omit proper covering of the *'aura*[5] to the one who has no garment, to permit substituting the ablution with sand (*tayammum*)[6] for the usual ablution when a person can find no water, to permit replacing the *Fātiḥa* by some pious phrase (*dhikr*) for the one who is unable to say it; and to allow replacing standing (in the ritual prayer) by sitting or reclining for the one unable to stand, and to allow substituting bowing the head for one unable to prostrate himself and bow.

2) The second principle is that it is necessary to preserve in the substitute something which recalls the original, and which feels like its substitute and its replacement. The secret of this lies in effecting the goal that is intended by legislating dispensations. This occurs when the attachment to the first action remains, so that the soul should be like one waiting (to do this act). Therefore in wiping the leather socks (as part of the ablution), cleanliness (of the feet) at the time of putting on the socks was made a condition,[7] and a time was set at which it runs out, and investigating the direction of the Qibla was made a condition.[8]

3) The third principle is that not every oppressive situation leads to a special dispensation, for the aspects of hardship are many, and concessions in all of these would lead to neglect of worship and obedience. Going to the limit of this would eliminate hardship and the

---

[4] The direction of prayer.

[5] The parts of the body which must be covered to ensure modesty.

[6] *Tayammum* is an alternate form of ablution performed when water is not available. In lieu of water clean soil, stones, clay brick dust, etc., may be used.

[7] The "*masḥ al-khuffain*" or wiping the inner shoe or leather sock may be substituted for washing the feet up to the ankles during the ritual ablutions before the prayers. The feet must be washed, however, at least once every twenty-four hours.

[8] If someone doesn't know the direction of Mecca, he must make an honest effort to find out before praying in an unconfirmed direction.

bearing of suffering, which is the way to recognize obedience to the law and the straightening out of the lower soul. Thus (the divine) wisdom requires that the injunction only concern matters which occur often and cause great affliction, especially in the case of a group in whose language the Qur'ān was revealed and with reference to whose customs the divine law was specified. It is necessary not to overlook the fact that the act of worship has special properties, whenever this is possible—hence a shortening (of prayers) during journeys was legislated, but not for those in arduous professions, and not for farmers and laborers; while for the traveler in comfort the same dispensations were permitted that were permitted in the case of the uncomfortable one.

Making up (*qadā'*) includes making up by a rationally calculated substitution and making up by a substitution which is not rationally calculable. Since what is essential in the act of worship is the heart's following the order of God and the soul's taking up the glorification of God, anyone who acts with neither a purpose nor a firm decision, or who is of a type whose purposefulness has not yet matured or who is not capable of holding his soul to the necessary respect—has it within his right to be excused, and not to be constrained by all of the constraints. In reference to this one must interpret the Prophet's saying, "The pen was raised from three,"[9] and God knows better.

---

[9] The sleeper, the child, and the feeble-minded were not blamed for falling short in their actions. Bukhārī Ṭalāq 11, Ḥudūd 22, Abū Dāwūd, Tirmidhī, Nasā'ī, Ibn Mājah, Dārimī, Ibn Ḥanbal.

CHAPTER 64

The Establishment of the Supports of Civilization (Irtifāqāt) *and the Reformation of Conventions*

We have already mentioned in what preceded, explicitly or implicitly, that human beings are naturally predisposed to the Second and Third Irtifāq, and that through these they are distinguished from the other species of animals, and that it is impossible that people should abandon these two or neglect them. In much of this they need a wise man (*ḥakīm*) who is knowledgeable concerning the need and the way of properly achieving it, who is led to the universal beneficial purpose, either as derived from thought and vision, or from a soul constituted with the angelic force inherent in it, so that he is prepared for the revelation of knowledge from the Highest Council. This latter way is the more perfect of the two types and the more reliable of the two ways. Conventions among the supports of civilization have a role analogous to that of the heart in the body. There may become interpolated into the conventions harmful elements due to the leadership of a group who lack the control of the comprehensive intellect (*al-ʿaql al-kullī*), so that they perform savage, lustful, or devilish acts and they promote these so that most people imitate them. From another similar aspect, there is a great need for a powerful man, aided by the Unseen, who is inclined to the universal beneficial purpose, in order to change their conventions to what is correct through a regulation to which in most instances only those supported by the Holy Spirit are guided.

Then, if you have comprehended some of this knowledge, you should know that the basic element in the mission of the prophets, although it was essentially and primarily to teach aspects of worship, still included besides this the desire to extirpate corrupt conventions and to encourage people to aspects of the supports of civilization. This is the Prophet's saying, may the peace and blessings of God be upon him, "I was sent to eradicate musical instruments,"[1] and his saying, "I was sent to bring the noble virtues to fulfillment."[2]

---

[1] Ibn Ḥanbal V:257, 268. The hadith has a variant wording "*baʿathanī* . . . *wa amaranī bi-maḥq al-maʿāzif.*"

[2] *Muwaṭṭa'* Ḥusn al-akhlāq. Ghazzālī in *Iḥyā'* from Ibn Ḥanbal and al-Baihaqqī.

You should know that there is no pleasure of God, may He be Exalted, in the neglect of the second and third stages of the supports of civilization, and not one of the Prophets, may peace be upon them, commanded this. The matter is not as a group of people thought who fled to the mountains and totally abandoned mixing with people whether for virtuous or sinful ends, and they became like the wild beasts. Therefore the Prophet, may the peace and blessings of God be upon him, refuted the one who wanted to renounce marriage and said, "I was not sent to bring monasticism but rather I was sent with the tolerant, monotheistic, (Ḥanīfī) religion."[3] Rather the prophets, may peace be upon them, ordered moderation in the supports of civilization, and that they should not use them to achieve the state of those deeply absorbed in luxury such as the kings of the Persians, nor that they should be reduced to the state of the inhabitants of the mountain peaks who become like wild beasts.

From this one may draw two contradictory analogies. One of them is that a life of ease and comfort is good and that through it the temperament is made sound and the morals upright, and through this emerge the commendable qualities which distinguish a man from others of his genus; while stupidity, weakness, and so on, arise through poor management.

The second of them is that a life of ease and comfort is repugnant due to its requiring conflict, competition, hard labor, drudgery, turning away from the Unseen World, and a neglect of planning for the next life. Therefore what is pleasing (to God) is the middle way and maintaining the supports of civilization while including with them the remembrances of God, the proper behaviors, and taking the opportunities for turning attention to Jabarūt. Without exception, what the prophets brought from God, may He be exalted, on this subject, took into account the customs of eating and drinking, dress, architecture, and methods of beautification, the marriage practices, the behavior of the husband and wife, the ways of selling and buying, the manner of deterring from disobedience, and the way of giving legal judgments, and so on. If whatever was necessary with respect to the comprehensive outlook (*al-ra'y al-kullī*) was compatible with this, then there was no sense in altering

---

[3] Variant (with no mention of monasticism) in Ibn Ḥanbal V:266, VI:166, 233. The inveighing against monasticism is mentioned in separate hadiths in Ibn Ḥanbal and Abū Dāwūd.

any part of it, nor of renouncing it in favor of something else. Rather it was necessary to spur the people on in what they were already practicing, to hold true to their opinion about this, and to guide them to the beneficial purposes of this. If this practice was not compatible with the universal outlook, and circumstances required that something be altered or eliminated, since it caused some of them to harm others or it led to their being absorbed in the pleasures of worldly life and their being distracted from righteousness, or it was a diversion which would lead to the neglect of the beneficial purposes of this world and the next, and so on—one must not proceed to what is totally contrary to their habit. Rather these must be changed to something similar to what they are already doing, or to what resembles something well-known as a practice of the righteous ones who were acknowledged as good models for that group of people. In summary, it should be changed to something which their intellects would not reject if it were proposed to them, but rather they would become satisfied that it was right. It is for this reason that the divine laws of the prophets, may peace be upon them, differ from one another.

The one firmly-grounded in knowledge knows that the divine law did not decree in marriage, divorce, social transactions, decoration, clothing, judgment, punishments and the division of spoils, something which people had no knowledge of, or which they would hesitate to carry out if it were required of them. Indeed, it only undertook the straightening out of the crooked and the restoration of the ailing. Usury (*ribā*) was prevalent among them, so they were forbidden to practice this. They used to sell fruit before its quality was evident, and they used to argue and remonstrate about defects which had afflicted it (once it had ripened), and then they were forbidden this type of sale. The blood money in the time of ʿAbd Al-Muṭṭalib was ten camels, then when he saw that they were not deterred from killing he raised it to one hundred, and the Prophet, may the peace and blessings of God be upon him, kept it at that. The first determining of the shares of spoils that occurred was by the order of Abū Ṭālib and the leader of the group used to receive one-quarter from every raid, then the Prophet, may the peace and blessings of God be upon him, made the practice to be taking one-fifth from every booty. Qabādh and his son Anushirvān[4] used to

[4] The pre-Islamic kings of Persia.

impose the *kharāj* and the *ʿushr* on the people, so the divine law imposed something like that. The Children of Israel used to stone adulterers and cut off the hands of thieves and take a life for a life, and the Qur'ān was revealed containing this also. Examples like this are very numerous and are not concealed from someone who investigates. Rather if you are astute and comprehensive regarding all facets of the rulings, you will also come to know that the prophets did not introduce acts of worship that people were not already practicing, or at least something like them. They did, however, repudiate the distortions of the Age of Ignorance, and firmly established the times and pillars which had become confused and they publicized among the people whatever had become obscure.

You should know that the Persians and the Byzantines, since they had passed on hereditary rulership for many generations and had become engrossed in the pleasures of this world and forgotten the next world, and the devil had gained mastery over them—had become deeply involved in life's comforts, and they prided themselves on these. Scholars traveled to them from the far horizons to discover from them the fine arts of living and its comforts. Thus they continued to practice them, and some of them became more excessive in them than others and they showed off to one another in this regard, until it was said that they used to rebuke those of their leaders who wore a girdle or a crown whose value was less than a hundred thousand dirhems, or who did not have a lofty palace, a bath tub, a bathing pool, and gardens, who did not have swift riding animals and handsome slaves, who was not expansive in dispersing food, and who did not wear beautiful clothes. The citation of these would take a long time, and what you see with your own eyes of the conditions of kings of your countries makes recounting these unnecessary for you. All of this had penetrated the foundations of their way of life, and could not be removed from their hearts even if they were cut to pieces. Due to this an incurable disease was engendered in all of the parts of the city, and a great calamity. Not one remained of their markets or villages, nor their rich or their poor who had not been overwhelmed, dominated, and weakened by it, and in whom it had provoked sorrows and anxieties without limits. This is because these things are not obtained without significant expenditure of wealth, and such wealth is only acquired through multiplying the taxes on the peasants, merchants, and their like, and oppressing them, so that if they refuse, they fight them and torture them, and if they obey they became like donkeys and cattle which are used in irrigating, threshing and harvesting, and which are only procured in order to fulfill their masters' needs. Therefore these people were not given an hour free from hardship until they got to the point where they did not give any of their attention to the felicity of the world to come, and they had become incapable of this. Sometimes there would be a wide region in which not one person assigned importance to his religion, and there would only be found people who had acquired the propensities for these foods, clothes, buildings, and so on, and they abandoned the principles of the professions on which the order of the world is built. The masses who surrounded them began to affect the ways of the leaders in these things, or else they would not have a privileged position with them nor would they receive any attention from them, and most of the people became entirely dependent on the Caliph, begging from him. Sometimes they did this because they were among the fighters and the managers of the city. They used to take on the customs of the leaders, and the intention was not to perform what was needed but to subsist in the manner of their ancestors. Sometimes (they begged) on the basis of their being poets on whom kings customarily bestow gifts, and sometimes on the basis that they were ascetics and Sufis and that it would be shameful of the Caliph not to take note of their situation. Thus some of these groups came to oppress the others, and their means of acquiring livelihood became conditional on companionship with the kings, friendship with them, good conversation with them, flattering them, and this became the art in which their thoughts were absorbed and their time was wasted. As these occupations multiplied despicable attitudes took form in the souls of people and they turned from the righteous virtues.

If you want to know the truth of this illness, look at a people who do not have a Caliphate,[5] and who are not absorbed in the pleasures of food and clothing—and you will find every one of them in control of his own affairs, and he will not be weighted down with heavy taxes, and such people will have time free for religious and community concerns. Then imagine their condition if the Caliphate had existed among them with its accompanying notables, and that they had reduced the subjects to servitude, and had

[5] In the later sense of a hereditary monarchy or similar administrative structure.

power over them. When this disaster increased and the sickness worsened, God and the nearer angels became wrathful towards them. It was God's pleasure, may He be Exalted, that He should cure this illness through cutting off its material aspect. Thus He sent an unlettered Prophet, may the peace and blessings of God be upon him, who had not mixed with the Persians and the Byzantines and who did not practice their customs, and He made him a measure through which to recognize the righteous guidance that pleases God from that which incurs His wrath. He made him speak censuring the customs of the Persians and the repugnance of being drowned in worldly life and being content with it, and He inspired in his heart to forbid them the chief things to which the Persians were accustomed and in which they took pride, such as wearing silk, purple and red robes, the use of golden and silver vessels, uncut gold jewelry, cloaks worked with pictures, embellishing their houses and other things. God decreed the removal of their empire in favor of (the Prophet's) empire, and their leadership in favor of the Prophet's leadership, and that Khusrau would be destroyed and there should be no Khusrau after him, and that Caesar would be destroyed and no Caesar should follow him.[6]

You should know that among people in the time of the Jāhiliyya there were mutual wrongdoings which oppressed and were difficult for the people, and these were only put to an end by removing their chief concerns in this matter, such as the blood-feuding in which one man used to kill another so that the representative of the murdered man would kill a brother of the murderer or his son, then this would be repeated and one from the other side would be killed and the affair would keep going on in this way. Thus the Prophet said, "All blood-revenge is abolished under these my feet and the first blood-revenge which I abolish is the blood of Rabīʿa."[7] (Another dispute in the Jāhiliyya concerned) inheritances which the leaders of the people used to decide in various ways, and the people did not refrain from things like extortion and usury, so they

[6] Based on a hadith in *Mishkāt*, p. 1130. Bukhārī Īmān 31, Khums 8. Khan trans. IV:225, Manāqib 25; Muslim, Tirmidhī, Ibn Ḥanbal transmitted it. Khusrau is the personification of Sāsānid royalty and a cherished symbol of past glory and of fate that overtakes even mighty rulers.

[7] The son of his cousin Rabīʿa ibn al-Ḥārith had been killed. *Mishkāt*, p. 546. Muslim. Farewell Pilgrimage. Siddiqi trans. II:615. Tirmidhī Tafsīr 9:6. Ibn Mājah Manāsik 76. Dārimī Manāsik 34.

strayed into this. Then came another generation,[8] and thus they argued[9] advancing various proofs; so the Prophet cut off the disputes among them when he said that everything which occurred during the Islamic era would be according to the order of the Qur'ān, and everything which had been divided in the Jāhiliyya, or that a man had gotten control of in some way in the Jāhiliyya, would be left as it was and not declared void. In the case of usury, one of them would loan wealth and impose interest and then he would oppress the borrower and combine the original sum and the interest together as the principal, and impose an increment on it and so on until it had become a "tremendous sum".[10] Thus (the Prophet) abolished *ribā*, and decreed the (use of) interest-free capital. "Wrong not and you shall not be wronged,"[11] and so on in other cases which would not have been discontinued but for the Prophet, may the peace and blessings of God be upon him.

You should know that sometimes a custom may be legislated for the people in order to cut off their rancors, such as beginning from the right-hand side in serving water, and so on. For there may be people who are quarrelsome with one another, and the superiority (of one of them) may not be agreed upon in order to let him commence. Arguments between them will not be suspended unless you have a regulation like this. Thus conventions like the leadership of prayer (at funerals) by the head of the household, and like the owner of a mount sitting in front of his companion while riding it, and so on, were legislated, and God knows better.

---

[8] After the coming of Islam.
[9] About inheritance, interest, etc.
[10] Qur'ān 3:14.
[11] Qur'ān 2:279.

CHAPTER 65

Rules which Bring into Effect Other Rules

God the exalted said,

> And We did not send any before you (as our Messengers) any except men whom We had inspired, so ask the People of Remembrance if you do not know. With clear proofs and writings; and We have revealed to you the Remembrance that you might explain to humankind that which has been revealed for them, so that perhaps they might reflect.[1]

You should know that God, may He be exalted, sent his Prophet, may the peace and blessings of God be upon him, to explain to the people what He had revealed to him about the ways of worship, so that they would take them up; and the categories of sins, so that they would avoid them; and whichever among the supports of civilization would make Him pleased with them, so that they would follow them. Part of this explanation was that he (the Prophet) should teach them what inspiration required or indicated, and so on.

A great body of the hadiths of the Prophet, may the peace and blessings of God be upon him, should be explained according to these principles, and here we mention the most important ones.

1) Among them are that God, may He be exalted, put into effect His practice in such a way that He arranged the causes leading to their results so that the beneficial purpose intended by His extensive wisdom and His complete mercy would be put in order. Therefore this required that altering what God has created should be considered an evil, an effort to corrupt, and a cause for hatred showering down upon such a person from the Highest Council. Thus, since God created humans in such a way that they do not come into existence at most times and instances from the earth, as do worms,[2]

---

[1] Qur'ān 16:43–44.

[2] The belief at that time was that worms were spontaneously generated in the earth.

and since the wisdom of God decreed the remaining of the human species, and moreover the spreading and increase of human individuals in the world, He therefore placed in them the faculties of procreation and made them desire to have descendants. He made carnal desire for each other prevalent among them, in order that God would accomplish through this a matter which His excellent wisdom required. When God informed the Prophet, may the peace and blessings of God be upon him, of this secret, and disclosed to him the true state of the situation, this required that he forbid cutting off this means and neglecting these decreed faculties, or employing them in some way other than their proper one, and therefore the strongest prohibition was applied to castration and sodomy, while coitus interruptus was considered reprehensible.

You should know that when human individuals are of sound temperament and their constituents are structured according to the determinants of the species, they emerge in a known form having an upright stature, bare skin, and so on, and this is the decree of the species, its requirement and its effect on individuals. In (the world) of the highest good the survival of the various species and the appearance of their forms on the earth is sought and decreed. Therefore the Prophet, may the peace and blessings of God be upon him, used to order that dogs be killed, then he forbade that saying, "They are a community among the communities."[3] That is, the species has a decree with God, and banishing its bodily forms from the earth would not be pleasing (to God). This decree leads to the decree of the manifestation of the determinants of the species in its individuals, and therefore opposing that decree and trying to deny it is repugnant and incompatible with the universal beneficial purpose. This principle is contravened by doing things to the body which are not decreed by the determinants of the species such as castration, making a gap between the front teeth, plucking the facial hair, and so on. As for putting on kohl and arranging the hair, these aid the manifestation of the intended determinants (of the species) and are compatible with it. When God, may He be exalted, established a divine law for human beings through which to organize their union and improve their condition, and there was in Malakūt a motive to manifest it; the situation (of this divine law) was like

[3] Abū Dāwūd III:108 #2845 Kitāb al-Ṣaid, Tirmidhī, Nasā'ī, Ibn Mājah, Dārimī, Ibn Ḥanbal.

the condition of the species in seeking the manifestation of their embodiments[4] on the earth. Therefore trying to efface it is loathed by the Highest Council and inconsistent with their decree and the aspiration of their concentrated resolve, and likewise for the supports of civilization on which the sects of people agree, both the Arabs and the non-Arabs, the furthest of them and the nearest, for they (the supports) are like the natural state.

Then, since God, may He be exalted, legislated oaths and giving evidence to clarify the true state of the situation, this required that bearing false witness and lying under oath would be loathsome to God and His angels.

2) Among these principles are that when He inspired the Prophet, may the peace and blessings of God be upon him, with a certain ruling of the divine law and disclosed to him its wisdom and its cause, then it was up to him to take this beneficial purpose and appoint for it a reason for legislation (*ʿilla*) and to make that rule refer to it, and this is the drawing of analogies by the Prophet. However the drawing of analogies (*qiyās*) on the part of his community is their recognizing the reason for the legislation of the rule specified in the text, and their basing the ruling upon it wherever it comes into operation. An example is the litanies (*adhkār*) for which the Prophet, may the peace and blessings of God be upon him, set the times of morning, evening and just before sleeping; for since he was made aware of the wisdom behind the legislation of the prayers, he could exercise individual reasoning about these litanies.

3) Among them are that when the Prophet, may the peace and blessing of God be upon him, understood from a Qur'ānic verse the reason for the sequence of the wording, and that no one else besides him would be able to understand this meaning from it due to the subtlety of its approach or the possible competing interpretations of it, then it was up to him to make the ruling according to what he understood, as in the case of God's, may He be exalted, saying, "Lo! (the mountains) Al-Ṣafā and Al-Marwah are among the emblems of Allah."[5]

---

[4] "The prototypes (of the species) in their *mithālī* forms stand before their Lord and plead that He emanate their outer forms in the physical world and manifest their rules as completely and abundantly as possible." *Lamāḥāt*, p. 50, no. 43.

[5] Qur'ān 2:158. "It is therefore no sin for him who is on pilgrimage to the house (of God) or visits it to go around them (as the pagans do.)"

From this the Prophet, may the peace and blessings of God be upon him, understood the precedence of Al-Ṣafā over Al-Marwah,[6] due to a correspondence of the expression to what was legislated for them, as might occur in the correspondence (of the word order in an answer) to that of the question, and so on, so he said, "Begin with what Allah begins with."[7]

(Another example of the Prophet's interpretation) is that of God's, may He be exalted, saying, "Adore not the sun nor the moon; but adore Allah who created them,"[8] and His, may He be exalted, saying, "But when it set, he said, I love not things that set,"[9] from which the Prophet, may the peace and blessings of God be upon him, understood the recommendation that they should worship God, may He be exalted, at solar and lunar eclipses.[10]

From His saying, may He be exalted, "Unto God belong the East and the West,"[11] the Prophet, may the peace and blessings of God be upon him, understood that facing the Qibla was a duty which could be waived at a time of excuse, and from it he derived the rule about someone looking for the direction of the Qibla on a dark night, so that he may make a mistake, and pray facing a different one; and the rule about the rider on a mount who prays the supererogatory prayer outside of the city.[12]

4) Among these principles is that when God, may He be exalted, charges someone with something related to his social interaction with people, this requires that the people be ordered to accede to him in this. Thus when the judges were commanded to administer the legally prescribed punishments this required that the disobedient ones be commanded to obey them in these. Likewise when the official was ordered to collect the *zakāt* from the people they were ordered that the collector should not depart from them until he was satisfied, and when the women were commanded to cover themselves, the men were ordered to avert their glances from them.

---

[6] That is, that Al-Ṣafā should be started with when performing the ritual of *sa'y* during the pilgrimage.

[7] *Mishkāt*, p. 544. Muwaṭṭa' Ḥajj 126.

[8] Qur'ān 41:37.

[9] This refers to Abraham, who worshipped first a star, then the moon, and sun, before turning to Allāh. Qur'ān 6:77.

[10] Special prayers are performed at these times. See Bukhārī Khan trans. II:83-99.

[11] Qur'ān 2:115.

[12] He can pray in the direction the animal is going. Bukhārī al-Taqṣīr 7, Khan trans. II:112.

5) Among these principles is that if a thing is forbidden, this requires that its opposite be commanded as obligatory or recommended according to the requirements of the situation, and if He commands a thing, this requires that He forbid its opposite. Thus when He commanded the Friday congregational prayer and making efforts to perform it, this required that He forbid being occupied with selling and professional activities at that time.[13]

6) Among the principles are that if a thing is conclusively commanded this requires that its antecedents and motivating factors be made attractive, but if a thing is forbidden conclusively, then it is required that what expedites it be prevented, and that its motivating factors be removed. Since worshipping an idol is a sin, and meddling with images and idols leads to this as has occurred among communities in the past, it is necessary to restrain the hands of the image-makers. Since drinking wine is a sin, it is necessary to restrain the hands of the grape-pressers, and to forbid being seated at a table where wine is served. Since killing in a civil war is a sin, it is necessary to forbid selling arms during civil strife.

Parallel to this topic in the civic policy is that if they get information of the heinous deed of concealing poison in food and drink, that they make pacts with the sellers of medicines that they will not sell poison except in an amount which is not usually lethal if drunk; and that if they learn of the treachery of a group they will make it a condition that they do not ride horses or bear arms. Likewise is the topic of acts of worship. Since prayer is the most important of the gateways to the good, it is necessary to spur people on to the congregational prayer, for this will aid them in performing it; and it is necessary to encourage them to (perform) the call to prayer in order to gather the assembly at one time and in one place; and it is necessary to encourage the building of mosques, and to perfume them and make them clean. Since recognizing the first day of Ramaḍān when there are clouds, and so on, depends on the calculation of the month of Sha'bān, computing the new moon of Sha'bān was categorized as recommended (*mustāḥabb*). Similar to it in the policy of the city is that if they see that shooting arrows has a great benefit, they will order an increase in the manufacturing of bows and arrows and the trade in them.

7) Among the principles is that if something is commanded or

---

[13] A reference to Qur'ān 62:9.

forbidden this requires that the status of the obedient ones be respected and that the disobedient ones be disdained. Since it is desired that the recital of the Qur'ān become widespread and performed diligently; it is necessary that it be made a practice that no one shall function as their leader in prayer except the one who is the best reciter, and that the reciters should be respected in the gatherings. Since false accusation is a sin it is necessary that the slanderer be dropped from the rank of having his witness accepted. In this vein is interpreted the report about being forbidden to be the first to address a heretic or a godless sinner by way of greeting or conversation. The parallel to these in the policy of the city is increasing the reward for archers and giving them precedence in acknowledgments and grants.

8) Among the principles is that if people are commanded to perform a thing or forbidden it, it is proper in this case that they should be ordered to make a decision to undertake the former and to avoid the latter, and that they should make up their minds to preserve the motive compatible with the act. Therefore censure has been mentioned for harboring the concealed intention not to pay a loan or a dowry.

9) Among them is that if a thing has in it the possibility of a damaging influence it should be considered reprehensible, as in the Prophet's saying, "(The one who wakes up from sleep) should not plunge his hand into a vessel, for he does not know where his hand has been during the night."[14]

In summary, God, may He be exalted, taught His Prophet the rulings of the acts of worship and the supports of civilization, and then the Prophet, may the peace and blessings of God be upon him, explained in this fashion and derived from them significant rules under each and every topic. The topic explained in this chapter together with the topic which follows if God, may He be exalted, wills, were both learned by the legal scholars of the community from the teachings of the Prophet, may the peace and blessings of God be upon him, and they comprehended these two (i.e., the basic rules and their derivatives) through careful consideration. Then whatever they set down in their compilations and their books branched out from these two, and God knows better.

---

[14] That is, he should first wash his hands. Therefore he must perform the ritual ablution. Bukhārī Wuḍū' 260, Muslim, Abū Dāwūd, Nasā'ī, Tirmidhī, Ibn Mājah, Muwaṭṭa', Ibn Ḥanbal. Siddiqi trans. *Ṣaḥīḥ Muslim*, I:166.

CHAPTER 66

Rendering the Ambiguous Precise, Solving the Problematic, and Deriving Rulings from All-Embracing Principles and So On

You should know that many of the things on whose designations the rulings are based are known through a pattern and a classification, and are not known by the comprehensive, exclusive definition[1] which discloses right away the status of each and every individual case as being included in this or not. For example, theft. God, may He be exalted, said, "As for the thieves, both male and female, cut off their hands."[2]

The legal punishment (*ḥadd*) is applied to the designation, "the thief," and it is known that what occurred in the story of the Banū al-Ubayraq[3] and Ṭu'aima,[4] and the Makhzūmiyya woman[5] was theft. And it is known that taking the possessions of another has various classifications, among them theft, highway robbery, snatching, betrayal (of a trust), picking up lost property, extortion and neglect. In situations like this the Prophet, may the peace and blessings of God be upon him, was sometimes asked about a case, i.e., whether its being considered a type of theft was a question of language or of situation, so he found it necessary to explain the true nature of theft, distinguishing what was the common factor so as to make clear the case of each individual instance separately. The way to distinguish was to consider the essential features of the designations that are not included under theft and through which the two categories can be differentiated; and to consider what are the es-

---

[1] "*Al-ḥadd al-jāmi'al-māni'*"—in the terminology of logic, an exhaustive, exclusive definition.

[2] Qur'ān 5:38.

[3] Tirmidhī IV:310-313 #5027 Tafsīr Ṣūra 4:22. The Bānū Ubayraq stole a person's armor and then tried to blame a Jew. Muḥammad saw through this and pronounced the Jew innocent. This is supposed to be a rare hadith.

[4] Ṭu'aima ibn Ubayraq. Instigated the plot to blame an innocent person for theft.

[5] This refers to Fāṭima bint Al-Aswad who stole. Uṣama ibn Zaid asked for clemency but the Prophet did not grant it and said, "If it were Fāṭima bint Muḥammad (his own daughter) who stole, I would cut off her hand." Bukhārī, Faḍā'il 18, Muslim Ḥudūd.

sential features of theft which the people familiar with the common language understand from this word. Thus theft can be defined by conceptual factors which distinguish it. For example, it is known that "highway robbery" and "plundering" and things like them are designations which refer to the use of force in relation to ones who are wronged, and by the choice of a place or time at which they cannot be rescued by others; and that "snatching" refers to stealing something from right in front of the people and within their sight or hearing; and that "betrayal" discloses that previously there had been partnership and trust, and "keeping whatever is picked up" (*ḥifẓ al-iltiqāṭ*) refers to taking something which is not in anyone's possession. "Taking by force" refers to public domination over the wronged individual based on a dispute, or based on the assumption that the issue would not be brought before the rulers, and that the true state of things would not be made manifest to them, or that they would not judge fairly due to bribery or some similar thing. "Pilfering" is said about a paltry thing which is customarily given freely or as charity, such as water and firewood, while "theft" refers to secretly taking. Therefore the Prophet, may the peace and blessings of God be upon him, precisely defined theft as (taking something) worth a quarter of a dinar or three dirhams, in order to distinguish it from what was insignificant, and said, "Neither (the hand of) the betrayer, nor the plunderer, nor the pilferer will be cut off."[6] And he said, "There will be no cutting off (the hand in the case of) stealing fruit which is hanging (on a tree), nor in what is protected by the mountain,"[7] indicating the condition that the thing has to be guarded.

(Another potentially ambiguous case regards) the high degree of luxury, for it is a source of corruption which is imprecisely defined; nor is there a criterion distinguishing the instances when it occurs by outer signs on the basis of which to blame all and sundry, while it is not doubted by anyone that luxury is certainly characterized by these outer signs. It is known that the habit of the Persians was to acquire swift riding animals, lofty buildings, fine clothing, extravagant jewelry and these sorts of excessive luxuries,

---

[6] *Mishkāt*, p. 767. Abū Dāwūd III:138 #4391, 4392 Ḥudūd 14, Tirmidhī, Nasā'ī, Ibn Mājah, Dārimī, Ibn Ḥanbal transmitted it.

[7] That is, the theft of sheep or goats stolen while not in the fold. *Mishkāt*, p. 766, Nasā'ī 11, Muwaṭṭa' Ḥudūd 22.

and it is known that opulence differs with the differences among people, so that the opulence of one group may be austere for another, and what is excellent in one region may be considered insignificant in another region. The supports of civilization may be (achieved) through the outstanding level or the lowly, and the latter is not through opulence. Achieving the benefits of civilized life through the outstanding (level) may be done without the intent of excellence, or in those instances where this (goal of excellence) does not predominate over a person in most cases; this will not be designated as "opulence" in common parlance Thus the divine law applied an absolute warning against the causes of harm due to luxury and specified the things that it found were only incorporated into civilization by people for the sake of opulence, and due to which the opulence would become a widespread custom among them. He (the Prophet) saw that the people of that age, the Byzantines and the Persians, were agreed on these things, so he designated them (these things) as an anticipated source (maẓinna) of excessive luxury and forbade them, and he did not take into account the rare benefits of other civilizations nor the customs of the far off regions. Thus the forbidding of silk and gold and silver vessels was subsumed under this category. Furthermore, he (the Prophet) found the true nature of luxurious living to be choosing the excellent in every support of civilization (*irtifāq*) and turning away from the lowly, and that full luxury was choosing the excellent and turning from the lowly in any one area. He found there to be dealings in which nothing was intended other than choosing the excellent and rejecting the lowly of one type (of thing); except, that is, for little items to which the divine laws did not attach importance. Thus he forbade such dealings because they were like an embodiment of the meaning of luxury and representations of it, and forbidding them was a natural requirement of his dislike of luxury. Since the anticipated sources for the thing were forbidden because of this (dislike), it was necessary that initially its embodiment and its representations be forbidden. The forbidding of selling money or food for the same commodity at different values[8] is derived from these principles, but the purchasing of the excellent thing for a high price was not forbidden because the price is paid for the very thing sold,

---

[8] For example, one cannot exchange four pounds of lentils for five pounds of lentils.

not for its attribute when there is a difference in kind. Neither was purchasing a slave girl for two slave girls forbidden, nor a robe for two robes, because these things possess inherent values so that spending a higher price is due to the particularities of the individual item and the excellence is inherent in these particular characteristics. Therefore the taking into account of "excellence" is not obvious at first glance.

What we have set out discloses many of the points connected with this topic such as the reason why selling one animal for another (of the same type) is deemed repugnant, and other things, so it should carefully be considered. There may be two ambiguous things which are not distinguished due to an undisclosed reason which only the Prophet, may the peace and blessings of God be upon him, and those well-versed in knowledge among his community understood, so that the situation required recognizing an outward sign (*'alāma*) for each of these and basing the ruling of righteousness and sin upon their signs. Examples of (the rulings) differentiating between two things are "marriage" versus "fornication," for the true nature of marriage is establishing the beneficial purpose on which is based the order of the world through the cooperation between the husband and wife, seeking descendants, keeping pure their sexual organs, and so on; and these are pleasing to God and desirable. The true nature of fornication is the lower soul's submitting to its excesses and desires in following its lust, tearing the mantle of modesty and its bonds, and abandoning the ascent to the universal beneficial purpose and the universal order, and this is hated and forbidden by God. These two things (marriage and fornication) are similar in most forms, for they share the satisfaction of lust, the elimination of the pain of carnal appetite, the inclination to women, and so on. Thus circumstances require that each be distinguished from the other by an outward sign, and by the act of requesting and refusing. Therefore the Prophet, may the peace and blessings of God be upon him, particularized marriage by certain things. Among them that it should be (arranged) with women, not men, since the seeking of descendants is only through women, and that there should be an intention, a consultation, and an announcement; and he made the presence of witnesses and the guardians a condition, and the consent of the woman. Among them also is the reconciliation of the soul to cooperation, and this does not occur in most cases unless it is forever and binding, not temporary, so that secret marriage and temporary marriage (*mut'a*)[9] were forbidden, and homosexuality was also forbidden.

Sometimes some pious act resembles what is a prelude to another, so the situation required that they be differentiated, such as the standing (in prayer) being legislated to separate bowing from the waist and bowing the head which is a prelude to the prostration. Sometimes the thing does not particularly augment the benefit, such as the sitting up between the two prostrations, and sometimes the condition or the pillar is in reality a hidden matter and an act of the heart, so an index (*amāra*) is assigned to it in the form of physical movements, or some sayings, and this is made a pillar fixing the hidden, such as the declaration of the intention. The devotion of an action to God is an unseen matter so that facing the Qibla and proclaiming "God is great" are assigned as a anticipated source of this, and are made essential for the prayers. If a textual report is forthcoming in a certain way, or if the situation requires establishing a category (of case) on which the ruling can be based then in certain matters ambiguity may occur. In this case, the proper course is to go back to the custom of the Arabs in interpreting this wording or in order to identify the comprehensive limit forbidding that type (of case). In this way, a textual report was forthcoming concerning the commencement of fasting in the month of Ramaḍān, and then ambiguity arose in the case of cloudiness, so that the ruling was according to what the Arabs did in completing the count of Sha'bān as thirty days, for the month may have thirty days, or twenty-nine. This is the Prophet's saying, may the peace and blessings of God be upon him, "We are an unlettered people who do not write or calculate. The month is thus."[10] Likewise in the case of the report related about reducing the prayer during a journey; ambiguity occurred in some cases, so that the Companions ruled that a journey was going out from one's home to a place you don't reach on the same day nor during the first part of its night, and this required that the distance should be that traveled on a day and some considerable part of the next day, and this is fixed at four *burūd*.[11]

You should know that the basic issue in the Prophet's, may the

---

[9] *Mut'a* is a form of temporary marriage permitted in Shī'a Islam.

[10] I.e., a complete thirty days. Bukhārī Ṣaum 13, Muslim, Abū Dāwūd, Nasā'ī, Ibn Ḥanbal. *Mishkāt*, p. 420.

[11] A distance of twelve miles.

peace and blessings of God be upon him, being singled out among his community by a ruling was that ruling's going back to a anticipated source (*maẓinna*) of a thing rather than its true nature, and this is the saying of Ṭā'ūs[12] about doing two *rak'at*s after the afternoon prayer, "They were only forbidden lest the people take them as proper."[13] The Prophet, may the peace and blessings of God be upon him, recognized the true situation, so there was no consideration in his case of the symbol once he knew the reality. For example marrying more than four women is anticipated source of abandoning righteousness in conjugal relations and the neglect of the wives' affairs, and this is suspect in the case of the rest of the people. The Prophet, may the peace and blessings of God be upon him, knew what pleased God in the affair of conjugal relations, so he commanded according to this, not according to the anticipated source (*maẓinna*).[14] (The Prophet's exemption from a ruling) may go back to the fulfillment of custom, not to the refinement of the lower soul, such as the Prophet's forbidding of conditional sales, then he bought a camel from Jābir on the condition that Jābir could ride it back to Medina.[15] Or (he was unique in a ruling because) it might lead to something else in relation to someone who was not at all free from sin and this is the saying of 'Ā'isha, may God be pleased with her, concerning kissing on the part of one who is fasting, "Who among you can control the domination of passion as the Prophet of God used to control it?"[16] Or it may be that his elevated soul required a certain type of piety, so that he was commanded with it because this soul longed for more attentiveness to God, and for a further divestment of the mantle of negligence just as the strong man longs to eat much food; for example the Prophet's praying the extra night prayer, the prayer after sunrise, and the forenoon prayers, according to a report, and God knows better.

---

[12] D. 101 A. H. (719/720 C. E.).

[13] Dārimī I:115 Muqaddima, Ibn Ḥanbal III:115.

[14] The Prophet alone was allowed to have more then four wives.

[15] As recounted in a hadith Bukhārī Khan trans. III:550.

[16] Bukhārī Ḥaiḍ 5, Ṣaum 23, trans. *Ṣaḥīḥ Muslim*, I:537, Abū Dāwūd, Tirmidhī, Ibn Mājah, Ibn Ḥanbal.

CHAPTER 67

The Facilitation (of Religious Duties—Taisīr*)*

God, may He be exalted, said, "It was by the mercy of Allah that you were lenient with them (O Muḥammad), for if you had been stern and fierce of heart they would have dispersed from around you."[1] And he said, "Allah desires for you ease; He desires not hardship for you."[2]

And the Messenger of God, may the peace and blessings of God be upon him, said to Abū Mūṣa[3], and Muʿādh ibn Jabal[4], may God be pleased with them, when he sent them to Yemen, "Be easy (with the people) and don't make things difficult, bring good news and don't estrange (them), accede and don't oppose,"[5] and he said, "you have been sent to make things easy (for the people) and not to make things difficult."[6]

Facilitation may be achieved in many ways:

1) By not making a thing which is hard for people a pillar or a condition of worship, and the basis of this is the Prophet's saying, "Were it not that I would have been hard on my community, I would have ordered them to clean their teeth with a *siwāk* before every prayer."[7]

2) That some portion of the acts of worship should be made conventions which they can take pride in, and these should be included in what they have been doing out of their own volition, such as (celebrating) the two *ʿĪds*[8] and the Friday congregational prayer. This is the Prophet's saying, "So that the Jews will know

---

[1] Qur'ān 3:159.

[2] Qur'ān 2:185.

[3] al-Ashʿarī d. ca 42. A.H. 662/3 C.E.

[4] Died. ca 18 A. H. 639 C. E.

[5] Bukhārī Wuḍū' 58, Adab 80, Abū Dāwūd, Tirmidhī, Nasāʾī, Ibn Ḥanbal. Previously cited in Ch. 54.

[6] *Mishkāt*, p. 791. Bukhārī Jihād 164, Maghāzī 60, Adab 80, Aḥkām 22. Muslim Khan. Trans. IV:171. *Mishkāt*, p. 791.

[7] *Mishkāt*, p. 79. Bukhārī Jumaʿ 8, Tamanna 9, Ṣaum 27, Muslim, Abū Dāwūd, Tirmidhī, Nasāʾī, Ibn Mājah, Muwaṭṭa', Ibn Ḥanbal.

[8] The Islamic festivals at the end of Ramaḍān and at the time of the pilgrimage.

that in our religion there is room for pleasure,"[9] for beautifying themselves in great assemblies and competitiveness in whatever they pride themselves on is a habit of humankind.

3) That such acts of worship will be made customary practices for them which they desire due to their physical nature, so that nature will motivate to that which reason calls for, and the two desires will reinforce one another. Therefore, keeping the mosques clean and fragrant and washing and scenting oneself on Friday were made practices, and melodious recitation of the Qur'ān was recommended, as well as having someone with a nice voice perform the call to prayer.

4) That they should be relieved of what is burdensome as well as that which they find hateful by nature. Therefore the leadership in prayer of a slave, or a Bedouin, or one of unknown lineage was declared repugnant, for following such a man would torment the people.

5) Those things should be retained which the nature of most of them demands, and the omission of which would produce disquiet in their souls, such as the Sultan being the most worthy of leading the prayers, the head of the household being most worthy of leading the prayers, that the one who marries a new wife should spend seven nights with her, or three (if she was previously married), and then he should again divide his time equally among his wives.

6) Imparting knowledge, giving sermons, and ordering the good and forbidding evil, should be made a practice among them; in order that their hearts will become filled with this, so that they will follow the divine law without discomfort, and thus the Prophet of God, may the peace and blessings of God be upon him, used to preach to them frequently.

7) That the Prophet, may the peace and blessings of God be upon him, performed the acts which he had commanded them or in which he had given them dispensations in order that they might take an example from his practices.

8) That the Prophet prayed to God, may He be exalted, to make the group refined and perfected.

9) That a spiritual tranquillity (*sakīna*)[10] descended to them from

---

[9] Ibn Ḥanbal VI:116, 233.

[10] *Sakīna* is a calm and dignity reflecting a spiritual influence. It may be experienced when the Qur'ān is recited, after prayer, etc., according to the hadiths.

the Lord through the intermediacy of the Prophet, so that they became absolutely still and calm in his presence.

10) That the one who wants something which is not his right would be spited by being deprived of it, as the murderer may not inherit, and the one who forces a divorce will find it not legally valid, so that this would serve as a check on those who used compulsion since their goal would not be achieved.

11) That something difficult should only be legislated for them gradually and this is the saying of ʿĀ'isha, may God be pleased with her, "The first thing that was revealed thereof was a chapter from *al-Mufaṣṣal*[11] in which Paradise and Hell were mentioned. Once the people had rallied to Islam the permitted and forbidden were revealed. If the first thing to be revealed had been, 'Do not drink wine,' they would have said, 'We will never give up wine,' and if 'Do not fornicate,' had been revealed, they would have said, 'We will never give up fornication.'"[12]

12) That the Prophet, may the peace and blessings of God be upon him, did not do what would be distasteful to them, and therefore he left aside some recommended things, and this is his saying to ʿĀ'isha, "If not for the recent proximity of unbelief among your people, I would have torn down the Kaʿba and built it on the foundation of Abraham, may peace be upon him."[13]

13) That the law-giver commanded types of pious acts such as the ablution, the full bath, prayer, the alms tax, fasting, the pilgrimage, and others, and did not leave them entrusted to their reasoning but made them precise through pillars, conditions, proper behaviors, and so on. Then beyond this he did not fix the pillars, conditions, and manners extremely precisely but left them up to their reasoning and to what they had understood from these words, and to what they had become accustomed in this area.

Thus, for example, he explained that no prayer is valid without the opening chapter of the Qur'ān,[14] while he did not explain the pronunciation of the letters on which it depends, and which letters

---

[11] The last part of the Qur'ān.

[12] Bukhārī Faḍā'il al-Qur'ān 6, Khan trans. IV:483. The Cairo text has "verses" (*suwar*) but the original text of the hadith uses the word "verse"—*sūra*.

[13] Bukhārī Ḥajj 42, Muslim, Nasā'ī, Dārimī, Ibn Ḥanbal. Trans. *Ṣāḥīḥ Muslim*. II:671. Khan trans. I:95.

[14] As in the hadith recounted in Dārimī I:283, "*man lam yaqra' bi-umm al-kitāb fa lā ṣalāt lahu.*"

are doubled, vocalized, and unvocalized. He explained that facing the Qibla is a condition for the prayer, but he did not explain a rule through which we could ascertain the direction of prayer. He explained that the minimum taxable amount of property for the alms tax is two hundred dirhems; but did not explain what is the weight of a dirhem, and when he was asked about things like this he did not go beyond what they could comprehend, and he did not tell them things not already found among their customs.

Thus he said concerning the question about the first crescent moon of Ramaḍān, "Then if it is obscured from you, complete the number (of days) of Sha'bān at thirty,"[15] and he said concerning the water on the ground from which wild beasts and livestock drink in an open desert space, "If the water reaches an amount to fill two jars (twelve hundred pints) then it should not be considered impure,"[16] and the basis of this had been customary among them, as we have explained.

The secret in this is that every one of these things could only be made clear through facts which were equally apparent, concealed or imprecise, and this would require an explanation and so on. This is a great source of trouble insofar as every setting of times is a restriction on them in a general way, for when the appointed times are many the scope will be completely restricted. Insofar as the divine law was imposed on all, the nearest and furthest, there is a great deal of trouble in keeping to these limits in their details. In addition the people, if they give strict care to undertaking whatever establishes piety, will not feel the benefits of the piety, nor will they pay attention to the spiritual aspects of these (acts). For example, you see many of the Qur'ān reciters not contemplating the meaning of the Qur'ān due to being caught up in the pronunciation of the words. Therefore there is nothing more suitable for the beneficial purpose than that the matter be left up to them once it had been basically established, and God knows better.

14) That the law-giver only addressed them according to the measure of intelligence placed at the basis of their nature before they took an interest in the fine points of theosophy (*ḥikma*), theology, and the roots of jurisprudence.

---

[15] *Mishkāt*, p. 420. Bukhārī ṣaum 5, 11, Muslim, Abū Dāwūd, Tirmidhī, Nasā'ī, Ibn Mājah, Dārimī, Muwaṭṭa', Ibn Ḥanbal.

[16] For ablution. Abū Dāwūd I:75 #65 Ṭahāra 33, Tirmidhī, Nasā'ī, Ibn Mājah, Dārimī, Ibn Ḥanbal. *Mishkāt*, p. 96.

Thus, God was confirming for Himself "a location" (*jiha*) when He said, "The Merciful, who sits upon the throne."[17] The Prophet, may the peace and blessings of God be upon him, said to a black woman, "Where is God?" and she pointed to the sky; then he said, "She is a believer."[18]

He did not impose on the people the memorization of the formulae of astronomy or geometry in order to determine the direction of the Qibla, the times of prayers, and the festivals, and he indicated this by his saying, "The Qibla is everywhere between the East and the West,"[19] when he faced the Ka'ba, concerning one aspect of the matter, and he said, "The Hajj is the day on which you make the pilgrimage; and the *'Īd al-Fiṭr* is the day on which you break the fast,"[20] and God knows better.

---

[17] Qur'ān 20:5. This Qur'ānic verse occasioned theological debates about the anthropomorphic interpretation of God. Shāh Walī Allāh's point is that this speculation was not intended in the scripture.

[18] Nasā'ī Sahw, Muwaṭṭa' 8.

[19] *Mishkāt*, p. 144. Tirmidhī Mawāqīt 139, Ṣalāt 139, Nasā'ī, Ibn Mājah, Muwaṭṭa'.

[20] Meaning that people should not become overly pre-occupied with determining the details behind religious practices. Tirmidhī Ṣaum 11, Ibn Mājah.

CHAPTER 68

The Inner Meanings of Encouragement (to Righteousness) and Deterrence (from Sin)

Among the blessings of God, Blessed and Exalted, upon His worshippers is that He revealed to His prophets, upon whom be peace, the reward or punishment that would accrue from actions, so that they might inform the people of it, whereby their hearts would become filled with either desire or fear. Thus they would bind themselves by the divine laws through a motive proceeding from themselves, just (as they are bound naturally) to the rest of what wards off harm, or brings benefit. This is His, may He be exalted, saying, "For indeed it (prayer) is a great burden except for the humble who think that they will meet their Lord and that they are returning to Him."[1]

Therefore, in this there are universal principles upon which the particular dimensions of incitement to good and deterrence from sin are based. The legal experts among the Companions knew them in a general way, although they didn't attain a detailed knowledge of them. One of the things which proves what we have mentioned is the hadith report that the Prophet, may the peace and blessings of God be upon him, said, "In one of you having intercourse (with his wife) there is a virtuous act (*ṣadaqa*)." So they said, "One of us satisfies his lust and in this there will be a reward for him?" And he said, "Don't you see that if he devotes it (his lust) to something forbidden it will be considered a sin for him?"[2] The Companions did not stop short at this question to the exclusion of others. Its rationale was only ambiguous for them because of what they knew about the relationship of acts to their rewards, and that the relationship goes back to a principle whose meaning could be arrived at by reason (*maʿqūl al-maʿnā*). For if this had not been so, there would have been no reason for their question, or for the

---

[1] Qur'ān 2:45

[2] *Mishkāt*, p. 404. Transmitted by Muslim Zakāt 52, Abū Dāwūd, Ibn Ḥanbal.

response of the Prophet, may the peace and blessings of God be upon him, with regard to a clear principle (*aṣl*).[3] This dictum of mine is like what the legal scholars say about the hadith, "If your father had a debt should you pay it or not?" He answered "Yes." The Prophet said, "Then the debt of God is even more worthy of being paid off."[4] They say that this hadith indicates that the rulings are connected with universal principles.[5]

The outcome of the question (of the Companions) is that good deeds (*ṣadaqāt*) come under the (topic of the) refinement of the soul, such as saying, "Praise to God," uttering "There is no God but Allah," and "God is great," or establishing the best interest with regard to the administration of the city, while evil deeds belong to whatever opposes these two things. The satisfaction of sexual lust is the following of the animalistic impulse and no beneficial purpose in addition to the habitual can be anticipated in this or anything else like it, which could be referred to knowing the universal principle, thus it is absurd to refer the question back to it (the principle).

The import of the answer, then, is that a man's intercourse with his wife keeps their sexual organs pure, and in this there is a deliverance from satisfying lust improperly.

There are various methods to encourage to good and to deter from sin, and there is an inner dimension to each method. We will alert you to the most important of these methods.

1) By explaining the effect of the action on refining the soul either through one of the two forces (the animalistic) being defeated or through the predominance and manifestation of the other one of the two forces.[6] The language of the law-giver expresses this by the "recording of good deeds and the erasing of evil ones" as in his saying, may the peace and blessings of God be upon him, "The one who says 'there is no God but God, unique, no partner to Him, His is the dominion and His is the praise, and He is in all things omnipotent,' one hundred times in a day, has done

---

[3] That is, a reward being given for not committing a sin is the principle *(aṣl)* on which the Prophet's indicating a reward was based.

[4] Nasā'ī Ḥajj 11, Ibn Ḥanbal IV:5.

[5] That is, fasting or pilgrimage could be performed on behalf of the deceased, therefore the universal principle is one of making up for indebtedness. See a discussion of this hadith in R. Brunschvig, "Raisonnement par analogie" *Studia Islamica* 34 (1971):67.

[6] I.e., the angelic.

the equivalent of setting free ten slaves, and one hundred good deeds are recorded for him, and one hundred bad deeds are effaced for him, and he is protected from the devil on that day until evening, and no one does better than what he does except the one who does it more times."[7] We have mentioned the secret of this previously.[8]

2) By explaining its effectiveness in protecting from the devil and others, such as his, may the peace and blessings of God be upon him, saying, "And he was protected from the devil until evening."[9] And his saying, "And the devils will not be able to do it,"[10] or by explaining its effectiveness in increasing sustenance and manifesting blessing and so on. The secret in one portion of this is that someone had sought well-being from God and this is the reason for his prayer being answered, and this is the Prophet's saying, may the peace and blessings of God be upon him, reporting from God, Blessed and exalted, "If he asks to take refuge in Me, I will give him refuge, and if he asks Me (for something) verily I will give it to him."[11] The secret in another portion is that being immersed in the remembrance of God, turning to Jabarūt, and drawing grace from Malakūt, cuts off the relationship to the devils, for the(ir) effectiveness is through the relationship. Another portion is that the angels pray for the one who is in this situation so that this comes into play through many channels, sometimes in attracting something beneficial and sometimes in warding off something harmful.

3) By explaining the effect of an act in the next world, and the secret of this is disclosed through two premises.

a) The first premise is that a thing is not judged to be a cause for reward or punishment in the next life unless it has a relevance to one of the two causes of requital. Either it impinges upon the four virtues on which happiness and the refinement of the soul are based by either confirming or negating (them); and these (four) are: cleanness (*naḍāfa*), humility before the Lord of the worlds,

---

[7] *Mishkāt*, p. 487. Bukhārī Bada' al-khalq 11, Du'āt 159, Tirmidhī, Ibn Mājah, Muwaṭṭa', Ibn Ḥanbal.

[8] In chapters 6 and 13, for example.

[9] The conclusion of the hadith cited in the preceding paragraph.

[10] This refers to the hadith, "Read the Qur'ānic chapter, "The Cow," for doing this is a blessing and abandoning this is an affliction, and the devils will not be able to do it." *Mishkāt*,. p. 449. Muslim Musāfirūn 252, Dārimī, Ibn Ḥanbal.

[11] Bukhārī Riqāq, 38.

magnanimity of soul, and striving to establish justice among people.[12] Or it may have a role, either positive or negative, in implementing what the Highest Council has agreed to set in motion by way of strengthening the divine laws and aiding the prophets, may peace be upon them.

The meaning of "relationship" is that the act is a anticipated source (*maẓinna*) of the existence of this meaning or customarily adheres to it, or is a means to it. For example, a person's praying two *rakʿat*s without becoming distracted is an anticipated source of humility, remembrance of the majesty of God, and progress above the depths of the animalistic. Performing the ablution is a means to cleanness, which affects the soul. Spending weighty sums of money with which one is usually miserly, forgiving someone who has done wrong, and the forgoing of dispute by the one who is in the right, are similarly anticipated sources (*maẓānn*) for magnanimity of soul and they are closely connected with it. Feeding the hungry, giving water to the thirsty, and trying to stop an outbreak of fighting between men is likewise an anticipated source for the reformation of the world and a means to it. Loving the Arabs is a way to taking on their style and this is a means for inclining one to join the monotheistic (Ḥanīfī) religion since it was instantiated through their customs and determined by the command of the divine law brought by Muḥammad. Similarly remaining punctual in breaking the fast distances one from the interpolations of (other) religions and their distortions.[13] The various groups of people, such as the wise men (*ḥukamā'*), artisans, and doctors continue to base rulings on their anticipated sources, and the Arabs continue to practice this in their discourses and speeches, and we have mentioned something of this.

Or (the relationship of the act to an anticipated source may be) that the act is difficult, obscure, or disagreeable by nature so that a person doesn't want to do it, and the only one who will attempt it is the one of real sincerity, thus the act testifies to his sincerity; like (after having gone a great distance) drinking deeply of the water of Zamzam, and like the love of ʿAlī, may God be pleased with him, for he was forceful in carrying out God's commands,

---

[12] The four virtues discussed previously in Ch. 32.

[13] As in the hadith reported in Abū Dāwūd. II:763 #2353. "The religion will continue to prevail as long as people hasten to break the fast, because the Jews and Christians delay in doing so."

and like love for the "Helpers" of Medina. Thus the Maʿad and the Yemeni Arabs used to hate one another until Islam reconciled them, and their being reconciled was a way of recognizing that the friendliness of Islam had entered the heart. Other examples are ascending the mountain and staying awake in order to keep a vigil over the armies of the Muslims, for this identifies the truth of one's commitment to promote the word of God and one's love for his religion.

b) The second premise (concerning an act's effect in the next life) is that when a man dies, and has recourse to his soul[14] and to the attitudes by which it has become colored, which are either suitable for it or incompatible for it—it is inevitable that the image of the painful or the enjoyable will manifest in close adherence to what is found there. In this there is no consideration of intellectual adherence, but rather of another type of adherence that draws one inner state of the soul (*ḥadīth al-nafs*) to another, and according to this meanings are depicted in dreams; just as the caller to prayer's prohibiting the people from sexual intercourse and eating (in Ramaḍān), appears (in a dream) in the form of a seal on the sexual organs and mouths. Therefore, in the World of Images (*ʿālam al-mithāl*) there are relationships on which the rulings are based, so that Gabriel manifested himself in the shape of Diḥya[15] rather than someone else, on account of a (special spiritual) meaning; and the fire did not appear to Moses except due to a (special spiritual) meaning. The one who recognizes these relationships knows in which form the requital of an act will be, just as the one who knows how to interpret dreams knows which meaning has manifested in the form of what he saw.

In summary, in this way the Prophet, may the peace and blessings of God be upon him, knew that the one who kept knowledge concealed and refrained from giving instruction when it was needed would be punished by a bridle of fire,[16] since the soul was pained by being restrained and the bridle is an embodiment and image of

---

[14] To the higher aspect of the spirit in the World of Images.

[15] Diḥya al-Kalbī, a Companion of the Prophet, who was a wealthy merchant who was so handsome that Gabriel is said to have taken on his features. See Khan trans. Bukhārī IV:531, and EI II:274 a.

[16] This is a hadith that whoever is asked about some knowledge which he has and keeps it hidden, will be bridled (*iljām*) on Judgment Day with a bridle of fire. Abū Dāwūd III:321 #3658 ʿIlm, 9, Tirmidhī ʿIlm, 3, Ibn Mājah and Ibn Ḥanbal also transmitted it.

restraint. The one who loves wealth and whose mind is always attached to it, will have his neck encircled by a bald snake,[17] and the one who is preoccupied with hoarding dirhems, dinars, and livestock, and is hesitant to spending them for God's purpose will be punished by these very things according to what is decided among the Highest Council as a way of punishment.[18] The one who has harmed himself (a suicide) by poison or a weapon and thus disobeyed God's command will be punished in that form, the one who clothed a poor man will be dressed in the silk brocade of Paradise on the Day of Judgment, and the one who manumits a Muslim and releases him from the hardships of slavery which surrounded him, in exchange for each limb of the slave, frees one of his own limbs from Hellfire.[19]

Among these ways of encouragement to good or deterrence from sin is the symbolizing of the act by something which is fixed in human minds as attractive or repulsive either from the aspect of the divine law or by custom. In this (symbol) there must necessarily be a shared element relating the two, even if in only one aspect; as the one who remained doing private devotions in the mosque from the morning prayer until the rising of the sun was compared to the one who performed the Hajj and the *ʿUmra*,[20] and the one who takes back a gift was compared to a dog who returns to his vomit.[21] Its (the symbol's) being ascribed to beloved or despised people and an invocation being made for the one who does the action or against him, alerts to the status of the action generally without giving attention to the reason why it is desirable or repugnant. Examples of this are the saying of the law-giver, "This is the prayer of the hypocrite and the one who does this is not one of us,"[22] and that a certain act was an act of the devils or the angels, or "may God have mercy on the man who does such and such an act," and expressions like these.

---

[17] Based on hadiths in Bukhārī Zakāt 3, Tafsīr 3:14, 9:6, etc. Muslim. Nasā'ī, Ibn Mājah, Dārimī, Muwaṭṭa', Ibn Ḥanbal.

[18] Such hadiths were referred to previously in Ch. 46.

[19] As in the hadith reported in Bukhārī ʿItq 1, Khan trans. III:419.

[20] Tirmidhī Jumʿa 60.

[21] Bukhārī Hibba 30, Muslim, Abū Dāwūd, Nasā'ī, Ibn Mājah, Ibn Ḥanbal.

[22] The hadith continues, "who sits watching the sun until it turns yellow and it comes between the horns of the devil, then he stands to pray and quickly prays four *rakʿats* in which he scarcely mentions God." That is, he performs the afternoon prayer very late. *Mishkāt* p. 122. Transmitted by Muslim Masājid 195, Tirmidhī, Nasā'ī.

Among the ways of incitement to good and deterrence from sin, is the status of the act in terms of its being connected to the pleasure of God or His anger, and a cause for the inclination of the prayer of the angels for a person or against him, like the saying of the lawgiver, "God loves so-and-so and hates so-and-so,"[23] and the Prophet's saying, may the peace and blessings of God be upon him, "God, may He be exalted, and His angels bless those in the right flanks of the row,"[24] and we have mentioned the inner meaning of this, and God knows better.

---

[23] As in the hadith cited in the chapter on the Highest Host.

[24] For the congregational prayer. *Mishkāt*, p. 225. Abū Dāwūd I:181 #676 Ṣalāt 25, Ibn Mājah Iqāma 55.

CHAPTER 69

The Ranks of the Community with Regard to Arriving at the Desired Perfection or Its Opposite

The basis of this topic is God's, may He be exalted, saying in the chapter of the Qur'ān called, "The Event."

> You will be of three types, those on the right hand; what of those on the right hand? And those on the left hand, what of those on the left hand? And those who excel, those who excel. They are those who will be brought near.[1]

And His saying, may He be exalted,

> Then we gave the Holy Book as an inheritance to those whom We chose among our servants, but among them are those who wrong themselves, and those who are in the middle, and those who excel in good deeds, by God's permission. And this is the great grace.[2]

You have come to know that the highest levels of souls are the souls of "the instructed ones" (*mufahhamūn*) and we have already mentioned them. Following "the instructed ones" is a group called "the ones who excel" (*sābiqūn*), and these are of two types; one type are the ones (whose angelic and animalistic sides combine in a relationship) of inner integration (*iṣṭilāḥ*) and exaltedness, whose capacity is like that of "the instructed ones" in receiving these perfections, except that happiness has not attained its full extent with them. Thus their capacity is like that of the sleeper who needs someone to wake him up, so that when the tidings of the prophets wake them, they turn to what suits their capacity among these branches of knowledge with a suitability latent in the interior of their souls. Thus they become like those independent scholars in a

---

[1] Qur'ān 56:7-11. The *aṣḥab al-shimāl* (the People of the Left or ill-omened hand), *aṣḥab al-yamīn* (People of the Right Hand), and the *sābiqūn* (those who excel or outstrip). This Qur'ānic verse which refers to the ranks of people in heaven and hell, is used by the Sufis as an allegory for the ranks of spiritual attainment.

[2] Qur'ān 35:32.

school of law (*mujtahidūn fī-l-madhhab*)[3] and their inspiration is learned from the comprehensive universal inspiration which is directed toward their souls due to their being pervaded with an inclination toward the Holy Enclave; this is the thing common to most of them (the ones who excel), and the prophets expounded it.

The other type who are the people of contention of the two sides (*tajādhub*) and a highness of nature, and the guidance to success from God drove them toward spiritual exercises and means of turning toward God with which to conquer their animalistic side. Thus God bestowed on them an intellectual and a practical perfection, and they achieved insight into their own state and divine visions appeared to them, and guidance and illumination, such as the great ones of the Sufi orders.

The "the ones who excel" combine two things: one of them is that they devote their strength to concentrating on God and drawing near to Him; and the second of them is that their innate constitution is strong so that the desired acquired habits are represented to them as they are, without their having to consider their embodiments, and they only need the bodily forms to express these traits and as a means to attain the traits through them. Among them are "the ones who retire to lead a solitary religious life," (*mufarridūn*) who concentrate their attention on the Unseen, and the remembrance of God (*dhikr*) throws off their burdens from them. (Then there are) "the truthful ones" (*ṣiddīqūn*) distinguished from rest of the people by the strictness of their obeying God and their pure devotion to Him; and "the martyrs" (*shuhadā'*) who were brought forth for humanity, and the tendency of the Highest Council was infused into them to curse the unbelievers, be pleased with the faithful, command the good and forbid evil, and to promote the religion by helping the Prophet, may the peace and blessings of God be upon him. When the Day of Resurrection comes, they will be resurrected arguing with the unbelievers and giving evidence against them, and they are like the limbs of the Prophet, may the peace and blessings of God be upon him, in his mission, in completing the thing intended by the mission. Therefore, they must be preferred to others and raised above them.

---

[3] Those who use their own interpretations within the scope of possibilities offered by previous rulings within their Islamic legal school.

Then, there are "those firmly grounded in knowledge,"[4] who possess intelligence and skill in reasoning. When they heard reported from the Prophet, may the peace and blessings of God be upon him, knowledge and wisdom, this suited their capacity so that there flowed within them the correct understanding of the meanings of God's Book.

'Alī, may God be pleased with him, referred to this when he said, "Or an understanding (of the Qur'ān) which was bestowed upon a Muslim . . ."[5] (Then there are) "the worshippers" (*'ibād*) who are those who perceive the benefits of worship with their own eyes and their souls become colored by its lights, and they (the benefits) enter in the innermost depth of their hearts so that they worship God with great insight.

"The ascetics" (*zuhhād*) are those who are certain of the Resurrection and the pleasures which are There, so that beside this they consider the pleasures of the world contemptible, and people, for them, become like camel droppings.

"Those prepared for the deputyship (Caliphate) of the Prophets," may peace be upon them, are those (*ṣābiqūn*) who worship God, may He be exalted, through the virtue of justice; so that they devote themselves to what God, may He be exalted, ordered. (There are also) "the possessors of good character," by whom I mean the people of magnanimity in generosity, modesty, and forgiveness of the one who wrongs (them). There are "those resembling the angels" who associate with them, as it is mentioned that certain of the Companions used to be greeted by the angels.

Each of these groups has an innate capacity whose perfection requires being awakened through the reports of the prophets, may peace be upon them; and an acquired capacity made ready by taking up the divine laws, so that through these two capacities their perfection is achieved. Those among the "instructed ones" (*mufahhamūn*) who are not sent on a mission to people, are counted in the divine laws as being among the "ones who excel," and following "the ones who excel" is a group called the "people of the right side" and they are of various types.

One type are those whose souls are near in origin to the "ones who excel," but who have not succeeded in perfecting what they

---

[4] "Al-rāsikhūn fī-l-'ilm." Qur'ān 3:5.

[5] Bukhārī 'Ilm 39.

were created for, so that they fall short in sufficing with the level of bodily forms without the spirits, although they are not complete strangers to that aspect.

Another type are those who are people of contention (*tajādhub*) of the two sides whose souls are weakly angelic and strongly animalistic. They succeed through rigorous spiritual exercises, which produce in them (equivalents) of what the Lower Council has. Or they are people of a weak animalistic side who are apathetic regarding to the remembrance of God so that there filter down on them partial inspirations, partial worship, and partial purification.

There is a group who are people of inner integration (*iṣṭilāḥ*) with very weak angelic sides, who grasp onto arduous spiritual practices if their animalistic side is strong, or reciting perpetual litanies if both sides are weak. This will not produce in them any mystical disclosure (*inkishāf*), but rather acts and attitudes which are embodiments of good character traits will enter the roots of their souls. Most of them do not have complete sincerity and being totally free of the requirement of nature and custom as a condition of their acts, so that they give alms with an intention mixed with some discomfort to the nature, and some hope of reward, and they pray in accordance with the practice of their group and in hope of reward' and they are prevented from fornication and drinking wine by the fear of God and the fear of other people, or they are not able to achieve their amorous desires nor to have money to spend for playthings. This (attitude to life) is accepted from them on the condition that their hearts are too weak for absolute sincerity and that their souls can only cling to the acts themselves, not to what are the expressions of the acquired virtues. It was believed according to the wisdom of earlier times that in modesty there could be both goodness and weakness; then the Prophet, may the peace and blessings of God be upon him, said, "all modesty is good," thereby supporting what we mentioned above.

On many of them the angelic twinkling shines for short times, but it does not become one of their permanent traits; still they are not total strangers to it, such as those who ask for forgiveness (from God) and blame themselves, and those who remember God in seclusion and their eyes flow with tears, and like those whose souls cannot control evil either due to weakness in the innate disposition, for indeed their hearts are like those of birds, or due to a disassociating factor affecting their temperament like one who has a gastric ailment, and people struck by afflictions whose difficulties are a penance for their sins.

In summary, the "people of the right side" lack one of the two virtues of the "ones who excel" and achieve the other, and below them is the group called the "people of the heights" (*aṣḥāb al-aʿrāf*)[6] who are of two types.

One type, is a group of who are sound in body and disposition, but who have not been reached by the message of Islam at all; or it has reached them, but only in such a way that the proof was not established, and confusion about it did not cease. Thus they grew up neither indulging in despicable traits nor in destructive actions, nor did they turn attention to the direction of God either to deny or confirm, and most of their concern is preoccupied with the immediate worldly supports of civilized life. Therefore, when they die these ones return to a state of unconsciousness, neither to punishment nor to reward. Finally their animalistic side is abolished, then some of the angelic twinkling shines on them.

Then, (there is a second group of the people of the heights), whose minds are deficient, such as most children, insane people, peasants, and slaves; and people often claim that the (religiousness of the) mentally deficient doesn't matter, for if they might not observe some conventions, still they (are excused) for lacking rationality. Therefore the extent of faith which is sufficient for them is as the Prophet of God, may the peace and blessings of God be upon him, found the faith of the black slave woman whom he asked, "Where is God?" and she pointed to the sky.[7] What is wanted from these people is that they imitate the Muslims, so that the unanimity will not be divided.

As for those people who arose absorbed in vices, and who addressed the Divine Majesty in a way other than that by which He must be considered, these are the "people of ignorance" (*Jāhiliyya*) who will be punished in various ways. Then following them is a group who are called "the hypocrites" (*munāfiqūn*) who are dissimulators in practice, and there are types of them whose happiness does not reach the commanded perfection as it should be. Either the veil of nature dominates them so that they are consumed by a depraved habit such as greed for food and women, and mali-

---

[6] *Al-aʿrāf* are the heights between heaven and hell. Mentioned in Qur'ān Chapter 7.
[7] He said, "She is a believer." Hadith previously cited in Chapter 67.

ciousness, and their obedience does not unburden them of their sins; or the veil of convention dominates them, so that they can scarcely permit themselves to abandon the conventions of the Ignorant Age, or to emigrate from brothers and homelands. Or the veil of misinterpretation (of God) dominates them such as in the case of the anthropomorphists, and those who secretly associate others with God in worship or in asking for help, claiming that the polytheism hated by God is other than what they practice, and that this type of "shirk" is something which the religion did not have a specific textual ruling on, and that it was not clearly explained.

Among them (the *munāfiqūn*) are people of weakness and abomination, and people of clowning and dim-wittedness whom the love for God and the Prophet has not availed in freeing themselves from acts of disobedience such as the tale of the man who used to drink wine, while he loved God and his Prophet, according to the testimony of the Prophet,[8] may the peace and blessings of God be upon him.

Then there is a group called "the depraved ones" (*fāsiqūn*) and they are those in whom evil actions predominate more than despicable traits. Among them are those of intense animalness who are carried away by the demands of their predatory and brutal traits, and among them are those of corrupt temperaments and listless thoughts such as the sick man who wants to eat clay and burnt bread, so that they are propelled toward devilishness.

After them come "the unbelievers" (*kuffār*) and they are the defiant disobedient ones who refuse to say, "There is no God but Allah," despite the maturity of their intelligence and despite true religious information having reached them, or they contradict the will of God to put into effect the command of the prophets, may peace be upon them. Thus they turn (people) away from the path of God, and are satisfied with the life of the world, and are not attentive to what comes after it. They are cursed eternally, and are imprisoned forever, and among them are the people of the Ignorant Age and some hypocrites who said they believed with their tongues while their hearts remained absolutely unbelieving, and God knows better.

---

[8] As recorded in a hadith, see Bukhārī, Khan trans. VIII:507-508 Ḥudūd 6.

CHAPTER 70

The Need for a Religion which Abrogates the Other Religions

Study thoroughly the religions present on the face of the earth. Do you see any inconsistency in what we have told you in the preceding chapters? Certainly not, by God! Rather, all of the religions believe in the truthfulness of the founder of the religion and respect him, and hold that he is perfect and has no equal, due to what they beheld in him concerning the establishment of acts of worship, the appearance of miracles, the answering of prayers, the establishment of the penalties, the divine laws, and the restraints without which the religion would not be orderly. Then, following this, there are things among those which we have already mentioned or similar ones, which facilitate obedience. Every group has a practice and a divine law in which they follow the custom of their ancestors, and among these the behavior of the bearers of the religion and its leaders is preferred. Then their foundations are strengthened and their pillars reinforced so that their people protect them and fight against what is other than them, and expend their wealth and their lives for their sake, and this is only due to firm management and sure beneficial purposes which ordinary people do not understand.

When every nation became distinguished by a religion, and turned to their own practices and ways, they defended themselves by their tongues from what was different than these, and fought for them with their swords. Injustice occurred among them, either due to the rise of someone who was not worthy of holding the religion to these rules, or due to the mixture of the divine laws with heretical innovations so that these become interpolated into them, or due to the indifference of the transmitters of the religion, whereby they neglected much of what was necessary, and nothing remained except a trace, which cannot speak any more about Umm Aufa.[1] Every

---

[1] A proverb based on a line of poetry from the pre-Islamic poet, Zuhair. The trace presumably refers to something left at the abandoned campsite of the poet's beloved, such as ashes and camel dung. See the *Mu'allaqa* Zuhair in al-Zauzānī, *Sharḥ al-Mu'allaqāt as-Sābi'a*. (Beirut: Dār Ṣādir li-l-ṭibā'a wā-l-nashr, 1963), 73.

religion reproached its sister religion, repudiated it, and fought against it, and the truth was obscured. Therefore circumstances required a leader-guide who would deal with the religions as a rightly-guided caliph deals with the oppressive kings.

There is a lesson for you about the mixing of religions in the discourse of the translator[2] of *Kalilah and Dimna* from the Indian language into Persian; that he wanted to achieve what was correct (in translation) but was only able to do a little of this, and in what the historians say about the condition of the Ignorant Age and the clashing of its religions.

This prophetic leader (*imām*) who unites the communities into one religion needs to have other principles besides the ones mentioned previously.

Among them is that he calls a nation to the right path and purifies them, and improves their condition. Then he takes them as if they were his limbs, and struggles against the people of the earth to bring them to the right path, and he spreads them (his people) to the far horizons. This is God's saying, may He be exalted, "You are the best community that has been raised up for humanity."[3] This is because this religious leader (Imām), by himself, cannot feasibly fight against countless communities, and since it is thus, it is necessary that the material of his divine law be something like a natural religion (*madhhab ṭabiʿī*) for the people of the moderate regions, both Arabs and non-Arabs. Therefore in what his people already have of knowledge and the supports of civilization, he makes more allowances for them than for others. Therefore he must induce all of the people to follow this divine law because there is no way to entrust the matter (of the divine law) to every group or to the leaders of every age, since the benefit of the legislation would not be achieved through this at all. There is no way to consider the condition of every people and deal with each one of them, so that for each a divine law would be made; since encompassing their customs and condition according to the differences of their cities and the disparity in their religions is something impossible. All of the transmitters have been unable to agree on the unanimous transmission of one divine law, so what would tran-

---

[2] Ibn al-Muqaffa c. 720–756 translated this book of fables into Arabic from a Pahlavi version of the Sanskrit *Panchatantra*.

[3] Qur'ān 3:110.

spire in the case of differing divine laws. Most of the time, the submission of the rest of the people only occurs after preparations and periods of time to which the lifetime of the prophet does not extend, as had occurred in the case of the presently existing divine laws. The Jews, Christians and Muslims in their beginnings did not believe except for a small group, then later they became victorious. Thus there is nothing better and easier than that he should take into account in the divine laws, the penalties, and the supports of civilization—the customs of the people to whom he has been sent, and that he should not put every restriction on the others who will come later, by making them keep all of these. Taking up this divine law was easy for the first ones due to the testimony of their hearts, and their customs, and it was easy for the later ones due to attraction to the behavior of the leaders of the religion and the caliphs, thus it is like the natural thing for every people in every age, ancient and modern.

The sound regions which produce moderate temperaments were gathered under two great kings at the time (of the Prophet). One of them was Khusrau, and he used to rule over Iraq, Yemen, Khurasan, and the territory adjacent to them. He was the king of Transoxiana, and India was under his command, and annual taxes were collected from them for him. The second was Caesar, and he used to rule Syria, Byzantium and the territory adjacent to them, and the kings of Egypt and the Maghreb and North Africa were under his rule and taxes were collected from them on his behalf.

Destroying the empire of these two kings and taking over their domain was like conquering all of the earth, and their habits of luxury were current in all the countries under their command, and changing these habits, and keeping people from them led, in sum, to a warning to all the countries about this, even if their situations differed afterward. Al-Hormuzān[4] mentioned something of this when ʿUmar, may God be pleased with him, sought his advice concerning making wars on the Persians. As for those regions which are far from having a balanced temperament, they were not considered in the universal beneficial purpose, and therefore the Prophet said, "Leave the Turks as long as they leave you (alone),"[5] and,

---

[4] A Persian general who became an advisor to ʿUmar after his defeat by the Muslims.
[5] Abū Dāwūd IV:112 #4302 Malāḥim 8.

"Don't bother the Abyssinians as long as they don't bother you."[6]

In summary, when God, may He be exalted, wanted to straighten the religion which was crooked and to bring out for humanity a community to command them with the good, to forbid them from evil, and to change their corrupt actions; this depended on the cessation of the power of these two, and this was facilitated by opposition to their condition since this way of life was in force, or almost so, in all of the healthy regions. Therefore God decreed the end of their power, and the Prophet, may the peace and blessings of God be upon him, informed that Khusrau would be destroyed and there would be no Khusrau after him, and that Caesar would be destroyed, and that there would be no Caesar after him.[7] Truth was revealed refuting falsehood all over the earth, through the triumph over the falsehood of the (Jāhiliyya) Arabs by the Prophet and his Companions and the triumph over the falsehood of these two kingdoms by the Arabs, and the triumph over the rest of the countries by these two, and God has the convincing proof.

Among the principles upon which the Imām must act is that his teaching them religion should include the establishment of the universal caliphate, and that he should make the caliphs after him be from among the people of his city or of his clan, who were brought up according to these customs and practices; for "the application of kohl to the eye is not the same as the kohl itself."[8] Religious fervor among them then will be combined with family loyalty, and their high rank and the eminence of their stature will exalt the dignity of the head of the religion and his eminence in stature. This is the saying of the Prophet, "The Imāms should be from the Quraish."[9] He enjoined the caliphs to establish the religion and to propagate it, and this is the saying of Abū Bakr al-Ṣiddiq, may God be pleased with him, "You will remain in it (your religion) as long as your Imāms are correct with you."[10] Among the principles of the Imām's action is that he must make his religion predominate over all other religions, and that he not leave

[6] Abū Dāwūd IV:114 #4309 Malāhim 11. *Mishkāt*, p. 1134.

[7] Following a hadith. *Mishkāt*, p. 1130. Bukhārī Īmān 3, 31, Muslim, Tirmidhī, Ibn Ḥanbal.

[8] This phrase is an Arabic proverb indicating that the genuine article cannot be feigned, and is the last hemstitch of a line of al-Mutanabbī, *Diwān Abī al-Ṭayyib al-Mutanabbī*, ed. ʿAzzām, (Cairo, 1944), 331.

[9] Ibn Ḥanbal III:129, 183.

[10] Bukhārī Manāqib al-Anṣār 26, Dārimī Muqaddima 63–67.

anyone unless religion has gained ascendancy over him, whether he is a respected notable, or a humble insignificant one. Thus, people are transformed into three groups: those who submit to the religion, outwardly and inwardly; those who submit outwardly—despite their defiance they cannot deviate from it; and the despicable unbelievers whom he makes use of in harvesting, threshing, and other works, as he makes use of the animals to plow and carry burdens. A way of curbing them is necessary, and they pay the *jizya* tax and are humbled.

There are causes for the predominance of a religion over others. Among them is publicizing of the practices of that religion over the practices of the other religions, for the practices of a religion are a manifest thing particularized for it which distinguish its adherents from those of the other religions, such as circumcision, respecting mosques, the call to prayer, the Friday prayer, and the other congregational assemblies.

Among them is that he should restrict people so that they do not publicly practice the rituals of those other religions. Among them is that the Muslims are neither made equal to the unbelievers in punishment and blood-money, nor in marriages, nor in leadership positions, in order that this will drive these others to the true faith.

Among them is that he should hold the people accountable for the external forms of pious and sinful actions, and that he should require them to follow these with a great compulsion, that he should not intimate to them much about the true spirits (of these actions), that he should not give them a choice concerning any of the divine laws, and that he should keep hidden the knowledge of the inner meanings of the divine laws which is the source of the detailed rulings and only bestow it on the one firmly grounded in knowledge. This is because most of those on whom the laws are imposed do not recognize the beneficial purposes, nor are they able to recognize them unless these are precisely determined by regulating devices; and become tangible and taken up by every person involved. For if he permitted them to omit some of them, and explained that the basic goal is something other than these outer forms, he would widen for them the avenues of unqualified discussion and they would disagree excessively, and what God wanted for them would not be achieved, and God knows better.

Among the necessary principles is that the Imām, if his dominance is through the sword alone, cannot remove the thick veils over their hearts, so it is possible that shortly afterwards they will return to unbelief. Therefore, he must confirm by proofs which are demonstrative or rhetorical, and which benefit the mind of the common person that these other religions must not be followed, because they are not transmitted from an infallible one, or because they are not consistent with the laws of the religion, or because there is distortion and displacement in them. He must prove this in public and explain the justifications for the true religion in that it is easy and tolerant, that its limits are clear so that reason recognizes their value, and that it is as clear as day,[11] that its practices will be more beneficial for the common people, and that it is most similar to what remains among them from the behavior of the preceding prophets, may peace be upon them, and other things like this, and God knows better.

[11] Literally, "its night is its day."

CHAPTER 71

Fortifying the Religion Against Distortion

It is indispensable for the man possessed of supreme authority who brings from God a religion which supersedes other religions that he fortify his religion against being penetrated by distortion, and this is because he unites many nations having manifold propensities and dissimilar goals. Often these people are misled by their self-interest or their love for the religion which they followed previously, or their deficient understanding when reasoning about something, so they miss many beneficial purposes and they neglect what the religion had explicitly assigned, or they interpolate into it something extrinsic, and the religion becomes corrupted as occurred in many religions before us. Then, since a thorough investigation in recognizing the avenues of the interpolation of unsoundness is not possible, for they are limitless and undetermined—and what cannot be completely achieved cannot be totally abandoned—he must warn them most strongly of the causes of distortion in a comprehensive way. He must single out the issues which he surmises will contain or cause laxity or distortion, or which will provoke a permanent malady among humans. He must block the entrance of corruption into them in the most complete way, and he must legislate a thing which is contrary to the familiar ways of the corrupted religions in whichever elements are most prominent among them, such as the prayers.

Among the causes of distortion is laxity, the true nature of which is that after the disciples comes a generation which neglects the prayers until they are lost and follows their lusts,[1] not giving importance to the spreading of the religion by studying, teaching, and acting; and they neither command the good nor do they forbid the evil. Thus there will coalesce, before long, customs which oppose the religion, and the desire of people's physical natures will

[1] Refers to Qur'ān 19:59, where after mentioning the prophetic missions the Qur'ān says, "There came a later generation which had lost their prayers and followed their lusts."

conflict with the demand of the divine laws. Then other succeeding generations will follow, increasingly lax and careless, until they forget most of the religious knowledge, and careless neglect on the part of the leaders and the great ones of the nation is more harmful to them and further undermines them. This is the reason that the religions of Noah and Abraham disappeared, for scarcely anyone from them could be found who knew the basis of their religion correctly.

The origin of careless neglect lies in several things. Among them is the absence of transmitting from the founder of the religion and acting upon this, and this is the Prophet's saying, "It won't be long before a man with a full stomach upon his couch talks to you about this Qur'ān, saying 'what you find permitted (in the Qur'ān) allow, and what you find forbidden in it, forbid.' But what the Prophet of God forbade is what God forbade."[2] And his saying, "Indeed God does not take away knowledge, per se, by removing it from the people, but He takes away knowledge by taking away the people of knowledge, until, when no learned man remains the people will take ignorant men as leaders, and will ask them (about various cases) and they will give legal opinions without knowledge (of the religion and sharīʿa), thus they will go astray and will lead others astray."[3]

Among the sources of distortions are corrupt intentions leading to false exegesis such as when people try to please the kings' (wishes) to follow their own desires, according to God's, may He be exalted, saying, "Those who conceal what God revealed in the scripture and purchase a small gain with this, are those who will eat nothing in their bellies except fire."[4]

Among the sources of distortion is the spread of evil things and the abandoning, on the part of the ulema, of prohibiting them, and this is God's saying, may He be exalted, "If only there had been in the generations before you people of excellence forbidding corruption on the earth, as did a few of those whom We saved. The wrongdoers followed that which made them softened by pleasure, and they were criminals."[5]

---

[2] *Mishkāt* p. 43–44. Abū Dāwūd IV:200 #4604 Sunna 5, Ibn Mājah, Ibn Ḥanbal I:291. A warning about following the Qur'ān only and not the Sunna.

[3] *Mishkāt* p. 51. Bukhārī ʿIlm 34, Muslim, Tirmidhī, Ibn Mājah, Dārimī, Ibn Ḥanbal. Khan trans. I:80.

[4] Qur'ān 2:174.

[5] Qur'ān 11:116.

This is (also demonstrated by) what the Prophet, may the peace and blessings of God be upon him, said about the Children of Israel falling into acts of disobedience, "Their religious scholars forbade them these things, but they did not abstain from them, then they (the religious scholars) sat in their meetings and they ate in their company and they drank with them, so that God made their hearts beat in the same rhythm, and cursed them on the tongues of David and Jesus, son of Mary; and this was due to their disobedience and transgression."[6]

Among the causes of distortion is unnecessary hair-splitting, and the nature of this is that the law-giver orders something and forbids something else, so that a man from his religion hears this and understands this as his mind is able, so that he overextends the ruling to things which resemble the (original) thing only in certain aspects or in which certain parts of the reason for the legislation (*'illa*)[7] are found, or to portions of the thing, or to some of its anticipated sources or its motives. Whenever the matter becomes ambiguous for this person due to a conflict in the hadith reports, he sticks to the most stringent and makes it obligatory, and he takes everything which the Prophet, may the peace and blessings of God be upon him, used to do, as an act of worship, while the truth is that the Prophet used to do things according to the current custom. Thus he thinks that the command and the forbidding include these things so he publicly proclaims that God, may He be exalted, commanded a certain thing, and forbade some other thing. For example when the law-giver legislated fasting to subdue the lower soul[8] and forbade sexual intercourse during it, one group thought that eating before dawn was against the law because it is incompatible with subduing the lower soul, and they thought that the fasting person was forbidden to kiss his wife because this is one of the inducements to sexual intercourse, and because it is similar to intercourse in satisfying lust, and therefore the Prophet revealed the error of these statements and explained that they were distortions.

Among the causes of distortion is being excessively strict, and the true nature of this is choosing austerities in worship which the

---

[6] Ibn Ḥanbal I:391.

[7] This is a criticism of deriving wrong conclusions through analogy (*qiyās*).

[8] Islamic fasting is carried out during the daylight hours.

law-giver did not command such as continual fasting, praying the whole night long, retiring from the world, remaining celibate, and making the recommended practices and proper manners just as compulsory as the obligatory things. This is referred to in a hadith report of the Prophet, may the peace and blessings of God be upon him, that he forbade ʿAbd Allāh Ibn ʿUmar and ʿUthmān Ibn Maẓʿūn from the arduous practices which they had in mind,[9] and this is his saying, "No one makes religion too strict but that it overwhelms him."[10] Thus, if this hair-splitter or strict man becomes the teacher and the leader of a group they will think that this is the command of the divine law and pleasing to God, and this is the ailment of the ascetics among the Jews and the Christian monks.

Among the causes of distortion is "preference" (*istiḥsān*).[11] The true nature of this that a man might see that the law-giver fixed for every rule an anticipated source related to it, and he knows that the law-giver fixed the legislation, but he appropriates certain of the inner meanings of the legislation which we have previously mentioned, so that he legislates for the people according to what he thinks is the beneficial purpose. An example of this is that the Jews believed that the Law-Giver only commanded the punishments as a deterrent from disobeying and to reform them, and they thought that stoning to death would result in altercations and fighting, so that in it there would be a greater evil, and therefore they applied (the principle of) "preference" in sanctioning blackening the face and skin (of the adulterer with coal) instead of stoning. The Prophet, may the peace and blessings of God be upon him, explained that this was a distortion and a discarding of the order of God laid down in writing in the Torah in favor of their own opinions.[12] Ibn Sīrīn[13] said, "The first one to apply analogies was Iblīs (the devil) and the worship of the moon and sun only came about through

---

[9] For example, *Mishkāt*, p. 145–146. Where ʿUthmān ibn Maʿẓūn asked permission to castrate himself and not marry. Bukhārī Nikāḥ 8, Khan trans. VII:8, Ibn Ḥanbal I:175.

[10] Bukhārī Īmān 29, Nasāʾī, Ibn Ḥanbal.

[11] *Istiḥsān* is a legitimate deviation from *qiyās* in which a discretionary interpretation is used more freely with the goal being a ruling for the public interest. Imām Mālik called it, "*maṣlaḥa mursala*".'

[12] As in the hadith reported in *Mishkāt*, p. 758. Bukhārī Tauḥīd 51, Khan trans. IX:476. Dārimī, Muslim, Ibn Mājah, Ibn Ḥanbal.

[13] Ibn Sīrīn (d.728), the first Muslim interpreter of dreams and a respected transmitter of hadith. He is reported to have said that the cunning questions which he was asked on hadith were prompted by the devils.

analogy."[14] There is a report about Ḥasan (al-Baṣrī) that he recited this verse of the Qur'ān, "You created me from fire and You created him from earth,"[15] saying that Iblīs used an analogy and he was the first to use an analogy.[16] A report from Al-Sha'bī is "By God, if you take up the use of analogies, you will forbid the permitted; and permit what is forbidden."[17] Mu'ādh ibn Jabal said,

> The Qur'ān will be opened to the people so that woman, and child, and man will read it. Then the man will say, "I have read the Qur'ān, but I am not followed, by God I will implement it among them so that perhaps I will be followed." So he implemented it among them but they still did not follow him. So he said, "I will build a mosque in my house so that they will follow me." So he built a mosque in his house, but was not obeyed; so he said, "I have read the Qur'ān and was not followed, and I implemented it among them, and I was not followed, and I have built a mosque in my house and I was not followed, so now, by God, I will provide them with a discourse which they will not find in the book of God, nor have they heard it from the Prophet of God, may peace be upon him, so that perhaps I will be followed."
>
> Mu'ādh said, "O, you people, beware of what this person relates, for what he relates is misguidance."[18]

It is reported of 'Umar, may God be satisfied with him, that he said, "The lapsing of the learned man, and the arguing of the hypocrite; using the book of God, and the command of the leaders who have gone astray, will destroy Islam."[19] What is intended by all of these examples are those things not derived from the book of God and the practice of His Prophet.

Among the causes of distortion is following the consensus (*ijmā'*), and the true nature of this is that a group of the bearers of the religion, whom the masses believe hold the correct opinion most of the time or always, agree; so the people think that this is a conclusive proof of the confirmation of the ruling, while this is a consensus on a matter which has no basis in the Qur'ān or the sunna. This is not the consensus which the community agreed on

---

[14] This and several subsequent references are to al-Dārimī I:63–67 Muqaddima, in the chapter on "The Changing of the Age and What Changes with It."
[15] Qur'ān 7:12.
[16] Dārimī I:65.
[17] Dārimī I:65.
[18] Dārimī I:67.
[19] Dārimī I:71.

(*ijmāʿ al-umma*) for they agreed to call something "a consensus" which was founded on the Qur'ān or the sunna or was directly derived from one of them, and they did not allow something to be called "consensus" which was not based on one of these things. This is God's, may He be exalted, saying, "And if it is said to them, believe in what Allah has revealed, they will say, rather we follow the way upon which we found our forefathers."[20]

The Jews only persisted in denying the prophethood of Jesus and Muḥammad, may peace and blessings be upon them, because their ancestors had tested their life stories and did not find them fulfilling the conditions of prophethood, and the Christians have many divine laws which oppose both the Torah and the Gospel which they have no cause to hold onto except the consensus of their ancestors.

Among the causes of distortion is conforming to the legal decisions (*taqlīd*) of someone who is not entirely infallible, by whom I mean someone other than the Prophet whose infallibility is established. The essence of this is that when one of the learned scholars of the religion uses independent reasoning (*ijtihād*) in a problem, his followers may think that he hits the right answer assuredly or in most cases, so that due to it (his independent reasoning) they reject a sound hadith. This reliance on following (*taqlīd*)[21] is different from that which the blessed community agreed on, for they only agreed on the permissibility of following (*taqlīd*) the experts in independent reasoning[22] (*mujtahidūn*), knowing that the *mujtahid* may err or may be correct, and searching for the text of the report of the Prophet in the issue, with the intent that if a reliable hadith comes to the fore which opposes what is followed, then conforming with (the reasoner) should be abandoned and the hadith should be followed. The Prophet said in the interpretation of God's saying, may He be exalted, "They took their rabbis (*aḥbār*) and monks

[20] Qur'ān 2:170.

[21] Shāh Walī Allāh opposes blind or strict *taqlīd* which occurs when the followers of a school or teacher prefer his reasoning to a hadith. He however, believes strongly in "free *taqlīd*" which allows a person to choose the best opinion from among the four schools of Islamic law. 'Abdel 'Aal. Khalīl 'Abdel Ḥamīd, "God, the Universe. and Man in Islamic Thought: The Contribution of Shāh Walīullah of Delhi (1702–1762)" Ph. D. Dissertation. University of London, 1970, based on *al-Tafhīmāt al-Ilāhiyya*, 1:208–210 (no. 66).

[22] Shāh Walī Allāh holds that "affiliated *ijtihād*" or "relative *ijtihād*" based on an informed individual reasoning in choosing among the opinions of the four schools is allowable, but that "free *ijtihād*" disappeared forever in the fourth Hijra century.

as lords besides God,"[23] that it was not that they used to worship them, but rather if they said that a thing was permitted to people, then they would take it as such, and that if they forbade them something, then they would take it to be forbidden.

Among them is the intermingling of one religion with another religion until one is not distinguishable from the other. This occurs because when a person previously professes another religion, he retains an emotional attachment to the branches of knowledge of this group. Then when he joins the religion of Islam the inclination of his heart to what it had previously been attached remains, and consequently he will seek a device (to make room for those things) in this religion (Islam), even if this is weak or fabricated. Sometimes he justifies fabrication or transmitting the fabricated hadith because of this and this is the Prophet's saying, "The situation of the Children of Israel was balanced until there arose among them those of mixed descent and the children of prisoners of (other) nations so they spoke on the basis of their own opinions and went astray and led (others) astray."[24] Among those things which have become interpolated into our religion is the lore of the Children of Israel,[25] the exhortations of the orators of the Ignorant Age, the science of the Greeks, the religious propaganda of the Babylonians, the history of the Persians, astrology, geomancy, and rationalist theology (*kalām*), and this is the ultimate cause of the anger of the Prophet, may the peace and blessings of God be upon him, when there was read before him a copy of the Torah,"[26] and the ultimate cause of 'Umar's, may God be pleased with him, beating the man who was trying to get the books of Daniel,[27] and God knows better.

---

[23] Qur'ān 9:31.

[24] Ibn Mājah I Muqaddima 8. *Mishkāt* p. 44.

[25] It is an historical fact that Judeo-Christian religious lore had begun to find its way into Islam at a very early date chiefly through the activity of popular preachers (*quṣṣās*) who wanted to make their sermons as effective as possible. This movement was criticized by certain early traditions and sayings. There is, for example, a tradition that 'Umar once advocated the acceptance of certain Jewish traditions but was sternly forbidden by the Prophet to do so." Fazlur Rahman, *Islamic Methodology in History*, (Karachi: Central Institute of Islamic Research, 1965), 49. This hadith is cited in *Mishkāt*, p. 49.

[26] Dārimī p. 115.

[27] "When 'Umar was informed about a man who copied (or read) the Book of Daniel, he ordered that man to be brought into his presence, and beat him with his whip until he promised to burn books of this kind and not to read them." In al-Khaṭīb al-Baghdādī, *Taqyīd al-'Ilm*, ed. Youssef Eche (Institut Francais de Damas, 1949), 51. al-Muttaqī al-Hindī I: 332–333 #1626, 335–336 #1633. Discussed in M.J. Kister, "Ḥaddithū 'an Banī isrā'īla wa-lā ḥaraja" in *Israel Oriental Studies* (II, 1972): 215–239.

CHAPTER 72

The Causes for the Differing of the Religion of Our Prophet, May Peace Be upon Him, From the Religion of the Jews and Christians

You should know that when God (*al-Ḥaqq*), may He be exalted, sends a prophet to a group, and establishes the religion for them through him so that he doesn't leave in the religion any "crookedness or distortion."[1] Then reports from him are put into practice, and the disciples from his community transmit them properly for a period of time, then after that there succeeds a generation which distorts them, and is careless about them, so that they are not pure truth but become coupled with falsehood and this is the Prophet's saying, "There was no prophet sent by God who did not have disciples and companions from his community who took up his practice; and obeyed his command; then generations succeed them who say one thing but do another, and who do what they are not commanded to do."[2]

From this falsehood comes patent polytheism and obvious distortion for which they are held accountable in every situation. Also from this come covert polytheism (*ishrāk khafī*) and concealed distortion which God does not hold them accountable for until he sends another prophet to them, who sets up the proof and reveals the obscure, so that, he who survived might survive by a clear proof and he who perished might perish by a clear proof.[3]

When a Prophet is sent to them he takes everything back to its origin, and takes into consideration the previous divine laws of the religion. Thus, whatever are religious rituals and emblems of God, unadulterated with associationism, and practices of worship or means of supporting the civilization consistent with the regulations of the

---

[1] Qur'ān 20:107.

[2] The rest of the hadith is, "Whoever fights them with his hand is a believer and whoever fights them with his heart is a believer; and there is not beyond this even a mustard seed of faith." *Mishkāt*, p. 42. Muslim Īmān 80, Ibn Ḥanbal.

[3] An echo of Qur'ān 8:42.

religion, he allows to remain; and he emphasizes what had lapsed among them, and sets out for everything pillars and causes. Whatever had arisen due to distortion and carelessness he nullifies, and demonstrates that this not part of the religion (*dīn*). He changes whatever rulings are conditional upon the presumed sources of the beneficial purposes of a certain time (*maẓānn al-maṣāliḥ*) once those anticipated sources have diverged due to variations in customs, since the essential goal in the legislation of the rulings is the beneficial purposes (*maṣāliḥ*), and they (people) are concerned with the anticipated sources for their occurrence (*maẓānn*). Sometimes a thing is a source or an anticipated source of a beneficial purpose, but then it ceases to be such a source for it.

For example the cause (*ʿilla*) of a fever is basically the agitation of the humors, then the doctor attributes the fever to the source which he anticipates (*maẓinna*) to be its cause, such as walking in the sun, exhausting activity, and eating a certain food, and it is possible that the anticipated source of these things will cease to remain in force.

Thus the judgments (*aḥkām*) must change in accordance with this principle. The Prophet builds on whatever consensus the Highest Council has resolved upon with respect to human practice, habitual behaviors, and whatever people have based their sciences upon, and totally internalized.

The prophets before the time of our Prophet used to increase things and not decrease them, and only changed things slightly. Abraham, may peace be upon him, added to the religion (*milla*) of Noah things such as acts of worship, natural acts (*aʿmāl al-fiṭra*)[4] and circumcision; and Moses, may peace be upon him, added to the religion of Abraham things such as the forbidding of camel's meat, the necessity of honoring the Sabbath, stoning adulterers, and others. Our Prophet, may peace be upon him, added to these, subtracted from them, and made changes in them.

The one who examines the fine points of the divine law, if he scrutinizes these matters, (of change, addition or subtraction), will find that they have various aspects. Among them is that the Jewish religion was transmitted by the rabbis and the ascetics, and they distorted it in the well-known ways mentioned previously, so that when the Prophet (Muḥammad) came, he restored everything to

---

[4] Practices which Abraham legislated as conforming to the sound natural constitution.

its original state, and his divine law differed in relationship to the Jewish one which they had at that time, so they said that this was adding, taking away and changing, but there was no change in reality.

Among them is that the Prophet, may the peace and blessings of God be upon him, was sent on a mission which included another mission. The first was that to the sons of Ishmael and this is God's, may He be exalted, saying, "He who was sent to the unlettered ones as a messenger from among them,"[5] and His, may He be exalted, saying, "that you warn a nation whose fathers were not warned so they are heedless."[6]

It was required that the material of the divine law of this mission be based on whatever they already possessed in terms of emblems, practices of worship, and aspects of the supports of civilization; since the legislation was only meant to reform what they were practicing, not to oblige them to do things of which they knew nothing at all. An example of this is God's, may He be exalted, saying, "An Arabic Qur'ān in order that you might understand."[7] And His, may He be exalted, saying, "If we had made it a Qur'ān in a foreign language they would have said, 'Why have its verses not been well set out? What? Non-Arab and Arab?'"[8] and His, may He be exalted, saying, "We never send a messenger except with the language of his people."[9]

The second mission of the Prophet was to the entirety of the people of the earth in order to bring the fourth stage of supports of civilization, and this is because in the time of the Prophet, God had cursed several nations and decreed the extinction of their rule such as Persia and Byzantium, so He ordered the establishment of the Fourth Irtifāq and made his (the Prophet's) honor and his victory a way of arriving at the fulfillment of the desired command, and He gave him the keys of their treasuries. Thus he received because of this consummation other rules besides the rules of the Torah, such as the *kharāj* tax and the *jizya*, rules for the Jihād, and precautions against the penetrations of distortions.

Among the causes of difference are that he was sent after a great

---

[5] Qur'ān 62:2.
[6] Qur'ān 36:6.
[7] Qur'ān 12:2.
[8] Qur'ān 41:44.
[9] Qur'ān 14:4.

gap of time[10] during which the true religions had become effaced and distorted, and the people had become overcome by prejudice and obstinacy, so that they wouldn't abandon their false religion nor their customs from the Ignorant Age, except through a profound emphasis on opposing these customs, and this occasioned many of the differences.

[10] Refers to Qur'ān 5:19, "O People of the Book, Our messenger has come to you who clarifies things to you after a great gap (of cessation) of the messengers."

CHAPTER 73

The Causes of Abrogation

The basis of this is God's, may He be exalted, saying, "We do not abrogate any verse or cause it to be forgotten unless we bring one better than it."[1]

Know that abrogation is composed of two categories.

1) One of them is that whereby the Prophet, may the peace and blessings of God be upon him, examined the supports of civilization or the aspects of acts of obedience, and then fixed them with different types of regulation consistent with the ordinances of the divine legislation, and this is the independent reasoning (*ijtihād*) of the Prophet. Therefore God did not determine this for him, but rather revealed to him what He had decreed upon the ruling in this issue, either

a) through a Qur'ānic revelation in agreement with it or

b) by a change in his independent reasoning in this, and his decision on this.

An example of the first case is the Prophet's, may the peace and blessings of God be upon him, command to face Jerusalem (during the prayers); the Qur'ān was later revealed abrogating this.[2] An example of the second is that he had forbidden the people from making *nabīdh*[3] except in animal skins, then he allowed them to make *nabīdh* in any container saying, "Don't drink anything intoxicating."[4] This is because he saw that the intoxicating property was a hidden thing, and he established for it an overt anticipated source which was the making of *nabīdh* in vessels which were not porous such as ones made of clay, wood, or gourds, for these hastened the intoxicating properties of the *nabīdh* made in them. Therefore he established making *nabīdh* in a water skin as an anticipated source of something in which intoxicating properties would not be present

---

[1] Qur'ān 2:106.

[2] Qur'ān 2:142–145.

[3] *Nabīdh* is a drink, allowed to remain and ferment from dates and barley.

[4] Muslim Janā'iz 106, āḍāḥī 37, ashraba 63, Abū Dāwūd, Nasā'ī, Dārimī, Ibn Ḥanbal.

for up to three days. Then he changed his independent reasoning to base the ruling on the intoxicating property because the intoxicating properties of something can be recognized by its bubbling and emitting froth. Therefore he established whatever had in it the properties of intoxication, or the attributes of an intoxicating thing, as being a more appropriate anticipated source than assigning something which was extraneous to it. According to another interpretation, we may say that the Prophet (at first) saw that the people were enamored of intoxicants, and that if he were to forbid these to them, there would be a loophole for someone to drink it (an intoxicating thing) and make the excuse that he had thought that it was not intoxicating or that the indications of its being an intoxicant were not clear to him, or that the vessels had become contaminated with some intoxicating thing, so that the intoxicating quality emerged more rapidly in whatever was brewed in ones like them. Then when Islam became strong and they were content to abandon intoxicants, and those vessels had gone out of usage, the order was made to apply to the intoxication itself, according to this interpretation. This is an example of the varying of the ruling according to the variation of the anticipated sources (*maẓinnāt*). Concerning this division of abrogation there is the Prophet's saying, "My speech does not abrogate the speech of God; but the speech of God abrogates my speech, and the speech of God can abrogate a previously delivered speech of God."[5]

2) The second category of abrogation is that whenever a thing is thought to be conducive to good or to harm, it is ruled on in accordance therewith, then a time comes in which it is no longer an anticipated source for this and thus the rule will be changed. An example of this is that when the Prophet, may the peace and blessings of God be upon him, emigrated to Medina and help for them from their blood relatives was cut off, and there only existed brotherhood aid,[6] which the Prophet made a means for the necessary beneficial purpose—the Qur'ān was revealed basing inheritance on a relationship of "brotherhood." God explained the benefit of this when He said, "If you don't do it there will be sedition on

[5] Muslim, Ḥaid 82

[6] The Prophet created a special relationship of "brotherhood (*ikhā'*) between the immigrants from Mecca (*muhājirūn*) and the Medinans (*Anṣār*) whereby each Anṣārī gave half of his property to a Meccan (*Muhājir*).

the earth and a great evil."[7] Then when Islam became strong, and the immigrants were reunited with those related to them, the command was revoked in favor of the previous inheritance by kinship. Or it may be that a thing did not have a beneficial purpose in the prophetic mission which had not included the caliphate as was the condition before the Prophet, may the peace and blessings of God be upon him, and as was the case in his time before the emigration to Medina; although the beneficial purpose came to be found in the Prophetic mission including the Caliphate. An example is that God, may He be exalted, did not make taking booty lawful to those before us, but permitted it to us, and this is justified in the hadith reports from two aspects: the first is that God saw our weakness and permitted us booty, and the second is due to God's esteeming our Prophet above the rest of the prophets and his community above other communities.

The explanation of these two aspects is that prophets before the Prophet, may the peace and blessings of God be upon him, were sent to their particular nations, and these people were of limited numbers. Sometimes the duration of the Jihād among them was a year or two and so on, and their peoples were strong and able to combine the Jihād with trades such as agriculture and commerce so that they did not have a need for booty. God, may He be exalted, wanted that a worldly intention should not to be mixed with their actions so that the people would get the fullest rewards for them. He sent our Prophet, however, to all people, and they are uncountable and the duration of the Jihād against them is not limited. The Muslims were not able to combine the Jihād with occupations such as agriculture and trade, so they had a need for the permission to take booty. The Prophet's community, due to the universality of his call, included people of weak intentions, and concerning them is reported the hadith, "Indeed God will support this religion even by a profligate man."[8] These people would only fight for a worldly goal, and God's mercy in the matter of the Jihād encompassed them mightily, and God's anger was directed against their enemies most strongly. This is the Prophet's saying, "God looked at the people of the earth, and loathed the Arabs and the non-Arabs among them."[9] This required the termination of the

---

[7] Qur'ān 8:73.
[8] Bukhārī Jihād 182, Qadar 5, Maghāzī 38, Muslim, Dārimī, Ibn Ḥanbal.
[9] Muslim Jannat 63, Ibn Ḥanbal.

safeguarding of their property and lives in the most complete way, and the galling of their hearts by the disposal of their wealth as when the Prophet of God bestowed upon the Sacred Precincts (to be sacrificed) the camel of Abū Jahl[10] with a silver ring in its nose, so that this galled the unbelievers,[11] and when he ordered the cutting down of the date palms and burning them, this galled their owners.[12] Therefore the Qur'ān revealed the permissibility of booty for this community.

Another example is that fighting the unbelievers was not forbidden to this community at first, but at that time there was no army, neither was there a Caliphate. Then when the Prophet emigrated, and the Muslims collected and the Caliphate[13] appeared, and the Jihād with the enemies of God became possible, God, may He be exalted, revealed, "Permission is given to those who fought because they have been wronged; and Allah is able to give them victory."[14] About this division (of abrogation) is God's, may He be exalted, saying, "We do not abrogate any verse or make it forgotten, but that we bring a better one or one like it."[15] His saying, "A better one" refers to the Prophetic mission containing the Caliphate and His saying, "or one like it," refers to the changing of the ruling according to the variation of the anticipated sources, and God knows better.

---

[10] Abū Jahl was the Prophet's uncle and one of his leading opponents. This incident of the sacrifice of a camel occurred after the battle of al-Ḥudaybiyya.

[11] The hadith may be found in Abū Dāwūd Manāsik II:145. Ibn Ḥanbal.

[12] This refers to the cutting down and burning of palm trees of the Jewish tribe, the Banī Naḍir, who were accused of treachery and later deported from Medina. It is referred to in Qur'ān 59:5.

[13] In the sense of the political rule by the Prophet.

[14] Qur'ān 22:39. The verb "are Fought" (*yuqātalūna*) has been read "fight" (*yuqātilūna*) by some commentators.

[15] Qur'ān 2:106.

CHAPTER 74

The Explanation of what had been the Condition of the People of the Jāhiliyya which the Prophet Reformed

Should you want to investigate the meanings of the divine law of the Prophet of Allah, may the peace and blessings of God be upon him, then you must first examine the state of those unlettered ones to whom he was sent, which formed the material object of his legislation; and secondly, his manner of reforming their condition through the goals mentioned under the topics of legislation, facilitation, and the rulings of the religion.

Thus you should know that he, may the peace and blessings of God be upon him, was sent with the Ḥanīfī religion of Ishmael[1] to straighten their crookedness and bring their distortion to an end, and to spread their light, and this is His, may He be exalted, saying "The religion (*milla*) of your father Abraham."[2]

Since this was the state of affairs it was necessary that the principles of that religion were accepted and its practice affirmed; for since the Prophet had been sent to a people in which a remnant of the rightly-guided practice still remained, there wasn't any sense in changing this and altering it. Rather it was necessary to affirm it since it would be easier for them, and proven over against their objection. The Children of Ishmael had inherited the codes of their father Ishmael, and remained following that divine law until the time of ʿAmr ibn Luḥayy,[3] for he interpolated things into it according to his worthless opinion and thus was led astray, and led others astray. Part of his legislation was the worshipping of idols, allowing a bull to wander at will, and setting free a she-camel, so religion was thereby falsified, and the sound was mixed with the

[1] That is, the Arabs followed the tradition of Abraham through his son Ishmael.

[2] Qur'ān 22:78.

[3] ʿAmr ibn Luḥayy was considered to be the founder of polytheism in Arabia who changed the Abrahamic religion by introducing idols. It is said that he set animals free in their honor. This practice is denounced in Qur'ān 5:103. Bukhārī Manāqib 9, Muslim, Nasāʾī.

corrupt, and ignorance, polytheism and unbelief overcame them. Therefore, God sent our master Muḥammad, may the peace and blessings of God be upon him, to straighten their deviations and to reform their corruptions. Therefore he, may the peace and blessings of God be upon him, examined their divine law and whatever in it agreed with the codes of Ishmael, may peace be upon him, or was from the rituals of God, he retained. Whatever in it was distorted or corrupted or adopted the emblems of polytheism and unbelief he nullified and recorded its nullification. When a thing fell under the topic of customary behaviors, and so on, he explained its proper manners and reprehensible aspects in such a way as to avoid the disasters of conventions, and he forbade the corrupt customs and commanded the sound ones. Whatever issue of principle or practice had been abandoned during the interval, he reinstated as fresh and as lush as it had been before, so that through this the blessing of God was fulfilled and His religion was made straight. The people of the Jāhiliyya at the time of the Prophet, may the peace and blessings of God be upon him, used to accept the possibility of the mission of the prophets, belief in requital, believed in the principles of the types of piety, and put into practice the second and third stages of the supports of civilization.

What we have said is not contradicted by the existence of two groups among them, their emergence, and the fact that they were prominent and wide-spread.

1) One of them were the profligates (*fussāq*) and the heretics in religion (*zanādiqa*). The profligates carried out bestial or vicious acts in contravention of the religion due to their being overwhelmed by their lower souls and due to the paucity of religiousness among them, so that they transgressed the decree of the religion and thus testified to their own depravity. The heretics were naturally disposed to be defective in understanding, neither fully able to ascertain what the founder of the religion intended, nor to imitate him. They didn't accept the information he gave, "so that in their doubt they wavered,"[4] in fear of their chiefs, and people repudiated them and held them to be outside of the faith (*dīn*), having discarded religion. Since the matter was as we mentioned concerning their being repudiated and considered despicable, their apostasy did no harm.

---

[4] Qur'ān 9:45.

2) The second group were the ignorant and neglectful ones who did not even pay attention to religion at all, nor turn even a single glance towards it, and this type of person was mostly found among the Quraish[5] and those close to them due to the remoteness of their era from the time of the prophets, and this is His saying, may He be exalted, "That you might warn a people to whom a warner has not previously come."[6] Yet they were not so completely remote from that path that a proof could not be confirmed for them or that a compelling argument could not be directed to them, or that they could not be defeated (in a debate).

Among the principles agreed upon among them (the people of the Ignorant Age) was the belief that God, may He be exalted, had no partner in the creation of the heavens and the earth and the substances in them, and that He had no partner in managing the great affairs and that no one could reject His order nor frustrate His decree once it had become settled and decided, and this is His saying, may He be exalted, "If you asked them who created the heavens and the earth they would answer Allah,"[7] and His saying, "No; upon Him you will call,"[8] and His saying, "All upon whom you call for help lose their way except Allah."[9] But it was due to their deviance in religion that they held that there were personages among the angels and the spirits who could manage (the affairs of) the people of the earth except for the most major matters, by improving the status of a person who worshipped them in matters which involved his personal affairs, his children, and his wealth. They compared them to the situation of the kings in relation to the king of kings and to the situation of the intercessors and courtiers in relation to the Sultan administering the power. What gave rise to this were the pronouncements of the divine laws concerning the entrusting of affairs to the angels, and the answering of the prayers of those people who are closest (to God), so they supposed that this was an administration (of power) on their part like

---

[5] Those of this tribe at the time of the coming of Islam who did not accept the Prophet.

[6] Qur'ān 32:3.

[7] Qur'ān 31:25.

[8] Qur'ān 6:40–41. "If God's chastisement comes upon you, will you call upon any other than God, if you speak truly? No; upon Him you will call, and He will remove that for which you call upon Him if He will, and you will forget whatever partners you associated with him."

[9] Qur'ān 17:67. That is, these others fail in times of crisis or disaster.

the administration of kings, by analogy of the unseen to the visible world, and this was false.

Among (the accepted principles of the pre-Islamic world) was God's transcendence of what is not appropriate for His exalted state and the forbidding of heresy concerning His names, but among their heresies was their claim that God took the angels as daughters,[10] and that the angels were made intermediaries so that God could acquire from them knowledge which He did not have, in analogy to the king and his spies.

Among their beliefs was that God decreed all events before they take place, and this is the saying of Ḥasan al-Basrī, that the people of the Jāhiliyya continued to mention predestination in their speeches and poetry and the divine law only added a confirmation to this.

Among the Jāhiliyya beliefs was that there is a place at which the predestination of events is determined one after another, and that there is an effect at this plane of some sort due to the supplications of the nearest angels and the best humans. This then became represented in their minds as something resembling the intercession of the courtiers with kings.

Among their beliefs was that God obligated His worshippers with whatever He willed and thus He permitted and forbade, and that He requited the actions good for good, and bad for bad. They believed that God has angels closest to His presence who are the chief ones in His kingdom and that they manage the world by the permission and command of God and they, "do not disobey God in what He commands them and they do what they are ordered."[11] They believed that the angels do not eat, drink, defecate, or get married, and that they may appear to the most favored people and give them good tidings and warn them. They believed that God might send to His worshippers, in His grace and kindness, a man from among them, then He sent His revelation to him, and the angels descended to him. Thus they believed that He had made obedience to him incumbent upon them, and that they would find no way to avoid this, nor any escape from it. The mention of the Bearers of the Throne (of God) and the Highest Council is common in the Jāhiliyya poetry and it is reported from Ibn 'Abbās,

---

[10] The Jāhiliyya Arabs worshipped, in particular, three of these daughters of God, Al-Lāt, Al-'Uzzā, and Manāt, mentioned in Qur'ān 53:19–20.

[11] Qur'ān 66:6.

may God be pleased with him, that the Prophet, may the peace and blessings of God be upon him, said that Umayya ibn Abī aṣ-Ṣalt was correct in two verses of his poem, for he said:

> A man and a bull are by its right foot,
> And an eagle and a watchful lion are by the other one.[12]

And the Prophet, may the peace and blessings of God be upon him, said, that he was correct when he said,

> "The sun rises at the end of every night
> Her color becomes red, like a rose
> She refuses and doesn't rise in her own sweet time
> But because she has been punished and whipped.[13].
>
> for he said, "He spoke truly."[14]

The precise explanation of this is the people of the Jāhiliyya used to claim that the Bearers of the Throne were four angels: one in the form of a man, who was the intercessor for the descendants of Adam with God; the second in the form of a bull, which was the intercessor for the animals; the third in the form of an eagle, which was the intercessor for the birds; and the fourth in the form of the lion, which was the intercessor for the wild beasts. The divine law reported something close to this except that it called all of them "mountain goats,"[15] and this is according to their form as manifested in the World of Images. All of this was known among them despite what had entered into it in terms of drawing an analogy from the visible world to the Unseen and mixing the familiar (cultural norms) with matters of theology. If you doubt what we have said, then look at what God, may He be exalted, related in the great Qur'ān. He argued using the knowledge that remained with them, and disclosed the ambiguities and doubtful things which they had brought into it, especially in His saying, may He be exalted,

---

[12] *Dīwān Umayya Ibn Abī al-Ṣalt*, ed., ʿAbd al-Hafīẓ Asatli (Damascus: Maṭbaʿa al-taʿāwuniyyia bi-Dimāshq, 1974), 365. These angels who bear the Throne correspond to the fixed astrological signs Aquarius, Taurus, Scorpio, and Leo. al-Dārimī II, 296.

[13] *Dīwān Umayya Ibn Abī al-Ṣalt*, p. 366. That is, the Sun rises by the power of its Creator.

[14] al-Dārimī II, 296. Ibn al-Ḥanbal I, 256.

[15] As in the Qur'ānic verse 69:17, "bearing the Throne of your Lord on that day there will be eight above them." The reference to mountain goats is found in a hadith cited by Ibn Mājah Muqaddima 13, Abū Dāwūd, and Ibn Ḥanbal.

when they denied the revelation of the Qur'ān, "Say, who revealed the book which Moses brought?"[16] When the people said, "What kind of Prophet is this who eats food and walks in the markets?"[17] He revealed His saying, "Say (O Muḥammad), 'I am no novelty among the prophets,'"[18] and so forth.

So you are informed from this that the polytheists, although they were far from the straight path, still were in a position which allowed the proof to be established for them through what remained of the knowledge which they had possessed. Look at the speeches of their wise men—such as Quss ibn Sā'ida[19] and Zaid ibn 'Amr ibn Nufail,[20] and at the reports of those before 'Amr ibn Luḥayy, and you will find this there in detail. Indeed, if you investigate their reports with close scrutiny you will find that their great and wise men used to believe in the next life, the guardian angels, and other things, and that they affirmed the unity of God (*tauḥīd*) in a sense, so that Zaid ibn 'Amr ibn Nufail said in his poetry

> Your servants err and You are the Lord,
> In your palms are the destinies and the fates.

He said also

> Shall I worship one Lord or one thousand lords,
> When functions can be divided up?
> I have abandoned both Al-Lāt and Al-'Uzzā[21]
> That is what the perceptive man does."

The Prophet, may the peace and blessings of God be upon him, said about Umayya ibn Abī Aṣ-Ṣalt, "His poetry believes but his heart does not;."[22] These things are among what they inherited from the codes of Ishmael and what had penetrated among them from the People of the Book, and it was recognized among them that human perfection consisted of bowing one's head before the Lord and worshipping Him with one's fullest efforts.

(They also knew) that among the divisions of worship is purity, and taking a full bath after a major ritual impurity continued to be

---

[16] Qur'ān 6:91.
[17] Qur'ān 25:7.
[18] Qur'ān 46:9. That is, the other prophets had also done these things.
[19] A pre-Islamic Arabian poet who used to pray.
[20] The archbishop of Najrān.
[21] Two of the pre-Islamic goddesses.
[22] Ibn 'Asākir, *Ta'rīkh Dimāsqh*. A weak hadith transmitted by Ibn 'Abbās.

a usual practice among them and likewise circumcision and the other natural virtues. It is said in the Torah that God, may He be exalted, made circumcision a brand of Abraham and his descendants. The ablution (*wuḍū'*) was performed by the Magians, the Jews, and others, and the wise men among the Arabs used to do it. Prayer was also found among them. Abū Dharr, may God be pleased with him, prayed for three years before he became acquainted with the Prophet, may the peace and blessings of God be upon him, and Quss ibn Sā'ida al-Iyādī used to pray.[23] What was preserved of the prayer among the religions of the Jews and the Magians and what was left among the Arabs were the respectful acts, especially the prostration, and saying petitionary prayers and litanies; and the alms tax was also found among them. A customary practice among them was the entertainment of the guest and the traveler, supporting the whole family, giving alms to the poor, keeping the bonds of kinship, and helping those struck by the calamities of God; they were praised for these things; and they recognized that these constituted human fulfillment and happiness. Khadīja said (to Muḥammad), "By God, God will never debase you, since you honor the bonds of kinship, entertain the guest, support the family, and aid those struck by God's calamities,"[24] and Ibn al-Daghina said something like this to Abū Bakr al-Ṣiddīq.[25] They also used to practice fasting from dawn to dusk and the Quraish used to fast in 'Āshūra' during the Jāhiliyya and retreat for devotions to the mosque, and 'Umar once made a vow to do a night of devotions during the Jāhiliyya and he asked the Prophet for advice about that,[26] and 'Āṣ ibn Wā'il left a will that a certain slave should be freed.[27]

In summary, the people of the Jāhiliyya used to perform various acts of religious piety. As for the pilgrimage to the house of God and respecting its rites and honoring the sacred months,[28] the matter is too obvious to be obscure. They had among themselves types of incantations and talismans and they had interpolated polytheism

---

[23] The Archbishop of Najrān.

[24] Bukhārī I:4 Bad' al-Waḥī 3. Muslim Īmān 252.

[25] A tribal chief who pledged the protection of Abū Bakr to Quraish during the Makkan period. Bukhārī Kafāla 5, Khan trans. III:277. Manāqib al-Anṣār 45, Khan trans. V:158.

[26] Bukhārī Khan trans. VIII:450 I'tikāf 16, Nudhūr 29, Ibn Ḥanbal II:10.

[27] Dāwūd Waṣāya 16, Ibn Ḥanbal II:182, V:110–111.

[28] The Jāhiliyya Arabs did not make war on each other during four sacred months three of which were reserved for the pilgrimage.

into this. Their method of slaughtering continued to be cutting the neck and piercing the throat, and they didn't strangle the animal nor did they cut open its stomach. They preserved a remnant of the religion of Abraham, may peace be upon him, in abandoning astrology and the discussion of the intricacies of natural phenomena, except those things to which natural insight spontaneously forced them. The foundation of knowing about the future was the art of dream interpretation, and the good tidings of the prophets before them; then soothsaying entered amongst them and casting lots with divining arrows, and augury from the flight of birds, although they recognized that this was not part of the original religion. This is the saying of the Prophet, may the peace and blessings of God be upon him, when he saw (in the Kaʿba) a picture of Abraham and Ishmael, may peace be upon them, with divining arrows in their hands, "Indeed they (the Meccans) knew that the two of them never had recourse to divination."[29] The descendants of Ishmael had kept the codes of their father until ʿAmr ibn Luḥayy arose among them, and this was about three hundred years before the mission of the Prophet, may the peace and blessings of God be upon him.

The Jāhiliyya Arabs possessed confirmed practices, whose abandonment was reprimanded, in whatever they ate, drank, and wore, how to give feasts, their festivals, burying their dead, marriage, divorce, the waiting and mourning periods, their ways of selling, and their transactions. They continued to forbid marriage within the forbidden degrees such as those with daughters, mothers, sisters, and others; they had deterrents against crimes, such as retaliatory punishment, blood money, and oaths, and as well they had punitive measures against adultery and theft. From the Khusraus and Caesars there entered among them the branches of knowledge of the third and fourth stages of the supports of civilization, but along with them entered excesses and the oppressions of taking prisoners and plundering, the spread of fornication, improper marriages, and usury. The people had abandoned the prayer and the recitation of the name of God, and discarded them, so the Prophet, may the peace and blessings of God be upon him, was sent among them when their condition had reached this point. He considered all that the people were practicing, and whatever remained of the

[29] Bukhārī Ḥajj 54. The basic hadith has the words "bi-hā" with them, i.e., divining arrows, in it.

true religion (of Abraham) he retained, and emphasized that it should be acted upon. He made the acts of worship precise for them through legislating the causes and the times, conditions, pillars, proper behaviors, the things leading to their invalidation, the dispensations, the strict interpretations, the (timely) performance of and the making-up for missing an obligation. He specified the acts of disobedience (sins) for them while explaining to them the relevant pillars and conditions, and he legislated for these things *ḥadd* punishments, deterrent punishments (*mazājir*), and atonements. He made religion easy for them through explaining incitements to righteousness and deterrents from sin, and he blocked the means to sin and incited them to extra acts of religious service for the good, and so on, in the matters which were previously mentioned. He did his utmost to spread the monotheistic (*Ḥanīfī*) religion and to make it predominate over all other religions. He negated whatever distortions they had harbored, and made the greatest efforts to repudiate them. Whatever from these matters belonged to the proper order of the supports of civilization he decisively decreed, and commanded, and he forbade to them all their unsound conventions, and barred them from these. He established the greatest Caliphate, and with those who accompanied him he waged Jihād on those who opposed them until the command of God was fulfilled despite their unwillingness. It is reported in some of the hadiths that the Prophet, may the peace and blessings of God be upon him, said, "I have been sent with the tolerant, clear, Ḥanīfī religion."[30] He meant by "tolerant" that which did not have any arduous practices in it, such as the monks had devised. Rather in it (this religion) there would be a special dispensation for every valid excuse, making the action attainable for both the strong and the weak, the working person and the unemployed. And by "Ḥanīfī" (monotheistic) he meant what we have mentioned about it being the religion (*milla*) of Abraham, may the blessings of God be upon him, in which were established the rituals of God and in which the practices of polytheism were suppressed, and in which distortion and false conventions were annulled. By "the clear" he meant that the reasons for its laws and its wise principles and the goals on which they were based, are clear, with no doubt about them for the one who reflects, and is of sound reason and not vainglorious, and God knows better.

---

[30] Variants of this hadith are cited in Ibn Ḥanbal V:266, VI:116, 233. The word "*baiḍa*'"—"clear" is not found in them.

BOOK VII

THE SEVENTH INVESTIGATION: THE DERIVATION OF THE SHARĪʿA LAWS FROM THE REPORTS OF THE PROPHET, MAY THE PEACE AND BLESSINGS OF GOD BE UPON HIM

CHAPTER 75

The Explanation of the Categories of the Prophet's Knowledge, May the Peace and Blessings of God be Upon Him

Be informed that what was reported from the Prophet, may the peace and blessings of God be upon him, and recorded in the books of hadith falls under two categories.

1) The first category (of the Prophetic sciences) comprises those things which are a means for the propagation of the message, and God's, may He be exalted, saying applies to it: "What the Prophet gives you, take, and what he forbids, abstain from."[1] These are:

a) Knowledge about the next life and the wonders of Malakūt, and all of this is based on revelation.

b) The divine laws, and the determining of the acts of worship and the supports of civilization according to the ways of determination mentioned previously. Some of these depend on revelation, while others depend on independent reasoning (*ijtihād*). The independent reasoning of the Prophet, may the peace and blessings of God be upon him, is at the level of revelation, because God made him safeguarded from having his opinion settle on error. It is not required that his independent reasoning be inferred from revealed statements as has been thought, but rather in most cases God, may He be exalted, used to teach him the intentions behind the divine law and the principle of legislation, facilitation, and the rulings, then he explained the intentions that he learned through revelation according to this principle.

c) Practical wisdom and general beneficial purposes which he did not appoint for a specific time, nor did he set their limits; as when he explained good character and its opposite. These generally depended on individual reasoning (*ijtihād*) in the sense that God, may He be exalted, had informed him of the principles of the supports of civilization from which he then inferred an underlying reason (*ḥikma*) and made this a general principle.

---

[1] Qur'ān 59:7.

d) The virtues of deeds and the outstanding traits of those who do good deeds and I think that some of these depend on revelation and some on independent reasoning.

The explanation of these principles has already been given. We intend to comment on and explain the meanings (of the issues stemming from the principles involved in this category of the Prophet's sayings).

2) The second category (of the Prophet's sayings) comprises whatever does not come under the topic of propagating the message. Concerning this we have the Prophet's saying, "I am only a man, and when I order you to do something regarding your religion, accept it, and if I order you with something according to my own opinion, then I am only a human being,"[2] and his saying in the story of pollinating the palms, "I only made a conjecture, so don't blame me for that opinion, but if I tell you something about Allah, then accept (it) for I will never lie about God."[3]

Included in this section is medicine, and also topics like the Prophet's saying, may the peace and blessings of God be upon him, "Be keen to acquire (as best for Jihād) a black horse; with a whitish blaze on its forehead,"[4] and this was based on experience. Included (in this) are what the Prophet, may the peace and blessings of God be upon him, did as part of his daily routine, not as religious practice; and incidentally, not intentionally. Under this category also fall things he used to speak about just like other people, such as the hadith of Umm Zara' and the hadith of Khurāfa,[5] and this is the saying of Zaid ibn Thābit when a group of people came to him and said, "Tell us some sayings of the Prophet of God, may peace and blessings be upon him." He said, "I was his neighbor, and it happened that whenever revelation came to him, he used to send for me, and I wrote it down for him. If we were talking

---

[2] Bukhārī Ṣalāt 31, Maẓālim 16, Ḥaiḍ 10, Aḥkām 20, 29, 31, Muslim, Abū Dāwūd, Tirmidhī, Nasā'ī, Ibn Mājah, Muwaṭṭa', Ibn Ḥanbal. *Mishkāt*, p. 40.

[3] The Prophet at one point ordered the Medinans not to fecundate their palm trees but when they found that the crop was less abundant this order was rescinded. Muslim Faḍā'il 139, Ibn Ḥanbal. *Mishkāt*, p. 40.

[4] Tirmidhī III:120 #1747 Jihād 20, Ibn Mājah, Ibn Ḥanbal. *Mishkāt*, p. 823.

[5] The hadith of Umm Zara' was related by 'Ā'isha and involves a story of eleven women telling about their husbands, Siddiqi trans. *Ṣaḥīḥ* Muslim IV:1302. In the hadith of Khurāfa, the Prophet explains that Khurāfa was a marvelous story teller who was kidnapped by the jinn and then returned to tell about them, such tales are known by the people as "hadith Khurāfa". Ibn Ḥanbal VI:157.

about the life of this world he would talk about it with us, and if we spoke of the next life he would speak about it with us, and if we spoke about food he would speak about it with us, so I would have to report to you all of these things as hadiths from the Prophet of God." Included here are things from which he intended a specific beneficial effect at that time, and not as things incumbent upon the whole community, for example when a Caliph gives a command regarding mobilizing the armies and specifying the banner. This is the saying of ʿUmar, may God be pleased with him, "There was no reason for us to do *ramal*[6] except that we wanted to make a show of force to a people (the pagans) and now Allah has destroyed them."[7] Then he feared that for the ramal there might be some other cause.[8] Many of the rules can be traced back to a specific beneficial purpose such as the saying of the Prophet, may the peace and blessings of God be upon him, "Who ever kills someone (in the *Jihād*) can keep his arms and clothing."[9] Included under this are specialized rulings and decrees, and in these he, may the peace and blessings of God be upon him, used to accept evidence and oaths and this is his, may the peace and blessings of God be upon him, saying to ʿAlī, may God be pleased with him, "The one who is present sees something that the absent one does not."[10]

---

[6] *Ramal* is fast walking accompanied by movements of the arms and legs to show one's physical strength. When Muḥammad and his Companions came to Mecca for pilgrimage and the Quraish were still in control, the pagans said that they had been weakened by the fever of Yathrib so the Prophet ordered them to do "*ramal*" in the first three circumambulations of the Kaʿba.

[7] Bukhārī Ḥajj 57. Khan trans. II:393.

[8] So ʿUmar added, "Nevertheless, the Prophet did that (ramal), and we do not want to abandon it."

[9] That is, these articles would not be subject to having the one-fifth share of the Prophet assessed on them. Bukhārī Khums 18, Maghāzī 54, Muslim, Abū Dāwūd, Tirmidhī, Ibn Mājah, Muwaṭṭa', Ibn Ḥanbal.

[10] Cited in Chapter 40. Ibn Ḥanbal I:83.

CHAPTER 76

The Difference Between the Beneficial Purposes (Maṣāliḥ) *and the Religious Laws*

Be informed that the law-giver benefited us with two types of knowledge which are distinct in their rulings and dissimilar in their ranking.

1) One of the two types is the knowledge of the beneficial purposes and the causes of sin, by which I mean whatever he explained about refining the soul through the acquisition of virtues which aid in this world and in the next life, and by the elimination of their opposites. Therefore how to manage a household, the manner of livelihood, and the governing of the city were not quantified by specific amounts, nor did he render the ambiguous precise by fixed limits, nor did he distinguish the obscure issues through determined signs. Rather the Prophet, may the peace and blessings of God be upon him, encouraged praiseworthy things and urged abstention from vices, leaving his speech as it would be normally be understood by people in basing demanding or forbidding on the beneficial purposes themselves, not on anticipated sources (*maẓānn*) attached to them, or signs which would made them recognizable. For example, he praised cleverness and courage, and commanded friendship, showing affection, and adopting a middle course as a way of life; but he did not explain the definition of cleverness on which the call for it was based, nor did he explain any anticipated source for it on the basis of which the people could be taken to task (for disregarding it).

Every beneficial purpose to which the divine law incites us and every cause of evil from which it keeps us, unavoidably goes back to one of three principles.

1) The first is the refinement of the soul through the acquisition of the four virtues[1] beneficial for the next life and the rest of the virtues which are of benefit in this life. The second of them is the

[1] Purity, humility, magnanimity, and justice. Previously discussed in Ch. 32.

propagation of the word of God and the consolidation of the divine laws and making efforts to spread them. The third is the organization of the order of mankind and the improvement of their supports of civilization, and the refinement of their conventions.

The meaning of their (the beneficial purposes and the causes of evil) going back to these principles is that a thing has some role in these matters either by affirming or negating them, in that it is a ramification of one of their properties or opposes this ramification, or it is an anticipated source of their presence or absence, or it is bound up with them or with their opposite, or it is a path to them or diverts from them. God's pleasure is originally connected with these beneficial purposes, and His anger is also attached to these evil deeds, before His sending the prophets and after it equally. If it were not for the connection of His satisfaction and anger with these two types of behavior (good and evil), the prophets would not have been sent. This is because the divine laws and the punishments only came after the mission of the prophets, so that initially there was no grace from God in having these laws imposed and being held to them, but rather the beneficial purposes and the causes of evil were effective in requiring the refinement of the soul or its being corrupted, or in putting their affairs in order or ruining them, already, before the sending of the prophets. Thus the grace of God decreed that people should have (at the same time) an understanding of what was important for them, and that they should have imposed on them whatever was incumbent on them, and this could not be effected except through setting stipulations and laws, so the grace of God decreed this sort of thing as the occasion arose. This type of thing is rationally comprehensible, but among it is a portion which the intelligence of the common man can understand on its own and a portion which can only be understood by the minds of the most intelligent on whom emanates the light from the hearts of the prophets. The law alerted them, so that they were alerted, and pointed out to them, so that they understood. The person who masters the principles which we have mentioned will not waver on any aspect of them.

2) The second type of knowledge is knowledge of the divine laws, the punishments (*ḥudūd*), and the obligatory religious duties. I mean whatever the divine law explained by way of the appointed quantities, so that it fixed anticipated sources and precise, well-known signs for the beneficial purposes, and based the ruling on

them, and imposed them on the people. The types of pious actions were set by determining the pillars, conditions, and proper behaviors, and for every type a limit was set to be sought from the group which was compulsory, and a limit which was recommended without compulsion. In every pious act a number which was compulsory was chosen and another which was recommended, thus the obligation was directed to these very anticipated sources, and the rulings were based on these very signs, and this type of knowledge is a basis for the rules of religious regulation. Not every anticipated source of a beneficial purpose is made compulsory for people, but only those which are precise and tangible or possess an obvious description which both the elite and the masses can recognize. Sometimes there are incidental reasons for the compulsory and the forbidden which result in their being recorded among the Highest Council, and There above is confirmed the form of the compulsory and the forbidden; such as a certain question of an inquirer, and a particular desire of a group for something or their rejecting it. All of this is not rationally comprehensible in the sense that even if we knew the rules of determining amounts and (divine) legislation, we would not know that they had actually been recorded among the Highest Council and that the form of the compulsory had been established with the Holy Enclave, except through the explicit statement of the divine law.[2] This is one of the things that there is no way to understand except through the reports of God's sayings, and this is comparable to ice—we know that the cause for its coming into being is a coldness which affects the water, but we don't know whether the water in a large cup at this time has frozen or not except through seeing it ourselves or through being told by one who saw it. On this analogy we know that the minimum taxable holding for the alms tax must be set, and we know that two hundred dirhems and five *ausāq* is the proper amount for the minimum taxable holding, because the one who has this much is reckoned as being able to spare (the tax), and these two are things in confirmed usage among the people; but we do not know that God, may He be exalted, decided this share for us, and based on it His being satisfied and angry, except through the text of the divine law. Indeed how could we, and how many causes are there which we have no way of knowing except through a report,

---

[2] "*naṣṣ al-sharʿ*".

and this is the Prophet's saying: "The Muslim who harms Muslims most"[3] and his saying, "I feared that it would become prescribed for you."[4]

The learned scholars of repute have agreed that analogical reasoning (*qiyās*) does not apply in the topic of set amounts, and that the proper methodology of analogical reasoning is extending the ruling on the original case (*aṣl*) to the assimilated case (*farʿ*) through a shared reason for legislation (*ʿilla*),[5] not that the symbol or anticipated source of a beneficial purpose should be made a reason for legislation, nor that some thing related to it should be made a pillar or a condition.[6] They also agreed that analogical reasoning does not apply due to the existence of the beneficial purpose but because of a precise reason for legislation on which the rule may be based. Therefore an analogy is not drawn from what determines "hardship" for the traveler in his being allowed a dispensation in prayer and fasting, since the prevention of "hardship" is the beneficial purpose behind receiving a dispensation, and not the reason for the legislation to shorten the prayers and break the fast. Rather, the reason for the legislation (*ʿilla*) is being on a journey. Therefore, on the whole, the religious scholars do not differ about such cases, but most of them support bringing them out in more detail. This is because sometimes the beneficial purpose becomes confused with the reason for legislation and the act of legislation. Some of the legal scholars, when they went deeply into analogical reasoning became confused, and insisted on some of the determined amounts, and forbade exchanging them for things similar to them, while they were tolerant about others and allowed substitutions for them. An example of their fixing amounts is their setting the minimum share of cotton liable for the *zakāt* tax as five loads, and their making riding on a ship the anticipated source of seasickness

[3] The Muslim who harms Muslims most is the one who asked about a thing so that it was forbidden because of his questioning, Bukhārī Iʿtiṣām 3, Muslim, Abū Dāwūd, *Mishkāt*. p. 42. This hadith was previously mentioned in the introduction and Chapter 57.

[4] *Mishkāt*, p. 270. Transmitted by Bukhārī Tahajjud 4. Khan trans. II:128. This hadith refers to the Prophet's allowing people to pray the special night prayers at home during Ramaḍān, and was cited in the introduction and Chapter 60.

[5] This is a reference to the technical syllogistic method of juridical *qiyās*.

[6] The idea of "condition" (*sharṭ*) is a fairly technical one discussed in works of *uṣūl al-fiqh* and varying somewhat from one legal school to another. A discussion of the term may be found in al-Thanavī, Muḥammad, *al-Kashshāf fī Istilāḥāt al-Funūn*, (Calcutta, 1862), 754.

so that they based the dispensation to remain seated during prayer (while at sea) on this, while they set the amount of water at ten by ten.[7]

Whenever the law made the beneficial purpose understood in a case and we find this (same) beneficial purpose in another case; then we should recognize that the pleasure of God is connected with it in its own right, not by the particularities of that case, in contrast to the determined amounts, for there God's pleasure attaches to these determined amounts themselves. To elaborate this; when someone does not perform the prayer on time it is a sin, even if at that time he is occupied in the remembrance of God and other acts of worship. The person who does not pay the tax incumbent on him but spends more than that amount in good works, still commits a sin. Likewise to dress in silk and gold in privacy where it won't wound the hearts of the poor, nor lead people to be more worldly, and when luxury is not intended by it, still is a sin; likewise to drink wine with the intention of medicinal use, when there is neither immoral purpose nor abandonment of the prayers—is still a sin because the pleasure and anger of God are connected to those acts themselves, even if the essential goal was to avert people from sin and to bring them to the beneficial purpose. God knew, however, that the regulation of the community would not be possible at that time except through compelling these very acts and forbidding them, and thus he directed His pleasure and anger to them in themselves, and this was recorded among the Highest Council.

This is in contrast to someone who wears the finest wool which is better and more costly than silk, and who uses vessels of ruby, for he does not sin by those actions in themselves. However, if breaking the hearts of the poor, leading other people to do these things, or seeking luxurious living are the result, then he is far from the mercy of God because of these sins, and if not, then he is not.

Wherever you find the Companions and the Successors doing something which resembles setting specific quantities,[8] their intent was only to clarify the beneficial purpose and to incite to it, and to clarify the evil and to deter against it. They only brought out

[7] The size of a pond whose water may be considered pure for the ritual bath according to some Ḥanafī jurists. See *Ḥujjat Allah al-Bāligha* II:183–183 "Rules About Water."

[8] Such as in defining the length of a "journey" as being a specific distance.

this case by way of exemplification, and they did not mean to specify by it,[9] for they only had in mind conveying the meanings, although on the surface the matter is confusing.

Whenever the divine law allows exchanging a set thing for its price, such as the two year old camel for a price according to one report; then even if we accept this report, this is also a type of setting a quantity, and this is because it is not possible to do a minute investigation into quantification in so far as this would lead to hardship. Rather, sometimes an amount is set for something common to many things, such as the two year old camel itself, for sometimes one such camel will be superior to another. Sometimes setting an amount by value sets a limit which is universal, such as setting the threshold for which a person's hand would cut off for theft at something that has the value of a quarter of a dinar or three dirhems.

Be informed that making things obligatory and forbidden (as such) are two types of assessment and this is because often what appears to be a beneficial purpose or a cause of evil has many forms. One form is settled upon to be commanded or forbidden because it is exact or because it is something whose status was recognized among the preceding religions, or because people strongly desired it. Therefore the Prophet, may the peace and blessings of God be upon him, excused himself saying, "I feared that it would become prescribed for you."[10] And he said, "Were it not that I would be hard on my community I would have ordered them to clean their teeth (before every prayer)."[11] Since this is the situation, it is not legitimate to construe something whose ruling was not stipulated by the textual pronouncement on the basis of something that was.

As for being something being recommended or reprehensible,[12] in these there are further detailed aspects. Whichever recommended

---

[9] For example, when they set the distance traveled on "a journey" as being four *burūd*.

[10] Hadith cited above in this Chapter.

[11] *Mishkāt*, p. 79. Bukhārī Juma' 8, Tamannā 9, Ṣaum 27, Muslim, Abū Dāwūd, Tirmidhī, Nasā'ī, Ibn Mājah, Muwṭṭa', Ibn Ḥanbal. Hadith cited previously in Chapter 60.

[12] According to most schools of Islamic law there are five degrees of actions—compulsory, recommended (*mandūb* or *mustaḥabb*), neutral (*mubāḥ*), reprehensible (*makrūh*), and forbidden (*ḥarām*). This therefore refers to two of the categories between compulsory and forbidden.

thing the law-giver commanded per se, praised, and laid down as a customary practice for the people—has the same force as the obligatory. Any recommended thing in which the Prophet confined himself to explaining its beneficial purpose, or which he himself chose to do without establishing it as a practice (sunna), and without emphasizing its significance; remains in the state which it had before the legislation, and reward for it will be through the beneficial purpose found within it, not the performance of the action itself.[13] Likewise is the situation of the reprehensible according to this more detailed explanation.

When you have really grasped this introduction it will become apparent to you that most of the analogies of which this group[14] are proud and on account of which they have become arrogant towards the group of the People of the Hadith lead to an evil consequence for them without their realizing it.[15]

---

[13] This distinction in Islamic Law is recognized as that between a confirmed practice (*sunna mu'akkada*) and a non-confirmed practice (*sunna ghair mu'akkada*) of the Prophet.

[14] Meaning those jurists who rely more on reason rather than on literal readings of revealed texts.

[15] Thus Shāh Walī Allāh is criticizing the excessive use of *qiyās*, particularly if it is not strictly based on the Qur'ān and the hadith.

CHAPTER 77

The Way the Community Received the Divine Law from the Prophet, May the Peace and Blessings of God be Upon Him

Be informed that the community received the divine law from the Prophet in two ways.

1) The first of them is overt reception, which must be through a transmission which is either handed down from the beginning by a large number of continuous channels (*mutawātir*) or non-*mutawātir*.

a) The *mutawātir* report may be word for word such as the great Qur'ān or like a small portion of the hadiths such as the Prophet's, may the peace and blessings of God be upon him, saying, "You will see your Lord."[1] Or it may be a *mutawātir* report which has been handed down according to its meaning such as many of the rules of purity, prayer, the alms tax, fasting, the pilgrimage, buying and selling, marriage, and making war; about which Muslims do not differ with one another.

b) Next there are the non-*mutawātir* reports of which the highest level are the *mustafīḍ*, and these are things reported by three or more of the Companions, then the transmitters continued to increase until the fifth generation, and this is a type of report found in great numbers, and on it are based the chief issues of jurisprudence.

c) The next category are the reports (*khabar*) judged sound (*ṣaḥīḥ*) or fair (*ḥasan*) according to the hadith scholars known for their having committed many hadiths to memory, and the great ones among them.

d) Next there are reports about which there is some controversy (among hadith scholars), so that some accept them while others do not. Those hadiths among them which are supported by parallel

---

[1] *Bukhā'rī* Mawāqīt 16, Tauḥīd 24, Muslim, Ibn Mājah, Ibn Ḥanbal. Jarīr ibn āAbd Allāh said, "We were sitting with the Prophet of God, may the peace and blessings of God be upon him, and he looked at the moon on the night it was full and said, 'You shall see your Lord as you see this moon and you will not be wronged in seeing Him and if you are able don't miss praying before sunrise and before it sets.'"

transmissions, or by the opinion of most of the knowledgeable scholars, or by clear understanding, must be acted upon.

2) The second way of receiving the divine law is through indication (*dalāla*), and this is that the Companions observed the Prophet, may the peace and blessings of God be upon him, speaking and acting, and then they derived from that a ruling of "obligatory" and other rulings, so that they informed about that ruling, saying, "Such and such a thing is compulsory, while some other one is simply permissible." Then the Successors likewise accepted this from the Companions, and the third generation recorded their legal opinions and judgments and strengthened the matter.

The greatest ones in this type (receiving through indication) are 'Umar, 'Alī, Ibn Mas'ūd,[2] and Ibn 'Abbās,[3] may God be pleased with them, but it is known that the conduct of 'Umar, may God be pleased with him, was to seek counsel from the Companions and debate with them so that the ambiguous would become clear and assurance would come from that. Most of his legal opinions and judgments were followed in the East and the West of the earth, and this is the saying of Ibrahīm[4] when 'Umar died, "Nine-tenths of knowledge has departed," and the saying of Ibn Ma'sūd, "'Umar was such that if he set us on a road we found it to be smooth."[5] 'Alī, may God be pleased with him, usually did not consult, and most of his decrees were issued in Kufa, and usually only a few people reported them from him. Ibn Mas'ūd was in Kufa, and his legal opinions were generally reported only by the people of that region. Ibn 'Abbās used to use independent reasoning (*ijtihād*) after the era of the first Muslims and contradicted the latter in many rulings, and his companions among the people of Mecca followed him in this, so that most of the Muslims did not adopt those of his rulings which were not supported by other rulings. As for other than these four, whose transmission of hadiths was on the principle of using an indication, they didn't distinguish the pillar and the condition[6] from the manners and the practices of the Prophet, and

---

[2] Ibn Ma'sūd, (d. 652/53) one of the first Muslims, a Companion of the Prophet who rose from humble origins to a position of authority. He migrated to Iraq and hence many Kufan jurists transmitted his hadith.

[3] Ibn 'Abbās (d. 687/88) another Companion and Meccan authority.

[4] Ibrahīm al-Nakhā'ī, a Kufan authority.

[5] al-Dārimī, II: 344, Farā'iḍ.

[6] The pillar of a legal obligation is an aspect which is indispensable for its per-

they only had a little to say when there was conflict among hadith reports, or an incompatibility of indications, such as Ibn 'Umar,[7] 'Ā'isha, and Zaid Ibn Thābit, may God be pleased with them. The greatest in this type of receiving (through indication) among the Successors of Medina were the seven jurists,[8] especially al-Musayyab[9] in Medina, and in Mecca 'Aṭā ibn Abī Rabāḥ,[10] and in Kufa, Ibrahīm (al-Nakha'ī), Shuraiḥ,[11] and al-Sha'bī,[12] and in Basra, al-Ḥasan.[13] In each of these two ways (of receiving the divine law) there are gaps which are only restored by the other, and neither can manage without the other.

As for the first (overt reception), among its defects are changes which entered into the reporting based on the meaning (of the hadith),[14] and there is no guarantee against changes in the meaning. Among its faults are that the command may have been given with regard to a specific event, and the transmitter thought that it was a general ruling. Also among them are that the statement may have been made in the emphatic voice, in order that they would adhere to it, but the transmitter took it to be compulsory or forbidding, while this was not the case. Therefore, whoever had legal acumen and was present on the occasion, inferred from the circumstantial evidence the true state of the matter such as the saying of Zaid (ibn Thābit), may God be pleased with him, about the ban on the sharecropping contract and the sale of fruit before its proper ripening, i.e., that it is evident that this was in the nature of advice.

As for (the defects of) the second type (reception through an

---

formance to be fulfilled while a condition is an aspect of the situation which brings the legal obligation into effectiveness. For example, ritual purity is a necessary requirement (pillar) for the prayers, while a certain time of day being reached is a cause (*sabab*) for bringing the requirement to pray the appropriate prayer into effect These terms are discussed in the books on Islamic legal theory (*uṣūl al-fiqh*).

[7] Ibn 'Umar, (692/93) was the son of Caliph 'Umar and a Medinan authority.

[8] The seven early jurists of Medina who are particularly distinguished are Sa'īd ibn Musayyab, 'Urwah ibn Zubayr, Qāsim ibn Muḥammad ibn Abī Bakr al-Ṣiddiq, Abū Bakr ibn 'Abd al-Raḥmān Makrūmī, Kharijah Zaid ibn Thābit, 'Ubaid Allāh ibn 'Abd Allāh ibn 'Utba Ma'sūdī, Sulaimān ibn Yassār Hilālī.

[9] Ibn al-Musayyab (d. 712/13).

[10] D. 114/15 A. H. 732/33 C. E.

[11] Shuraiḥ, the oldest Iraqi authority after Ibn Mas'ūd, said to have been appointed judge of Kufa by 'Umar.

[12] D. 728.

[13] Ḥasan al-Baṣrī (d. 728).

[14] Rather than the literal wording.

indication), there entered into it the analogical reasonings of the Companions and the Successors, and their inferences from the Qur'ān and the practice of the Prophet, and their independent reasoning (*ijtihād*) was not always correct in all circumstances. Sometimes the hadith didn't reach one of them or reached him in such a way that it could not serve as a proof, so it was not acted upon. Then after that the true state of affairs was made know through the statement of another Companion, like the report of 'Umar and Ibn Mas'ūd, may God be pleased with them, about making the ablution with earth after ejaculation.[15]

Often the leaders of the Companions, may God be pleased with them, agreed upon something because reason indicated its benefit (*irtifāq*). On that account the Prophet stated, "You should follow my sunna and the sunna of the rightly-guided caliphs after me."[16] Therefore this (agreement of some of the Companions) is not one of the roots of legislation.

Thus, whoever has gone deeply into the reports and the wording of the hadith will escape from pitfalls. Since the situation is thus, it is necessary for the one dealing with jurisprudence to be proficient in both movements and well-versed in both schools,[17] and the best practices of the religion are those on which the majority of the transmitters and the bearers of knowledge have agreed and in which the two methods coincide, and God knows better.

---

[15] 'Umar ibn al-Khaṭṭāb held that the ablution with sand was not sufficient for someone in a state of major ritual impurity who did not find water. Then 'Āmmār reported in his presence that he had been with the Messenger of God (S) on a journey and had became ritually impure and did not find any water so that he rolled himself in the dirt and then mentioned this to the Prophet of God (S). The Prophet of God (S) said, "It would have been sufficient for you to have done thus," and he lightly struck his hands against the earth and rubbed both of them across his face and arms. Although 'Umar did not accept that this was sufficient, this report became the more accepted at a later period. Bukhārī Tayammum 8, Khan trans. I:208–209. Nasā'ī Ṭahāra 195, 199–201.

[16] Mishkāt, Faith p. 44. From Nasā'ī, Ibn Ḥanbal and al-Dārimī. Tirmidhī IV:150 #2816 'Ilm, 16 with the slight variation "he should follow my sunna . . ."

[17] Expertise in both the reports themselves and the methodology for interpreting their legal implications.

CHAPTER 78

The Ranks of the Books on Hadith

Be informed that there is no way for us to (obtain) knowledge of the divine laws and the rulings except through the report of the Prophet, in contrast to the case of the beneficial purposes, for these may be understood through experience, true reflection, surmisal, and so on. There is also no way for us to have knowledge of the sayings of the Prophet, may the peace and blessings of God be upon him, except by receiving reports which go back to him by successive links and transmission, whether they are in his, may the peace and blessings of God be upon him, words; or they are interrupted hadiths (*mauqūf*) whose transmission was verified by a group of the Companions and the Successors, in so far as they were remote from taking the initiative to decide on something like that if there were not a proof text or sign from the law-giver. An example of this is transmitting an indication (*dalāla*) from him, may the peace and blessings of God be upon him.[1]

In this our time there is no way to receive these reports except to follow the literature written in the science of hadith, for today there is not to be found any report which is reliable unless it is written down. The books about hadith are of different ranks and at various levels and it is necessary to exercise care in recognizing their ranks.

We hold that these collections are of four ranks with respect to accuracy and repute.

1) This is because the highest types of hadith, as you learned in what preceded—are those confirmed from the beginning by many reports (tawātur), and the community agreed to accept them and to act on them.

2) Next are those which are well-known and transmitted abundantly through numerous lines (*mustafīḍ*). There remains no doubt

[1] "*Dalāla*" or indication, in the Islamic theory of signification refers to how an utterance may signify through various modes and logical implications.

worth considering regarding them, and the majority of the legal scholars of the garrison towns[2] agreed on them, and in particular the learned scholars of the two sacred cities did not disagree with them, for Mecca and Medina were the location of the rightly-guided caliphs in the first generations and were frequented by the itinerant learned scholars generation after generation, so that it was unlikely that an obvious error would be accepted by them, or that it would become well-known and implemented within a large area and be reported by a large body of the Companions and the Successors.

3) The third group of hadith are the ones whose chain of reporters is sound or good, and is testified to by the scholars of the hadith, and they are not abandoned sayings which no one of the community holds to be true.

4) As for (the fourth group which are) weak, invented (*mauḍūʿ*), interrupted in chain, or transposed (*maqlūb*) in text, transmitted from unknown persons, or opposed to (a hadith) which the pious ancestors agreed on generation after generation; there is no way to uphold them. Thus, the accuracy resides in the compiler of the book having made a condition for himself the citing of what is sound and good, not transposed, anomalous (*shādhdh*),[3] or weak, unless he provides an accompanying explanation of its status; for citing the weak hadith together with an explanation of its status would not detract from the book.

The reputation (of books) consists of the hadiths cited in them being current among the hadith experts before their being recorded as well as after, and that the leading scholars (*imāms*) of hadith even before the compiler used to relate them by multiple chains of transmission and present them in their compilations of traditions and hadith collections. Scholars coming after the compiler should have been devoted to transmitting the book, memorizing it, clarifying its obscurities, elucidating its uncommon words, analyzing them grammatically, sorting out the lines of transmission of its hadiths, inferring its juristic significance and seeking information about the biographical circumstances of the transmitters, generation after generation, until our present time, so that there doesn't remain anything connected with it which is not investigated, except what

---

[2] The garrison towns (*amṣār*) where the early Muslims settled in the populated areas of Syria and Iraq and scholarship and legal schools developed in the second and third Islamic centuries.

[3] A report from a single authority which differs from what others report.

God wills. The hadith critics both before the compiler and after him agreed to cite its hadiths, judged them to be sound, were satisfied with the opinion of the author on them, and accepted his book with praise and appreciation. The leading figures of jurisprudence continued to derive rulings from its hadiths and relied on them, and devoted attention to them, and the general public also continues to believe in and honor them.

In summary, when these two virtues are perfectly combined in a book it is of the first rank, and so on and so forth; and if these are completely absent it won't be considered at all. Those at the highest degree of the first rank must reach the degree of multiple chains of transmission going back to the Prophet (*tawātur*), and those beneath must attain the degree of being well-known (*istifāḍa*), and next is the degree of definite soundness (*al-ṣiḥḥa al-qaṭ'iyya*), by which I mean that it is ascertained with full confidence through the science of hadith that it is "suitable for being acted upon" (*mufīd l-il-'amal*).[4] The second rank comes close to the degree of "istafāḍa" or "either certain or conjectural soundness," and thus, the matter descends.

Investigation has established that only three books belong to the first rank: *The Muwaṭṭa'*, the *Ṣaḥiḥ* of al-Bukhārī, and the *Ṣaḥīḥ Muslim*. Al-Shāfi'ī said, "The most sound book after the book of God is the *Muwaṭṭa'* of Mālik."[5] and the experts in hadith studies agreed that everything in it is sound (*ṣaḥīḥ*) according to Mālik's opinion and those who concurred with him. As for the opinion of other hadith scholars, no hadith interrupted just before reaching the Prophet[6] and no interrupted hadith (*munqaṭi'*) was included in it, unless its chain went back directly to the Prophet through some other line, so that it is definitely sound due to this reason. In the time of Mālik many *Muwaṭṭa*'s were written in which Mālik's hadiths were investigated and classified and their interrupted chains were taken directly back to the Prophet, may the peace and blessings of God be upon him, such as the books of Ibn Abī Dhi'b,[7] Ibn 'Uyaina,[8]

---

[4] I.e., It does not lead to absolute certainty (*mufīd l-il-'ilm*).

[5] Shāh Walī Allāh's cosmopolitan training in hadith while he studied in the Hijāz led him to accept the authority of Mālik's *Muwaṭṭa'*. Although Shāh Walī Allāh was of the Ḥanafi school of law, he allowed the competent person to choose according to reason (*ijtihād*) from among the rulings of any of the four schools.

[6] *Mursal*.

[7] Ibn Abī Dhi'b (d. 774/775), a Medinan scholar and traditionalist.

[8] Sufyān Ibn 'Uyaina (d. 813/14), a traditionalist.

al-Thaurī,[9] and Muʿammar,[10] and others who shared the same teachers with Mālik. More than one thousand men transmitted[11] this book from Mālik without an intermediary and people journeyed to Mālik in search of knowledge from the farthest regions of the land, as the Prophet had foretold in his hadith.[12] Among them were the most prominent legal scholars such as Al-Shāfiʿī, Muḥammad ibn al-Ḥasan (al-Shaibānī),[13] Ibn Wahb,[14] and Ibn al-Qāsim,[15] and among them were the most adept at hadith such as Yaḥyā ibn Saʿīd al-Qaṭṭān,[16] ʿAbd al-Raḥmān ibn Mahdī[17] and ʿAbd al-Razzāq,[18] and among them were the kings and princes such as (Hārūn) al-Rashīd and his two sons. The *Muwaṭṭa'* became famous in Mālik's lifetime until at last its fame reached all of the lands of Islam, and its reputation continues to grow with the passage of time and interest in it only becomes stronger. The legal scholars of the garrison towns based their law schools upon it, even the (school of the) People of Iraq[19] on some points. The learned scholars continued to establish the authenticity of its hadiths and to mention hadiths which support it through other transmissions or through similar meanings. They continue to explain its unusual expressions, solve its problematic features, investigate its juristic significance, and do research on the hadith transmitters cited in it, to the greatest extent possible. If you want the clear truth then compare the *Muwaṭṭa'* with the *Kitāb al-Āthār* of Muḥammad al-Shaibānī and the *Amālī* of Abū Yūṣuf[20] and you will find between it and them the distance between the East and the West; for have you heard one of the

---

[9] Sufyān al-Thaurī (d. 161 A. H. 777/8), a Kufan authority.

[10] Muʿammar ibn Rāshid al-Azdī, author of *Kitāb al-Jāmiʿ*.

[11] Hārūn al-Rashīd (d. 799) was the Abbasid Caliph, and his two sons were later the caliphs, al-Amīn and al-Ma'mūn.

[12] Shāh Walī Allāh refers to this hadith in his *Muṣaffā*, p. 3. The Prophet said, "Soon men will travel great distances in search of knowledge, but they will not find any scholar more knowledgeable than the one from Medina." Tirmidhī IV:152 #2821 ʿIlm 18, Ibn Ḥanbal II:299.

[13] Al-Shaibānī (d. 803/4) a Kufan legal scholar.

[14] ʿAbd Allāh ibn Wahb ibn Muslim al-Fihrī al-Qurashī (743–813). Mālikī hadith scholar from Egypt who studied in Medina. Wrote a Jāmiʿ work.

[15] Ibn al-Qāsim (d. 806) a Medinan pupil of Malik who laid the real systematic foundations of Mālikī law.

[16] Yaḥya ibn Saʿīd al-Qaṭṭān (d. 760).

[17] ʿAbd al-Raḥmān Ibn Mahdī (d. 813/14).

[18] ʿAbd al-Razzāq ibn Humām 743–826.

[19] A reference to the Iraqi legal school crystallized by Abū Ḥanīfa.

[20] Abū Yūṣuf (d. 799) was appointed chief Qaḍī by Hārūn al-Rashīd. He had been a pupil of Abū Ḥanīfa and wrote on fiscal and penal law.

hadith experts or the legal scholars taking an interest in or paying attention to those works?

As for the two *Ṣaḥīḥ*'s (of Muslim and Bukhārī), the hadith scholars agreed that all of the uninterrupted hadiths going back to the Prophet, may the peace and blessings of God be upon him, in them are certainly sound, and that the (contents of) the two books reached their compilers in multiply-transmitted connected chains, and that whoever holds them in contempt is an innovator in religion who is not following the path of the believers. If you wish the clear truth compare them with the book of Ibn Abī Shayba[21] and the book of al-Ṭaḥāwī[22] and the *Musnad* of al-Khāwarizmī[23] and between these two and the others you will find the distance between the East and the West. Al-Ḥākim[24] supplemented them with hadiths which met the standards of these two, but which had not been mentioned in them, and I traced back what he had added and found that he was correct from one aspect and incorrect from another. This is because he found hadiths reported from the people cited by the two Shaikhs (Muslim and Bukhārī), conforming to their criteria of soundness and reaching the Prophet, may the peace and blessings of God be upon him, through uninterrupted lines of transmission, thus he oriented his additions to both of them according to this aspect. But the two Shaikhs did not cite a hadith unless their teachers had discussed it among themselves and agreed to report it and declare it authentic, as Muslim indicated when he said, "I didn't mention here any of them (hadiths) except those that they had agreed on." However the major portion of (the hadiths) that (Ḥākim's) *al-Mustadrak* was unique (in citing) were obscure ones whose status was unrecognized in the time of the teachers of Muslim and al-Bukhārī, even if they become well-known afterward) or they were ones about which the hadith scholars disagreed concerning the transmitters. The two Shaikhs, like their teachers, were careful in investigating the evidences for the hadiths in terms of whether they reached back to the Prophet, or were inter-

---

[21] Abū Bakr ibn 'Abī Shayba (d. 849), his book is *Kitāb al-Muṣannaf.*

[22] Abū Ja'far Aḥmad ibn Muḥammad al-Ṭaḥāwī (d. 933).

[23] Abū'l-Mu'ayyad Muḥammad ibn Maḥmūd al-Khwārizmī (d. 1257), put together the so-called "Musnad of Abū Ḥanīfa".

[24] Al-Ḥākim al-Nīsābūrī (933–1014) wrote the book *al-Mustadrak 'alā al-Ṣaḥiḥain fi-l-ḥadīth. Mustadrak* is a genre of hadith work in which the author conforms to the standards of predecessors in selecting hadith but adds ones which they did not cite conforming to these same standards.

rupted and so on, until the situation was clarified. Al-Ḥākim usually relied on principles derived from the methodologies (of the hadith scholars) such as his saying, "Whatever reliable transmitters add (to a hadith) is accepted." When people disagreed as to whether the hadiths went back to the Prophet, were interrupted at the level of a Successor, stopped at some transmitter, or were transmitted directly to the level of the Prophet, and things other than this; (he held that) the one who had preserved the additional material to the hadith was more authoritative than the one who had not done so. The truth is that often those defects which arose occurred through the hadith memorizers due to the chains of transmitters being interrupted and their having made them go back to the Prophet, especially because of their desire to have them go back uninterruptedly, and the importance which they gave to this. The two Shaikhs do not cite much of what al-Ḥākim does, and God knows better. It is these three books which the Qāḍī ʿĪyāḍ[25] gives attention to in *al-Mashāriq*[26] by clarifying their obscurities and correcting their copying mistakes.

2) The second rank of books are books which don't attain the rank of the *Muwaṭṭa'* and the two *Ṣaḥīḥ*'s, but which are next after them. Their authors were well-known for their reliability, integrity, memory and being deeply immersed in the laws of hadith, and in their books they were not satisfied to be lax about whatever they had set as conditions for themselves. Therefore those after them accepted these books as sound, and the hadith scholars and legal experts paid attention to them, generation after generation, and these books achieved a reputation among people. Some people were devoted to explaining their unusual (*gharīb*)[27] hadiths, investigating the transmitters, and making inferences about their juristic significance, and upon these hadiths are based the general hadith sciences. They include the *Sunan* of Abū Dāwūd and the *Jāmiʿ* of Al-Tirmidhī and *al-Mujtabā* of al-Nasā'ī.[28] The hadiths of these books together with those of the first rank, are given attention by Razīn in *Tajrīd al-Ṣiḥāḥ*[29] and Ibn al-Athīr in *Jāmiʿ al-*

---

[25] ʿIyāḍ ibn Mūsā, 1088–1149. A Mālikī scholar.

[26] *Mashāriq al-Anwār ʿalā Ṣiḥāḥ al-Athār* (Cairo: Maṭbaʿa al-ʿAtīqa, 1973).

[27] A hadith resting on the authority of only one Companion.

[28] Al-Nasā'ī, Aḥmad ibn Shuʿaib (830–915). His book is called al-Sunan, *al-Mujtabā* is the condensed version of this work by al-Suyūṭī.

[29] Razīn ibn Mu'āwiya d. 1140 author of Tajrīd al-Ṣiḥāḥ al-Sitta.

Uṣūl, and the *Musnad* of Aḥmad (ibn Ḥanbal) is generally included in this rank, for Imām Aḥmad made (his book) a basis for discerning the sound and the faulty saying, "Don't accept what is not in it (my book)."

3) The third rank is comprised of those *Musnad*, *Jāmiʿ*, and *Muṣannaf* works[30] compiled before al-Bukhārī and Muslim, during their time, and after them which combine the sound, the good, the weak, and the recognized (*maʿrūf*) and the uncommon; the anomalous and the objectionable (*munkar*); the erroneous and the correct; and the confirmed and the transposed. They do not have the same reputation among the learned scholars, even if they are not given the designation of "absolute rejection." The learned religious scholars have not given much currency to those hadiths found uniquely in these works, and the hadith scholars have not carried out major investigations into their soundness or faultiness. Among these books are the ones for which no linguist has rendered the service of explaining its rare expressions, nor has any legal scholar worked to reconcile it with the opinions of the pious ancestors, nor has any hadith scholar explained its problematic (*mushkil*) hadiths, nor has any historian made mention of its transmitters. I do not refer to those recent authors who are hair-splitters; rather I am speaking of the early leaders (*imāms*) of the hadith scholars. Therefore their books have remained obscure, unknown, and undistinguished, such as the *Musnad* of Abū ʿAlī,[31] the *Muṣannaf* of ʿAbd al-Razzāq,[32] the *Muṣannaf*[33] of Abū Bakr ibn Abī-Shayba,[34] the *Musnad* of ʿAbd ibn Ḥumayd,[35] al-Ṭayālisī,[36] and the books of al-Baihaqqī,[37] al-Ṭaḥāwī,[38] and al-Ṭabarānī.[39] The goal of these authors was to collect whatever

---

[30] A *muṣannaf* is a collection of hadiths arranged in chapters (*bāb, abwāb*) according to different topics. A *musnad* is arranged according to the names of the transmitters, and a *jāmiʿ* is a hadith work which contains hadiths on the whole range of topics.

[31] Abū ʿAlī al-Bazzāz (d. 1034).

[32] Ibn Hammam al-Himyārī 743?–827.

[33] A *muṣannaf* is a collection of hadiths arranged in chapters (*bāb, abwāb*) according to different topics.

[34] Abū Bakr ibn Abī Shayba, 775/6-849, author of *Kitāb al-Muṣannaf fīl-aḥādīth wa-l-athār* (Hyderabad: Maṭbaʿa al-ʿUlūm al-Sharqiyya, 1966).

[35] A Ḥanbalī scholar, d. 863/4.

[36] D. 818.

[37] Al-Baihaqqī, Aḥmad ibn al-Husain (944–1066). Especially *al-Sunan* and *Shuʿab al-Imān* (Beirut: Dār al-Kutub al-ʿIlmiyya, 1990).

[38] Al-Ṭaḥāwī (d. 933) was a Ḥanafī scholar who wrote on hadith.

[39] Sulaymān ibn Aḥmad al-Ṭabarānī (d. 970/1) wrote three works, *Muʿjam Kabīr*, *Muʿjam Ausāt*, and *Muʿjam Ṣaghīr*, (al-Madīna, 1968), ed. ʿAbd al-Raḥmān Muḥammad ʿUthmān.

they found, not to extract the best, nor to refine them, nor to make them more accessible for use.

4) The fourth rank are those whose authors after long centuries sought to gather hadiths which were not found in the first and second ranks of books, and these were in obscure *Jāmiʿ* and *Musnad* works, then they brought them to light. These had been transmitted from those whose reports the hadith scholars did not record, such as many of the jabbering preachers, heretics, and unreliable persons, or they were traditions (*āthār*) of the Companions and the Successors, or reports of the Children of Israel,[40] or from the philosophers and preachers, which were mixed by other transmitters with hadiths of the Prophet, may the peace and blessings of God be upon him, mistakenly or deliberately. Or they were interpretations of the Qur'ān or the sound hadiths, so that a group of righteous people transmitted their sense not realizing the abstruseness of the science of transmission, and therefore they made these ideas into hadith going back to the Prophet. Or these were concepts understood from the indications of the Qur'ān and the sunna which they deliberately made into completely independent hadiths, or there might have existed an assorted group of various hadiths which they combined into one hadith, as a uniform narration. The place to find such hadiths is the *Kitāb al-Ḍuʿafā'* (The Book of the Unreliable Hadiths) of Ibn Ḥibbān,[41] *al-Kāmil* of Ibn ʿAdī,[42] and the books of al-Khaṭīb,[43] Abū Nuʿaym,[44] al-Jūzāqanī,[45] Ibn ʿAsākir,[46] Ibn al-Najjār,[47] and al-Dailamī,[48] and the *Musnad* of al-Khwārizmī[49]

---

[40] These "reports of the Children of Israel" or "Isrāʿīliyyāt", represent a corpus of legendary material from which stories and beliefs were incorporated into Islam, hence they are suspect.

[41] Ibn Ḥibbān (d. 965), a Shāfiʿī traditionist of Khurasān.

[42] Abū Aḥmad ʿAbd Allāh ibn ʿAdī (d. 975–6) wrote *al-Kāmil fī-Ḍuʿafā' al-Rijāl.*

[43] Al-Khaṭīb al-Baghdādī (d. 1070–71) was a Ḥanbalī who changed to Shāfiʿism.

[44] Abū Nuʿaym al-Isfahānī (948–1038) author of *Ḥilyat al-Auliyā'* and a *Kitāb al-Duʿafā'*

[45] al-Juzajānī, Ibrahīm ibn Yaʿqūb al-Jūzajānī (d. 870). A Syrian who wrote on the reliability of transmitters. A Kitāb al-Ḍuʿafā' by him is mentioned in Ibn Ḥajar's al-*Tahdhīb.* Juynboll *Muslim Tradition* (Cambridge: Cambridge University Press, 1983), 239, or perhaps Ḥusain b. Ibrahīm al-Jauzaqānī, d. 1148/49 who according to Azmi compiled the first collection of spurious hadith.

[46] In his *Ta'rīkh Dimashq.*

[47] Author of *Dhayl Ta'rīkh Baghdād* 1118–1245.

[48] In his *Musnad al-Firdaus*

[49] al-Khwārizmī, Muḥammad ibn Maḥmud, *Jamiʿ Masānīd al-Imām Abī Ḥanīfa.* (Hyderabad 1332).

is almost of this rank. The best of this rank are those which are weak but equivocal, and the worst are those which are fabricated and whose word order is transposed[50] and which merit the strongest denial. This rank is the subject of the book *al-Mauḍū'āt* (The Fabrications) of Ibn al-Jauzī.[51]

5) There is also a fifth rank including what is widespread among the legal scholars, the Sufis, the historians and so on, and it has no basis in these four ranks. This includes the interpolations of those insolent about their religion, who know its language and invent strong chains which cannot be invalidated, and eloquent sayings which seem to issue from the Prophet, may the peace and blessings of God be upon him, and thus they have provoked in Islam a great misfortune. However the brilliant minds among the People of the Hadith have adduced things like this on this basis of the recensions and the context in which these are used and they have torn away the veils and exposed the flaws.

As for the first and second ranks, it is upon them that the hadith scholars rely, and it is within the sanctuary of these two ranks that their grazing ground and pasture lies. As for the third rank, no one undertakes implementing them and holding them to be accurate except the rare brilliant ones who have memorized the names of the transmitters and the defects (in the *isnād*s) of the hadith, although indeed sometimes supporting evidence for the text or meaning of other hadiths can be taken from these. "God made for everything a measure."[52]

As for the fourth rank, being preoccupied with collecting them or deriving from them is a sort of hair-splitting of the latter day writers. If you want the truth, sects of innovators among the Rafidites[53] and the Mu'tazilites and the others were capable, with the least effort, to extract from these evidence for their points of view. However, referring to such hadiths for assistance during the disputations of the scholars of hadith is not correct, and God knows better.

---

[50] Maqlūb.
[51] Ibn Qayyim al-Jauzī d. 1201.
[52] Qur'ān 65:3.
[53] The Rafidites were an extremist Shi'a faction.

CHAPTER 79

The Manner of Understanding What is Meant by the Utterance (Kalām)

Be informed that the speaker's expression of what is in his mind and the hearer's understanding of this occur at successive levels in terms of clarity and obscurity.

The highest of them is what has explicitly stated within it the confirmation of the ruling which was designated by it; and the utterance was intended to convey this information, and cannot have any another meaning.

Following this (in clarity) is something which is lacking one of three qualifications

1) The ruling is established by a general designation taking in a group of designations inclusively or interchangeably such as "people," "Muslims," "nation," and "men," or by demonstrative pronouns when their referent is general, or by things described by a common description, or denied by the "*lā*" of absolute negation, for often the general entails a specific designation.

2) The utterance was not uttered to convey that information but this was entailed by what was said, such as "excellent Zaid came to me," in relation to his being excellent, or "O poor Zaid," in relation to establishing his poverty.

3) The utterance may also be interpreted in another sense such as the homonym, something which has a literal meaning and a commonly recognized metaphorical sense; and something which might be known through example or category, but is not known by a comprehensive exclusive definition. An example is the term "journey," for it is known that an example would be leaving Medina with the goal of Mecca, and it is known that among such movements may be pleasure trips or frequent comings and goings due to needs in which a person returns to the village on the same day, and there may be a journey; while the differentiating limit and the common ground between two individual instances is not known. Another example is the demonstrative pronoun or the pronoun,

when there is a mutual applicability in the referential context, or in which the relative pronoun may refer to both things.

Then following this level is what the utterance connotes beyond the explicit wording in it, and the major types of this are three.

1) The first is the import (*faḥwā*) of the utterance, and this occurs when the utterance conveys an understanding about something which it does not explicitly state through the sense bearing on the ruling. For example, (the Qur'ānic injunction) "Don't even say 'uff' to them (your mother and father)."[1] From this is understood the forbidding of striking the parents, a fortiori;[2] and for example, "Whoever eats during the day in Ramadan must make up for it."[3] From this is understood that what was meant is any breaking of the fast, and eating was only specified because this form is the first to come to mind.

2) The second is entailment (*iqtiḍā'*) and this is that it (the wording) conveys an understanding of meanings through a meaning being associated with (an expression) used in it, on the basis of customary, rational, or legal (usage). For example, "he set free" or "he sold" both require that previously someone owned something; "walking" entails the soundness of the legs; "praying" entails that a person be in a state of ritual purity.

3) (The third is) indirect allusion (*īmā'*). This is when the intent is conveyed through expressions conveying the related implications, so that the rhetoriticians mean by this combining the expression with the related implication which is additional to the basic meaning. Thus the utterance conveys its related implication, for example, restriction; through a description or condition indicting that the ruling is nullified if these are not present,[4] in those instances where neither the form of the question nor the expression of an obvious instance nor the explanation of the benefit of the ruling, is meant. Indirect allusion is also meant by what is understood as being in the nature of an exception, a statement of extent (until, or up to), or the specification of a number.[5]

---

[1] Qur'ān 17:23.

[2] Discussed in Wael Hallaq, "Non-Analogical Arguments in Sunni Juridical Qiyās", *Arabica* (36, 1989): 286–306.

[3] Hadith

[4] For example, qualifying a particular breed or type of animal rather than a whole species, when the ruling is not meant to be generalized.

[5] Statements which qualify a command by "except for such and such," "up to such and such a limit or extent," or which specify a certain number.

A condition for considering "indirect allusion" is that contradiction should occur according to the convention of the users of the language. For example, (if someone says), "Give me ten except for one," and then, "Give me only one," this will be judged a contradiction according to most people. As for that which is only be comprehended by those of utmost expertise in semantics, it should not be taken into account.

Then, following this level is whatever is inductively deduced by syllogism (*istadalla*) from the content of the utterance, and the majority of this falls within three categories.

1) Inclusion through generalizing, such as saying "a wolf has fangs" and that "everything having fangs is unlawful to eat," and its proof is through the conjunctive syllogism[6] and this is the Prophet's, may the peace and blessings of God be upon him, saying, "Nothing was revealed to me about donkeys except for this single and comprehensive verse, 'The man who does an atom's weight of good shall see it and the one who does an atom's weight of evil shall see it.'"[7] In this category is the inductive inference (*istidlāl*) of Ibn 'Abbās using God's, may He be Exalted, saying, "With their guidance, be guided,"[8] and His saying, may He be Exalted, "David thought that we had put him on trial but he asked forgiveness of his Lord, and he fell to the ground kneeling and was sincere in repentance,"[9] when he (Ibn 'Abbās) said, "your Prophet was ordered to follow him (David)."[10]

2) Deduction (*istidlāl*) through logical entailment (*mulāzama*) or logical exclusion (*munāfā*)—for example, "If the single (*witr*) prayer had been compulsory it would not have been performed while mounted; however it may be performed this way.[11] Its proof

---

[6] *Qiyās Iqtirānī*. A conjunctive syllogism is of the type that if the two initial premises are true then a third may be deduced. If all A's are B and all B's are C then all A's are C.

[7] Bukhārī Manāqib 28, Khan trans. IV:537. The verse is Qur'ān 99:7–8.

[8] Qur'ān 6:90. In the previous verses seventeen prophets were named and Muḥammad was told to be guided by them, therefore, Ibn 'Abbās' derivation was that people should be guided by Muḥammad.

[9] Qur'ān 38:25.

[10] Bukhārī Anbiyā' 39, Khan trans. IV:419; Tafsīr 6:5, 38:1, Ibn Ḥanbal. That is, the Prophet was ordered to follow David. This refers to the prostration which Muslims may perform at the recitation of the verses of the Qur'ān concerning David's prostration which occur in the Qur'ānic Chapter Ṣād 38:25.

[11] The non-compulsory prayers can be performed while mounted. Logical exclusion works by comparing the statements: i.e., "The *witr* prayer can be performed while mounted." "The obligatory prayers cannot be performed while mounted." Therefore by exclusion *witr* cannot be compulsory.

is through the conditional syllogism and of this type this God's, may He be Exalted, saying, "If there had been in them (heaven and earth) Gods other than Allah, they (heaven and earth) would have been corrupted."[12]

3) Analogical reasoning (*qiyās*) and this is correlating one case to another case through a rationale for legislation (*'illa*) which is shared by both of them such as that there is *ribā*[13] in chick peas just like there is in wheat.[14] And among them is his, may the peace and blessings of God be upon him, saying, "Do you think that if your father had a debt and you paid it off that it would be paid off in this way?" He answered "Yes." The Prophet said, "Then you can also make the pilgrimage on his behalf,"[15] and God knows better.

---

[12] Qur'ān 21:22.

[13] This interpretation of usury (*ribā*) here is that barter of foodstuffs against a like commodity is only permissible when offerings on both sides are exactly equal in weight or quantity and when delivery on both sides was immediate.

[14] Wheat is mentioned in this context in a hadith cited in Muslim Masāqā 83, Tirmidhī, Nasā'ī, Ibn Mājah, Ibn Ḥanbal.

[15] I.e., on the deceased father's behalf. Indicting that the idea of a "debt" is the common *'illa*. See R. Brunschvig, "Raisonnement par analogie" *Studia Islamica* 34 (1971):67. Nasā'ī Ḥajj 11, Ibn Ḥanbal IV:5. Previously cited in Ch. 68.

CHAPTER 80

The Manner of Understanding the Legal (Shar'ī) *Meanings From the Qur'ān and the Sunna*

Be informed that the wording indicating God's pleasure and anger is love and hatred; mercy and cursing; nearness and distance; and the attribution of the act to the ones with whom God is pleased or displeased; such as the believers and the hypocrites, the angels and the devils, and the people of Heaven and those of Hell, by commanded and forbidding, by the explanation of the requital resulting from the action, by the comparison to something customarily praised or blamed, and by the Prophet's, may the peace and blessings of God be upon him, giving importance to doing it or avoiding it despite the presence of motives for doing it.

As for the distinction between the degrees of pleasure and anger as being compulsory, recommended, forbidden and reprehensible; the clearest of them is what clarifies the condition of its opposite, as in "the man who does not contribute the alms tax will find his wealth represented to him,"[1] and the Prophet's, may the peace and blessings of God be upon him, saying, "And whoever does not do it, it does not matter."[2] Next is wording like, "it is incumbent" and "it is not lawful" and the thing being made a pillar of Islam or of unbelief, or giving great emphasis to doing something or not doing it, and (wording like), "This is not a manly attitude" or "This is not proper." Then there are the rulings of the Companions and the Successors concerning these cases, such as the saying of 'Umar, may God be pleased with him, "The prostration when reciting the Qur'ān is not obligatory,"[3] and the saying of 'Alī, may God be pleased with him, "The witr prayer is not obligatory." Next is the status of the intended thing, for example,

---

[1] On the Day of Judgment as two venomous snakes biting him and hanging around his neck. Bukhārī Zakāt 3, Tafsīr 3:14, Ibn Mājah, Muwaṭṭa', Ibn Ḥanbal.

[2] That is, there is no obligation to do it.

[3] When certain verses of the Qur'ān are read or recited which mention prostrating before God, Muslims are to prostrate once.

its being a completion of an act of worship, or a way to block something leading to sin, or its being something which comes under the topic of dignity and good manners.

As for recognizing the reason for legislation, the pillar, and the condition; the clearest among them is what is provided by a textual designation, for example, "Every intoxicating thing is forbidden,"[4] or "The one who does not recite the *Umm al-Kitāb*[5] has not prayed,"[6] or, "The prayer of one of you is not accepted unless he performs the ablution."[7] Or it is through an textual indication (*ishāra*) or an indirect allusion (*īmā'*) such as the saying of a man, "I slept with my wife during (the day in) Ramaḍān," and the Prophet, may the peace and blessings of God be upon him, said, "Then free a slave."[8] Or as the designation of the prayer as consisting of standing, bowing, and prostrating, gives the understanding that these are its pillars.

The saying of the Prophet, may the peace and blessings of God be upon him, "Leave them (my leather socks), for I put them on when they (my feet) were clean,"[9] gives the understanding of cleanliness being made a condition when putting on the socks.[10]

Therefore a ruling is frequently found accompanied by the presence of something else, or it is not found when this is absent, so that in the mind it will be established that this thing is a reason for legislation, a pillar, or a condition of something. (This is) similar to what unconsciously enters the mind of a Persian speaker in recognizing the proper applications of Arabic expressions according to the practice and usage of the Arabs and their employing these in situations conjoined with circumstances, in that he does not comprehend (the meaning by convention) but rather his standard is this very recognition (of the impinging contextual factors). Thus if we see that the law-giver whenever he prayed performed

---

[4] Muslim Ashraba 73–75, 64,69 Bukhārī Adab 80 aḥkām 22 maghāzī 60, Abū Dāwūd, Tirmidhī, Nasā'ī, Ibn Mājah, Ibn Ḥanbal.

[5] *Umm al-Kitāb*—"The mother of the book" refers to the Fātiḥa or opening chapter of the Qur'ān.

[6] Variant in Ibn Ḥanbal II:250, 457.

[7] Bukhārī Wuḍū', Ḥaiḍ, Muslim, Tirmidhī, Ibn Ḥanbal.

[8] From this it can be inferred that breaking the fast induces a penalty. Abū Dāwūd Ṣaum 19, Ibn Ḥanbal II:281.

[9] Bukhārī Wuḍū' 49, Libās 11, Muslim, Dārimī, Ibn Ḥanbal. This hadith is cited in *Mishkāt*, p. 103.

[10] That is, for it to be permissible to rub the leather socks rather than washing the feet during the ablution, the feet must have been washed before putting on the socks.

bowing and prostrations, and removed unclean things from himself, and that he did this repeatedly; we can firmly decide that this was what was intended. If you want the truth, then this what is absolutely reliable in recognizing the individual properties. Thus if we see people collecting wood, and from it making something to sit upon, and calling it "the seat," we can deduce from this its individual characteristics; and thus (we can) identify an anchoring point (*manāṭ*)[11] (for the ruling) depending (methodologically) on finding a relationship, or on probing and deleting.

As for recognizing the goals on which the rulings are founded, this is a delicate science which only the person of subtle understanding can penetrate and correctly comprehend. The legal scholars among the Companions had learned the principles of the acts of obedience and the sins from well-known things on which the religions present at that time agreed, such as the Arab polytheists, the Jews, and the Christians, so they did not have a need to know their rationale, nor to investigate what was connected with this.

As for the rules of legislation, facilitation, and the rulings of the religion, they learned them from observing the occasions of the commanding and forbidding, as the companions of a doctor come to know the purposes of the medication which he prescribes by virtue of their lengthy association and practice along with him; and the Companions were at the high level in knowing these. Among this is the saying of 'Umar to the man who wanted to combine the supererogatory prayer with the obligatory prayer, "Through this people have been destroyed before you," so the Prophet, may the peace and blessings of God be upon him, said, "God has granted you the true opinion, O Ibn al-Khaṭṭāb."[12] And there is the saying of Ibn 'Abbās explaining the reason for the command to take a full bath before the prayer on Fridays, and the saying of 'Umar,

---

[11] "*Takhrīj al-Manāṭ*". The jurist in dealing with the reason for legislation (*'illa*) faces three problems. 1) Verification of the basis of the ruling. 2) Revising the *'illa* by isolating non-essential properties. 3) identifying the *'illa* (*takhrīj al-manāṭ*) in a case where the ruling was given without an apparent reason. See Wael B. Hallaq, "The Development of Logical Structure in Sunni Legal Theory." *Der Islam*, Band 64 Heft 1 (1987): 42–67.

[12] Previously cited in the introduction of this book. This hadith signifies that there should be some pause between the obligatory prayer and the supererogatory (*nafl*) prayers. Abū Dāwūd I,:263 #1007 Ṣalāt 188, with the specification of "People of the Book." Ibn Ḥanbal V:368, has a variant of the hadith.

"My Lord concurred with my judgment in three things,"[13] and the saying of Zaid ibn Thābit about the forbidden (future) sales (of fruit), "The fruit used to be struck by blight, splitting, blistering from the heat, and so on,"[14] and the saying of ʿĀ'isha, "If the Prophet had seen what the women are doing these days, he would have forbidden them from coming to the mosques just as the women of the Children of Israel were forbidden."[15]

The clearest way (of understanding the rulings) is:

1) What is explained by the text of the Book and the Sunna. For example (from the Qur'ān), "And there is life for you in retaliation (*qiṣāṣ*), O men of understanding."[16] "God knew that you were deceiving yourselves, so He turned in mercy to you and forgave you."[17] "Now God has lightened your burden for He knows that there is weakness in you."[18] "If you don't do it there will be a crisis on the earth and a great corruption."[19] "So that if one makes an error the other will remind her."[20] And from the Prophet's sayings, "He does not know where his hand has spent the night,"[21] and "The devil spends the night in his nose."[22]

2) Then comes the level of meanings that the Prophet pointed out or alluded to indirectly, such as his saying, "Be on guard against the two causes for being cursed,"[23] and his saying, "The drawstring on action is the two eyes."[24]

---

[13] That is, ʿUmar had given opinions about praying at the station of Abraham, the observance of veiling by women, and the treatment of prisoners at Badr. Later revelation corroborated these In the version reported in Bukhārī trans. 1, 239–240 the third opinion was that given by ʿUmar to the wives of the Prophet concerning their behavior and then the Qur'ān was revealed warning them. Bukhārī Ṣalāt 32, Tafsīr 2:9, *Ṣaḥīḥ Muslim*, IV:1280, Dārimī, Ibn Ḥanbal.

[14] Bukhārī Buyūʿ 85, Ibn Ḥanbal V:190, Abū Dāwūd III:253 #3372 Buyūʿ 22.

[15] Bukhārī Adhān 163, Muslim Ṣalāt 144, Dārimī, Tirmidhī, Muwaṭṭa', Ibn Ḥanbal.

[16] Qur'ān 2:179. In that retaliation (*qiṣāṣ*) restrains random violence and blood-feuding.

[17] Qur'ān 2:187.

[18] Qur'ān 8:66.

[19] Qur'ān 5:73.

[20] Qur'ān 2:282.

[21] Therefore he should wash his hands. Bukhārī Wuḍū' 260, Muslim, Abū Dāwūd, Nasā'ī Tirmidhī, Ibn Mājah, Muwaṭṭa' Ibn Ḥanbal. *Mishkāt*, p. 82.

[22] Bukhārī. Bad' al-khalq, Muslim, Nasā'ī. Meaning that the person must rinse the nose during the ablution to eliminate accumulated dirt. *Mishkāt*, p. 82.

[23] These are being cursed for defecating under a shade tree or on a public thoroughfare. *Mishkāt*, p. 72–73. Muslim Ṭahāra 68.

[24] When a person sleeps the ties controlling action are loosened, therefore ablution must be performed after sleeping. Abū Dāwūd I:51–52 Ṭahāra 79. Ibn Mājah, Dārimī, Ibn Ḥanbal. *Mishkāt* p. 70.

3) The next level is that which a Companion with legal acumen mentioned.

4) Then comes the identification of the anchoring point (for a ruling)[25] in a way that it goes back to a goal which came to be considered, or something like it came to be considered, in a parallel case.

There is no haphazardness in the (religious) command, so it is necessary to investigate why certain amounts are specified while others like them are not, and to investigate why the specifications of general referents are excluded—whether it is because their goal is voided, or whether it is due to the existence of some impediment which was given preference when they conflicted—and God knows better.

[25] See footnote 11 above.

CHAPTER 81

Judging Among Divergent Hadiths

The basic principle is to implement every hadith unless inconsistency would preclude acting according to them all. In reality there can be no disagreement, except from our perspective. Thus if two opposing hadiths come to light under the topic of telling about an action of the Prophet, so that one Companion says that the Prophet, may the peace and blessings of God be upon him, did one thing and another says that he did something else, there is no conflict; for these two may both be considered permissible (*mubāḥ*) if they fall under the heading of habit, not of religious observance; or one of the actions may be recommended (*mustaḥabb*) while the other is permitted, since there resulted from one of the actions the effects of drawing nearer to God and not from the other; or both of the acts may be recommended or compulsory and one is sufficient to fulfill the other when both come under the heading of drawing nearer to God.

The memorizers of traditions among the Companions have made statements of this type about many of the practices of the Prophet such as his performing the *witr* prayer with eleven *rakʿat*s, or with nine or seven; and such as his praying the *tahajjud* prayer aloud or silently. It is according to this principle that it is necessary to judge the case of raising of the hands at the time of ritual prayer to the ears or to the shoulders; and to decide among the ways of making the *Tashahhud*[1] of ʿUmar, Ibn Maʿsūd and Ibn ʿAbbās; and concerning the *witr*, whether it is composed of a single *rakʿa* or three *rakʿas*, and concerning the invocations commencing the ritual prayer,[2] and the invocations of the morning and evening, and the other causes[3] and times.[4]

---

[1] *Tashahhud* is a portion of the ritual prayer in which the person declares, "I believe that there is no God but God and Muḥammad is his Prophet," together with other declarations on which the phraseology of the three reporters differed here.

[2] "*Adʿiya' al-istiftāḥ*". An invocation (*duʿā*), based on the practice of the Prophet, which is recited at the commencement of the ritual prayer.

[3] Causes are things which bring the rulings of the divine law into effect such as the time for prayer bring the necessity to pray into effect.

[4] Why and when the Prophet varied his practice.

The fact that there are two (apparently conflicting) hadiths may indicate a way to be rid of some hardship when a case had preceded which caused this (hardship), such as certain features of the expiatory offerings,[5] and the amount of the poll-tax for one who fights (against God and the Prophet), mentioned in a dictum (of the Prophet).[6]

Or there may be in this a hidden reason for legislation which categorizes as obligatory, or recommended, one of the acts at one time and the other at another, or requires a thing on one occasion and gives a dispensation in it at another, so it is necessary to study this deeply. Or one of them may be an ordinance as interpreted strictly (*'azīma*) and the other a dispensation; if the legal force was manifest in the first case and the consideration of hardship in the second.

If a proof of abrogation comes to light, this should be upheld. If one of them is the account of an action and the other takes a report back to the Prophet, then if the report is not absolutely definitive in forbidding or compelling, nor is the report absolutely sure in going back to the Prophet, then both may have a bearing in some aspect. But if the report is absolutely definitive, bearing on the Prophet's, may the peace and blessings of God be upon him, specifying acting upon it or its being abrogated, then the contexts of these two hadiths should be researched. If there are two reports and one of them is explicit in conveying a meaning lacking in the other, and their interpretation is close, then this has a bearing on one of them being the explanation of the other. However, if the interpretation is far-fetched it does not have any bearing on the other unless there is extremely powerful circumstantial evidence, or the interpretation was transmitted from a Companion of the Prophet who was also a man of legal acumen such as the saying of 'Abd Allāh ibn Salām about the hour (on Friday) when prayers are most likely to be answered, that it is just before sunset, and Abū Huraira (refuted this) by reporting that prayer was prohibited at this time. The Prophet had said, "No Muslim standing to pray asks Allāh for something at that time . . ."[7] Thus 'Abd

[5] These expiations for missed obligations may vary, for example, to free a slave, fast, feed the poor, and so on.

[6] The amount of *jizya* applied to various populations.

[7] "But that his prayer is answered." Ibn Mājah 1:360 Iqāma, 99, Bukhārī Juma' 37. The hour is said to be the last hour of the day on Friday so that the difference

Allāh ibn Salām said, "When a person is waiting for the prayer, it is as though he is in prayer," and this interpretation is far-fetched and would not have been accepted had it not be held by a Companion possessing legal acumen.

What confirms an exegesis as far-fetched is that if it were presented to sound minds without circumstantial evidence or its without having undergone belabored discussion, it would not be considered plausible, and if it is in opposition to a manifest allusion, a clear understanding, or a revealed textual source, then it is not permissible at all.

Under plausible interpretations come:

1) The restriction of a general statement when the customary practice has been to apply it to certain individual cases only in rulings similar to this one on these certain cases.

2) Generalizing in a situation where the customary practice has been not to be overly particular, such as in praising and blaming.

3) A generalization leading to the legislation being installed in a ruling after the principle of that ruling has been made clear; so that this comes to have the force of an unqualified proposition (*qaḍiyya muhmala*). An example is the Prophet's saying, "On whatever (land) the sky pours down (rain), the *zakāt* of one-tenth applies,"[8] and his saying, "There is no alms tax on whatever is below five *ausuq*."[9]

Among them (ways to deal with differing hadiths) is to reduce each hadith to one pattern if the anchoring reason (*manāṭ*) adduced (for the *'illa*) and the relevance (to the judgment) attest to this, and to interpret both of the hadiths as making an act reprehensible or indicating permissibility, in a general way, if this is possible; or interpreting emphasis as functioning as a deterrent factor if this has been preceded by a disputation.

His saying, "Carrion is forbidden (*ḥurrimat*) you,"[10] refers to

---

concerned whether the prayer meant was the sunset prayer which occurs just after sunset, or a supplication done before this prayer.

[8] In Arabic this phrase contains the general (*'āmm*) word "*mā*"—"*what* is watered by springs." Bukhārī Zakāt 55, Muslim, Abū Dāwūd, Tirmidhī, Nasā'ī, Ibn Mājah, Dārimī, Muwaṭṭa', Ibn Ḥanbal. Bukhārī Zakāt 55, Muslim, Abū Dāwūd, Tirmidhī, Nasā'ī, Ibn Mājah, Dārimī, Muwaṭṭa', Ibn Ḥanbal.

[9] Bukhārī Zakāt 4, 32, 42, 56, Sales 83. If these two hadith are combined there is a specification of the former by the latter which also contains the general word "*mā*" (what).

[10] Qur'ān 5:3.

eating it, while his saying, "Your mothers are forbidden to you (*ḥurrimat*),"[11] refers to marrying them.[12]

The Prophet's, may the peace and blessings of God be upon him, saying, "The evil eye is a reality (*ḥaqq*),"[13] means that its influence is proven while "The Prophet is true (*ḥaqq*),"[14] means he has truly been sent (by God).

The Prophet's, may the peace and blessings of God be upon him, saying, "Error and forgetfulness have been removed from my community,"[15] refers to the (removal of the) sins which occur in these states.

His sayings, "There is no prayer except in a state of purity,"[16] "there is no marriage without a guardian,"[17] and "acts are only through their intentions,"[18] refer to the effectiveness of these things not ensuing (without these factors) which the law-giver established for them.

(The Qurʾānic verse), "When you are going to perform the prayer then perform the ablutions,"[19] means if you are not already in a state of having made the ablution.

These are evident, and they are not to be interpreted, because the Arabs used each of the words in them in this sense and they meant what was suitable for that context, and this is their language in which they do not find any words diverging from what is apparent.

Then if two (hadiths) come under the scope of a fatwa (a legal recommendation) about an issue, or a judgment concerning a particular case, and if there comes to light a reason for legislation (*ʿilla*) which differentiates between the two, the judgment is to be made according to this rationale. An example is that a youth asked the Prophet about kissing in the case of someone who is fasting and he forbade it; and when an old man asked, he permitted it. If

---

[11] Qur'ān 4:23.

[12] As discussed in al-Ghazzālī, *Mustaṣfā* II:187–188.

[13] Bukhārī Ṭibb 36, Libās 86, Abū Dāwūd, Tirmidhī, Muwaṭṭa' Ibn Ḥanbal.

[14] "Muḥammad ḥaqq" in Bukhārī Tahajjud 1, Du'āt 9, Tauḥīd 8, 24, 35.

[15] Ibn Mājah Ṭalāq—with slightly different wording (*tajāwaz*) instead of "*rufiʿa*."

[16] This does not seem to be a hadith, per se, but is cited as synonymous with the hadith found in Bukhārī Wuḍū', and explained in Ibn Ḥajar, *Fatḥ al-Bārī* XII:329.

[17] Bukhārī Nikāḥ 36, Abū Dāwūd, Tirmidhī, Ibn Mājah, Dārimī, Ibn Ḥanbal.

[18] Bukhārī Bad' al-wahī 1, Īmān 41, etc., Muslim, Abū Dāwūd, Tirmidhī, Nasā'ī, Muwaṭṭa', Ibn Ḥanbal. Cited in the introduction.

[19] Qur'ān 5:6.

the context of one of them and not the other indicates the existence of a need or a persistent request on the part of the questioner, or its being an easing of the demand for perfection, or the repudiation of the person who was too strict on himself, then the decision will be made on the principle of (either) a strict interpretation (*ʿazīma*) or a dispensation (*rukhṣa*).[20]

If there are two hadiths which present a solution to a person in difficulty or indicate two punishments for a criminal or two monetary expiations for the one who broke his oath; then the soundness of both hadiths is conceivable, although their abrogation is also possible. On this principle is the judgment about the woman who has a prolonged flow of menstrual blood who was sometimes given an opinion that she should take a full bath between every two prayers, and sometimes that she should calculate menstruation as lasting the number of days usual for her, or its being the days when a copious flow of blood is apparent, according to a report. She can choose between the two, for the usual length of the menstrual flow and the color of the blood, both of them, are proper signs for indicating menstruation for the fasting person. (And the Prophet gave the order about) giving food (to the poor) in the name of the man who died and had not made up for missing his fast, according to a report.[21] The doubt of the person who thinks he has erred (in counting the *rakʿāt*s) in the prayer may be resolved in one of two ways; either by trying to find out what is correct, or by counting only what he is sure of, according to a report.[22] The judgment in establishing lineage may be done through comparing physical features or by casting lots, according to a report.[23]

If the proof of abrogation is forthcoming it must be applied, and abrogation can be known through the words of the Prophet, may the peace and blessings of God be upon him, such as his saying, "I had forbidden you to visit graves, but now you may visit them,"[24] or through recognizing the posteriority of one hadith

---

[20] *ʿAzīma* principles are ones considered in the ideal sense as a priori. Their original rigor is maintained and they constitute the law as intended by God. When considered with reference to the attenuating circumstances of life it is said that a *rukhṣa* (dispensation) is given.

[21] Ibn Mājah Ṣiyām 50, Tirmidhī.

[22] He may have performed two, three, or four, but he assumes that the minimum (two) have been performed, in order to be sure.

[23] For example in hadiths reported in Bukhārī Farā'iḍ 1, Nasā'ī Ṭalāq 50.

[24] *Mishkāt*, p. 369. Muslim transmitted it. Muwaṭṭa' Ḍaḥāyā 8.

to the other when it is not possible to combine them. If the lawgiver legislated a law, and then legislated in its place another without mentioning the first, the Companions with legal acumen recognized that the latter one abrogated the first. Or if hadiths disagreed and one of the Companions decreed that one of them abrogated the other, that one is presumed abrogated but is not conclusively abrogated. The opinion of the legal scholars, i. e., that whatever they found to be in opposition to the practice of their teachers was abrogated, is not conclusive. Abrogation in those things which they showed to be the changing of a ruling to another ruling, was in reality the coming to an end of the ruling due to the cessation of its reason for legislation (*ʿilla*), or due to the cessation of its being an anticipated means (*maẓinna*) of serving the intended purpose, or due to something arising which prevented its being a reason for legislation, or due to the emergence of the preference for another ruling on the part of the Prophet, may the peace and blessings of God be upon him, through a manifest revelation (*waḥi jalī*),[25] or through his individual reasoning—that is, if the previous ruling had also been arrived at through his independent reasoning. God, may He be Exalted, said, speaking about the Night Journey,[26] "The statement which comes from Me cannot be altered."[27]

When, however, it is impossible to reconcile (two contradictory traditions) or to explain them by each other, and when nothing is known of an abrogating factor, then contradiction is established. Then, if a preference for one of them emerges, either due to a quality in the chain of transmitters in its having had many transmitters, the legal acumen of the transmitter, the strength of the link to the Prophet, the clear evidence of its going back to the Prophet, the transmitter's being the one involved in the affair such as the one asking for the opinion or the one being addressed or the one carrying out the act; or due to a quality of the text in its being very emphatic and explicit, or in view of the fact that the ruling and its reason for being legislated are more in conformity with the rulings of the divine law, and it being a reason for legislation which is strongly relevant (for the rulings) and whose effectiveness is recognized, or due to an extrinsic factor in its being

[25] The Qur'ān, as opposed to the Prophet's own ideas (*wahī khafī*).
[26] The Night Journey of Muḥammad through the seven heavens.
[27] Qur'ān 50:29. That is, the hadith cannot abrogate the Qur'ān.

accepted by more of the knowledgeable people; then this ruling will be preferred, and if not then both of the hadiths will have no force. This is a hypothetical situation which is scarcely to be found.[28]

The saying of a Companion, "The Prophet commanded, forbade, judged and gave dispensations," then his saying "we were ordered and forbidden," then his saying, "such and such is the practice of the Prophet, and whosoever does this has disobeyed Abū Qasīm,"[29] then his saying, "this ruling of the Prophet, clearly going back to him," and the Companion's using methods of independent reasoning in conceptualizing the reason on which the ruling is based, or in determining whether the ruling was compulsory or recommended, or general or particular; and the Companion's saying that the Prophet used to do such and such a thing—is clear in indicating that the action was done repeatedly, and the statement of another Companion that he used to do something else does not contradict it. His statement, "I used to keep company with the Prophet and I did not see him forbidding such-and-such," and "we used to do it in his time," clearly represent the sanctioning (*taqrīr*) of an act[30] and do not constitute a ruling stipulated in either the Qur'ān or the sunna.

The wording of the hadiths may vary due to variations in the manner of transmission, and this is due to the process of transmitting the hadith according to its meaning, so if a hadith is brought forward and the reliable scholars are unanimous regarding its wording, then these are manifestly the words of the Prophet, may the peace and blessings of God be upon him. In this case it is possible to make an induction (*istidlāl*) on the basis of something coming early or being later (in the text) and its having a "*Waw*" or a "*Fa*,"[31] and so on, concerning expressions which are additional to the basic meaning. If the transmitters differ with a plausible difference and they were of nearly equal status in legal acumen, memory, and number, then the obviousness (that these are the Prophet's words) collapses, and only that meaning upon which they all concur can be deduced. The majority of transmitters attended to

[28] That contradiction between two hadiths could not be reconciled or solved in some way.

[29] The Prophet.

[30] When the Prophet witnessed a certain action and did not forbid it.

[31] These are the two most common conjunctive particles in Arabic. "*Waw*" beginning a phrase means "and" while "*Fa*" can mean "so that," "then," "but," etc.

the main ideas of the meaning, not to the peripheral factors, and if their status varies the saying of the reliable one, that of the majority, or that of the person most acquainted with the story should be adopted. If the saying of someone reliable indicates their greater precision, for example the transmitter's saying, "She said ('Ā'isha), 'He sprang up' and she did not say, 'he stood up', and she said, 'he poured water on his skin' and she did not say, 'he washed,'" then it should be accepted.[32]

If they disagree inordinately (on some points) and are otherwise close and there is no reason for preference, then the particulars in which they vary can be discounted.

If the hadith which does not go back to the Prophet uninterruptedly, but is broken at the level of the Companion (*mursal*) is combined with circumstantial evidence, for example, it is supported by a hadith stopping at one of the Companions, or by a hadith whose chain is weak, or is made to reach the Prophet by some other hadith the reporters of which are different, or by the opinion of most of the knowledgeable people, or by a sound analogy, or by an allusion in a revealed statement, or if it is known that it was transmitted by a reliable authority; then it (this hadith) may be advanced as an argument, but it is lower in status than a traceable hadith, and if (it is not supported) then it cannot be used.

Similar is the case of the hadith whose transmitter is an imprecise person, but against whom there is no charge, or whose transmitter is a person of unknown status. The preferable course is that it should be accepted if it is combined with evidence such as agreeing with an analogy or the practice of most of the knowledgeable people, and otherwise it should not be accepted.

If a reliable transmitter is unique in reporting an addition which the silence of the rest does not preclude, then this is accepted; such as in tracing back the mursal hadith, or adding a person in the chain of transmitters, or mentioning the situation in which the hadith originated, the cause of its being transmitted, its prolixity, and its citing an extra sentence which does not alter the meaning of the text. If this (silence of the others regarding the addition) is impossible, such as in the case of the addition which changes the meaning or an unusual thing which would not customarily have been omitted; then it is not accepted. If one of the Companions

---

[32] A hadith cited in Muslim Musāfirūn 129. Trans. of *Ṣaḥīḥ Muslim*, I:357.

construes a hadith in a certain sense and has undertaken independent reasoning (*ijtihād*) concerning this, then this is presumed correct, on the whole, as long as no proof can be furnished against it (this interpretation). If *ijtihād* was not used then this interpretation is strong, as in the case where an intelligent person who is knowledgeable about the language specifies (an interpretation) in the light of related circumstances and contextual evidence.

As for the disagreement of the reports of the Companions and the Successors, if it is easy to combine them through certain ways previously mentioned, then well and good, and if not, then two or more opinions may be found concerning the case under discussion, and which of them is the most correct must be investigated. Part of the well-kept knowledge is knowing the source of the schools of thought of the Companions, so make an effort and you will be granted a portion of this, and God knows better.

CHAPTER 82

The Causes of the Disagreement of the Companions and the Successors Concerning Applied Jurisprudence (al-Furūʿ)[1]

Know that in the noble time of the Messenger of God, may the peace and blessings of God be upon him, law had not yet been put into writing. The mode of investigating legal rulings at that time was not like the investigative method of today's jurists who make their greatest efforts in explaining the pillars, conditions, and principles governing[2] each matter as distinguished from other matters on the basis of its indicant (*dalīl*),[3] they posit hypothetical cases which they discuss and formulate definitions for whatever may be definable, and stipulate limits for whatever may be limited, and so on with the rest of their accomplishments.

As for the Messenger of God, may the peace and blessings of God be upon him, he used to perform ablution and the Companions would see his manner of ablution and imitate it without his explaining what was a (necessary) pillar (*rukn*) and what was his preferred mode of behavior (*adab*).[4] He used to pray, and they saw his prayers so they prayed just as they had seen him praying. He performed the pilgrimage and the people noted the way he performed the pilgrimage and did as he had done. This was his usual way, may the peace and blessings of God be upon him, and he did not explain whether there were six or four obligatory aspects to the ablution,[5] nor did he hypothesize that it would be possible that a person should do the ablution in any way other

[1] This chapter opens the author's treatise *al-Inṣāf fī Bayān Sabab al-Ikhtilāf* (Cairo: Maktaba al-Tablīghiyya, 1385) and continues to p. 10 of that work.

[2] The pillars are the necessary elements of a ruling. The conditions are attributes or elements without which a ruling cannot be fulfilled, for example, purity is a prerequisite of prayer, and being impure nullifies prayer.

[3] Indicants are features of the textual ruling which signify the reason for or force of an injunction.

[4] I.e., without his explaining what was obligatory (*farḍ*) or recommended as his practice (*mustaḥabb*).

[5] Ḥanafī's hold that they are four; Shāfiʿī's, six.

than in an uninterrupted sequence, so that he should rule on the soundness or invalidity of this, except on rare occasions. The Companions only rarely asked him about these things.

It is reported from Ibn ʿAbbās, may God be pleased with them (he and his father), that he said, "I never saw any group better than the Companions of the Messenger of God, may the peace and blessings of God be upon him. Up to the time of his death they only had asked him about thirteen issues, all of which were found in the Qur'ān. Among those issues were 'They will ask you about the fighting in the sacred month. Say, fighting in it is a great sin,'[6] and 'They will ask you about menstruation.'[7] He (Ibn ʿAbbās) said, "They only used to ask about what would be beneficial for them."[8] Ibn ʿUmar said, "Don't ask about things which have not yet arisen for I heard ʿUmar ibn al-Khaṭṭāb curse someone who asked about something hypothetical."[9] Al-Qāsim said, "You are asking about things which we didn't use to ask about and you are probing into things which we didn't use to probe into. You are asking about things which we didn't know about and if we had known them we would not have been permitted to keep them hidden." It is reported from ʿUmar ibn Isḥāq that he said, "Those of the Companions of the Prophet, may the peace and blessings of God be upon him, whom I have met outnumber those who had passed away before me and I never saw any group more easygoing in behavior and more lacking in severity than them." It is reported from ʿUbāda ibn Bisr al-Kindī, that he was asked about a woman who died while among a group of people where she had no guardian.[10] He said, "I have met many groups of people who weren't as severe as you and who didn't ask about (these) issues the way that you do." Al-Dārimī reported these accounts.[11]

The Prophet, may the peace and blessings of God be upon him, was asked by people to give legal opinions about things as they came up, so that he gave opinions concerning them, and cases

---

[6] Qur'ān 2:217.

[7] Qur'ān 2:222.

[8] Dārimī I:50.

[9] Dārimī I:50.

[10] To wash her corpse for burial.

[11] al-Dārimī, ʿAbd al-Raḥmān 797–869, from whose hadith collection *Sunan* (Beirut: Dār Iḥyā' al-Sunan al-Nabawiyya, 197–) Shāh Walī Allāh draws a number of examples in these pages. The ones cited above are drawn from hadiths in the chapter "The Repugnancy of Asking for Legal Opinions," 50–51.

were brought before him to adjudicate so that he judged them. He saw people doing something good so he praised it; or a bad thing, so he forbade it. Whenever he issued a legal opinion on something, passed a judgment, or forbade an action, this occurred in public situations. Similar (was the procedure of) the two Shaikhs (Abū Bakr and 'Umar), who when they didn't have any authoritative knowledge (*'ilm*) about an issue would ask the people for a hadith of the Messenger, may the peace and blessings of God be upon him.

Abū Bakr, may God be pleased with him, said, "I didn't hear the Messenger, may the peace and blessings of God be upon him, say anything about her," i.e., the grandmother, and therefore he asked the people. After he led the noon prayer he asked, "Did any of you ever hear the Prophet of God, may the peace and blessings of God be upon him, say anything about the grandmother?" Al-Mughīra ibn Shu'ba (669/70) said, "I have." Abū Bakr said "What did he say?" He replied, "The Prophet of God, may the peace and blessings of God be upon him, accorded her one-sixth (as a share of the inheritance). Abū Bakr then asked, "Does any one else besides you know of this?" Then Muḥammad ibn Salama said, "He has spoken truly." So Abū Bakr accorded the grandmother one-sixth as a share.[12]

(An example of the same type is) the story of 'Umar asking the people about the case of the compensation for causing the death of a foetus, then his having recourse to the report of Mughīra;[13] and his asking them about the plague, then his accepting the report of 'Abd Al-Raḥmān ibn 'Awf,[14] and likewise his having re-

---

[12] This hadith is reported in Dārimī II:359, Tirmidhī, Muwaṭṭa' and other collections.

[13] This was the compensation which Mughīra reported that the Prophet had established, i.e., "I heard him judging that a male or female slave should be given as compensation (*diya*)." al-Bukhārī (Khan trans. IX:2–33. See al-Shāfi'ī, *al-Risāla*, 263–264, for a discussion of this issue.

[14] During an expedition to Syria in 639 'Umar was informed that the country had been struck by a plague, and during consultation with other companions he did not obtain a definite answer concerning a precedent of whether he and his men should continued into the afflicted territory. Then 'Abd al-Raḥmān Ibn 'Awf informed him that the Prophet had said, "If you hear that a territory is afflicted by plague don't enter it, and if you are in a place where plague strikes, don't flee from it." Shāh Walī Allāh, *Muṣaffā* II:206–207. Bukhārī Ṭibb 30, Muslim, *Muwaṭṭa'* trans. Muhammad Rahimuddin (Lahore: Ashraf, 1980), 372–373. al-Shāfi'ī, *al-Risāla*, 264. This hadith is referred to in Ch. 41 of the present work.

course to 'Abd al-Rahmān ibn 'Awf's report in the story of the Magians;[15] and the joy of 'Abd Allāh ibn Mas'ūd at the report of Ma'qil ibn Yasār when it agreed with his opinion,[16] and the story of Abū Mūsā turning back from the door of 'Umar and 'Umar asking him about the hadith and Abū Sa'īd's bearing witness in his favor.[17] Examples of this are well-known and abundantly reported in the two *Ṣaḥīḥ*s and the *Sunan*.[18]

In summary, this was his noble habit, may the peace and blessings of God be upon him, and thus each Companion saw whatever God enabled him to see of his acts of worship, legal opinions, and judgments, then he committed them to memory, reflected upon them and recognized the reason for each thing due to the convergence of contextual evidence.[19] Thus the Companions interpreted some things as being permitted and some as being abrogated due to textual signs (*amārāt*) and contextual evidence which satisfied them. What was most salient for them was the sense of confidence and assurance and they scarcely ever resorted to methods of legal reasoning, just as you observe that Arabic speakers understand the meaning of a conversation among themselves as they become assured through declarations, signals and allusions, without their doing this consciously. The noble era of the Prophet came to a conclusion while the Companions were still proceeding in this manner.

Once the Companions had dispersed among different regions and each one had become the exemplar for some region, new legal problems proliferated and questions began to arise so that they

---

[15] 'Abd al-Raḥmān reported that the Prophet had treated the Magians of Hajar according to the same rules as the People of the Book. Therefore the *jizya* tax was levied on them. *Muṣaffā* II:167. Khan trans. *Bukhārī* IV:253. al-Shāfi'ī, *al-Risāla*, 265–6.

[16] Recounted in detail on the next page.

[17] According to a hadith Abū Mūsā al-Ash'arī came to 'Umar's door and knocked, then went away when it was not opened. 'Umar sent after him and asked, "why did you not come in?" Abū Mūsā replied that he had heard the Prophet say, "Ask permission to enter three times and if you are given it, enter, and if not, then leave." 'Umar said, "Who can confirm this?" Abū Mūsā went to the mosque and found that Abū Sa'īd al-Khudrī had heard the same statement. 'Umar assured Abū Mūsā that he had not doubted his veracity, but rather he had dome this as a deterrent to the circulation of false, unconfirmed hadiths. Khan trans. Bukhārī VIII:173; III:158. al-Shāfi'ī, *al-Risāla*, 267.

[18] The three most authoritative hadith collections of al-Bukhārī, Muslim, and Abū Dāwūd.

[19] "Qarā'in." For the nuances of this term see, Wael Hallaq, "Notes on the term *qarīna* in Islamic Legal Discourse", *Journal of the American Oriental Society*, 1989, 475–480.

were asked to give legal opinions about these. Each one answered on the basis of his recollection of the texts or resorted to legal inference. If he didn't find in his recollection or what he had deduced something which could serve to respond he would try to figure it out based on his opinion and ascertain the rationale for legislation on which the Messenger of God, may the peace and blessings of God be upon him, had based the ruling in his textual pronouncements. Thus he would search for the rationale of the ruling wherever he could find it and would spare no effort in order to remain consistent with his, upon him be blessings and peace, intent.

At this point disagreement of various types occurred among them.

1) Among them are that one Companion heard a ruling of a judgment or legal opinion while another one did not, so that the latter used his own opinion to do *ijtihād* about the case and this *ijtihād* could also turn out in various ways.

A) One of them is that his *ijtihād* might turn out to concur with the hadith. An example of this is what Al-Nasā'ī[20] and others recounted about Ibn Mas'ūd, may God be pleased with him. He was asked about a woman whose husband had passed away without settling her dowry portion.[21] Then he replied, "I did not see the Prophet of God, may the peace and blessings of God be upon him, making a judgment in such a case," but they kept on coming one after the other and asking him for a month, and remained insistent. Finally he performed *ijtihād* based on his own opinion, and ruled that she should receive the dowry of his other wives, neither less nor more, and that she should observe the waiting period, and that she would inherit. Then Ma'qil ibn Yasār stood up and testified that the Prophet, may the peace and blessings of God be upon him, had ruled similarly in the case of another woman, so Ibn Mas'ūd was overjoyed at this to a greater extent than he had been at anything else since accepting Islam.[22]

B) The second way is that there may occur a case disputed between two Companions and then the hadith comes to light in such

---

[20] al-Nasā'ī, Aḥmad ibn Shu'ayb, 830/31–915. Compiler of one of the standard six collections.

[21] That is, they had not yet had conjugal relations. Abū Dāwūd II:237 #2114 Nikāḥ 31, Tirmidhī, Nasā'ī, Ibn Mājah, Ibn Ḥanbal.

[22] Nasā'ī, Nikāh 68.

a way that it gets accorded the rank of being highly probable (*ghālib al-ẓann*) so that one of them retreats from his own reasoning in favor of the transmitted text. An example of this is what the Imāms report about Abū Ḥuraira, may God be pleased with him, who held that whoever got up in the morning in a state of ritual impurity could not keep the fast, until some of the wives of the Messenger of God, may the peace and blessings of God be upon him, informed him of a hadith which was contrary to his opinion, so he withdrew it.[23]

C) The third is that a Companion comes to know of a hadith but not at that level which brings with it the rank of being highly probable (*ghālib al-ẓann*) so that he does not abandon his *ijtihād*, but rather impugns the authenticity of the hadith. An example is what the masters of theoretical jurisprudence relate about Fāṭima bint Qais who testified before 'Umar ibn al-Khaṭṭāb that she had been divorced by the triple formula, and that the Prophet of God, may the peace and blessings of God be upon him, had neither granted her maintenance nor a dwelling. He rejected her testimony and said, "I will not abandon the Book of God[24] on the statement of a woman when I don't know if she is telling the truth or lying. A divorced woman will receive maintenance and a dwelling." 'Ā'isha, may God be pleased with her, said to Fāṭima, "Don't you fear God!" i.e., because of her statement, "neither a dwelling nor maintenance."[25]

Another example of this type is that the two shaikhs (al-Bukhārī and Muslim) reported that 'Umar ibn al-Khaṭṭāb held that the ablution with sand was not sufficient for someone in a state of major ritual impurity who did not find water. Then 'Āmmār reported in his presence that he had been with the Messenger of God, may the peace and blessings of God be upon him, on a journey and had become ritually impure and did not find any water so that he rolled himself in the dirt and then mentioned this to the Prophet of God. The Prophet of God, may the peace and blessings of God be upon

---

[23] al-Bukhārī III:81 Fasting. When asked about Abū Huraira's ruling 'Ā'isha and Umm Salama reported. "At times Allāh's Apostle used to get up in the morning in a state of *janāba* (ritual impurity after having had sexual relations). He would then take a bath and keep the fast." *Muṣaffā* I:239.

[24] For example, Qur'ān 65:6, "Lodge them where you are lodging" and 60:11, "Give to those wives who have gone away the like of what they have expended."

[25] *Ṣaḥīḥ Muslim*, trans. II:772 #3524. The wording is slightly different in the main hadith collections.

him, said, "It would have been sufficient for you to have done thus," and he lightly struck his hands against the earth and rubbed both of them across his face and arms.[26] However, 'Umar did not accept this and the proof of it according to him was not established due to a concealed defect which he saw in it. Later that hadith became abundantly transmitted by many chains after the first reporter, and the suspicion that it was defective faded into obscurity so that they implemented it.

D) The fourth of them is that the hadith hadn't reached the Companion at all. For example, what Muslim related about Ibn 'Umar who ordered women to unbind their hair when they were doing the ritual bath. 'Ā'isha heard this and said, "I'm amazed at Ibn 'Umar—he commands women to unbind their hair, why didn't he order them to shave their heads too! I used to take a bath together with the Prophet of God, may the peace and blessings of God be upon him, using the same vessel, and I did no more than pour water over my head three times."[27] Another example is what al-Zuhrī mentioned about Hind, i.e., that she had not heard about the Prophet allowing women to pray during their menstruation, so she used to cry because she could not pray.

2) Among (causes for disagreement among the Companions) are that they saw the Prophet carrying out an action, and some interpreted it as a means of drawing nearer to God (*qurba*) and others as being (merely) ethically indifferent. An example of this is what the experts in legal theory relate about the ruling on *taḥṣīb* (i.e., the stopping to sleep at a place called al-Abṭaḥ between Mina and Mecca when returning from Mina during the pilgrimage)—i.e., that the Prophet, may the peace and blessings of God be upon him, stopped there to rest. Abū Huraira and Ibn 'Umar held that this was done by way of performing an act of worship and thus they took it to be one of the normative Hajj practices. On the other hand, 'Ā'isha and Ibn 'Abbās held that it was coincidental and not one of the practices of the pilgrimage.[28]

---

[26] Bukhārī Khan trans. I:208–209 Tayammum 8, Nasā'ī Ṭahāra 195, 199–201.

[27] Muslim Ḥaiḍ 29, Ibn Mājah Ṭahāra 108, Ibn Ḥanbal.

[28] *Saḥīḥ Muslim*. Arabic II:951–952 #337–345, Siddiqi trans. II:659–660 Hadiths #3005–3015. Ibn Mājah, Ibn Ḥanbal. Shāh Walī Allāh prefers the latter in *Ḥujjat Allāh al-Bāligha* II (Cairo: Dār al-ṭab' wa-l-nashr, 1952–3), 548 ff., "Account of the Farewell Ḥajj."

Another example is that the majority held that walking with a fast gait (*ramal*) during the circumambulations of the Ka'ba was a normative practice of the Prophet. Ibn 'Abbās held that the Prophet, may the peace and blessings of God be upon him, had only done this coincidentally due to a coincidental occurrence which was the polytheists' saying that the fever of Yathrib had overcome the Muslims so that this was not a normative practice (sunna).[29]

3) Among them are disagreements based on misconstrual (*wahm*) and an example is that when the Messenger of God, may the peace and blessings of God be upon him, made the pilgrimage people saw him, so that some of them held that he had entered the state of *Iḥrām* for 'Umra and then later performed the Hajj (*tamattu'*), while others said that he had entered *Iḥrām* for *'Umra* and Hajj together (*qirān*) and some that he had entered the state of *Iḥrām* for the Hajj only (*ifrād*).[30]

Another example that Abū Dāwūd reported from Sa'īd ibn Khabīr is that he said, "I said to 'Abd Allāh ibn 'Abbās, 'O Ibn 'Abbās, I am surprised at the disagreement of the Companions of the Messenger of God, may the peace and blessings of God be upon him, concerning when the Prophet began to observe the *Iḥrām*. He replied, 'In fact I am the most knowledgeable person about this. It occurred due to the fact that the Prophet only performed the Hajj once and out of this arose their disagreement. The Prophet, may the peace and blessings of God be upon him, set out (from Medina) on the pilgrimage and when he had prayed one *rak'a* in the mosque of Dhū Ḥulaifa he entered the state of Iḥrām while he was sitting and made the exclamation 'Labbaika' when he had finished his two *rak'a*s, so that some groups of people heard him do this and I have preserved the recollection of his doing this. Then he mounted, and when his camel stood up bearing him he cried out, 'I am at your service O Lord,' and (other) people also saw that he

[29] Bukhārī Ḥajj 55, Muslim Ḥajj 247, Abū Dāwūd, Ibn Mājah. This hadith was referred to in Chapter 75.

[30] *Iḥrām* is the state of consecration maintained by the pilgrims during the greater (Ḥajj) pilgrimage to Mecca. There are three types of *iḥrām*. A pilgrim may declare the intention at the outset of remaining in the state of *iḥrām* throughout both the period of performing *'umra* and Hajj without any break of the sanctified state (*qirān*). The pilgrim may break the *iḥrām* state after the *'umra* and then resume it for the Hajj (*tamattu'*). The pilgrim may perform only a Hajj without any *'umra* (*ifrād*). This is discussed in *Ḥujjat Allāh al-Bāligha* II, Chapter "Account of the Farewell Ḥajj," pp. 548ff.

did this. This is due to the fact that people were coming to him group by group so that some heard him say 'Labbaika' when his camel stood up bearing him, so that they said, 'The Messenger of God, may the peace and blessings of God be upon him, began saying "Labbaika" when his camel rose up bearing him.' Then the Prophet, may the peace and blessings of God be upon him, set out on the journey, and when he had climbed the heights of Baida', he called out 'I am at your service, O Lord' and some of the people saw this so that they said, 'He began saying "Labbaika" when he had climbed the heights of al-Baida'.' By God's oath he entered the state of *Iḥrām* at his place of prayer and said 'Labbaika' both when his camel rose up with him, and also when he reached the heights of al-Baida'."[31]

4) Among the causes for disagreement are inattentiveness and forgetting. An example is that it is reported that Ibn 'Umar used to say that the Prophet, may the peace and blessings of God be upon him, had made *'Umra* during the month of Rajab, then 'Ā'isha heard this and judged that Ibn 'Umar had been inattentive.[32]

5) Among them are disagreements of judgment. An example is what Ibn 'Umar or 'Umar related from him, may the peace and blessings of God be upon him, about the dead person being tormented by the weeping of his family over him. Then 'Ā'isha adjudged that he had not construed the hadith properly. The Messenger of God, may the peace and blessings of God be upon him, was passing by the funeral of a Jewish woman whose family were weeping over her so that he said, "They are weeping over her while she is being tormented in her grave."[33] He (Ibn 'Umar) had supposed that the torment was causally related to the weeping, so that the ruling was generally applicable to the case of every dead person.

6) Among them are their disagreement over the rationale for legislation behind the ruling, for example, standing up as a funeral procession passes. One opinion is that it is out of respect to the angels so it should be generalized to funerals of both Believers and Unbelievers. Another says that it is due to the awe of death, so it should be done for a Believer or a Unbeliever. Al-Ḥasan ibn 'Āli, may God be pleased with both of them (he and his father),

---

[31] Abū Dāwūd II:150 #1770 Manāsik 21.
[32] Ibn Mājah Manāsik 47, Bukhārī 'Umra 3, Tirmidhī, Ibn Ḥanbal.
[33] Bukhārī Janā'iz 12, 36; Muslim, Tirmidhī, Nasā'ī.

said, "The funeral bier of a Jew was passing by the Messenger of God, may the peace and blessings of God be upon him, and he stood up because he disliked that it should be raised above his head,"[34] so that he considered that this applied only in the case of Unbelievers.

7) Among them are their disagreeing over resorting to two conflicting rulings. For example, the Prophet, may the peace and blessings of God be upon him, gave permission for temporary marriage during the year of Khaibar and the year of Auṭās, then he forbade it. Then Ibn ʿAbbās said, "The dispensation was based on an extenuating circumstance and the prohibition came due to the cessation of that circumstance, thus the prohibition remains in force."[35] The majority of the scholars held that the dispensation (*rukhṣa*) had made it allowable and that the prohibition has now abrogated this. Another example is the Prophet's, may the peace and blessings of God be upon him, forbidding facing the Qibla while performing *istinjā*ʾ[36] so that one group held this ruling to be generally applicable and not abrogated. Jābar saw him urinating while facing the Qibla a year before his death, so that he held that this abrogated the previous prohibition. Ibn ʿUmar saw him relieving himself with his back to the Qibla and his face toward Syria,[37] so that he used this to refute their opinions. Another group combined the two reports, so that Al-Shaʿbī and others held that the prohibition applied in the particular circumstance of being out of doors, but that if the person were in a toilet, then there would be no importance given to facing towards or having one's back turned to the Qibla. One group held that the Prophet's prohibition was definitive and universally in force, and that what he himself had done was possibly particular to his case only, so that it (the prohibition) could neither be considered abrogated nor limited to specific circumstances.[38]

---

[34] This story is found in Bukhārī, Muslim, Nasāʾī, and Ibn Ḥanbal, but none mention the phrase about his disliking the bier being raised above his head. Nadvī, *Arāʾ*, 14–5.

[35] That is, Ibn ʿAbbās, had according to some reports, considered temporary marriage permissible in times of need, but then changed his opinion. Tirmidhī, chapter forbidding temporary marriage. Bukhārī Khan trans. VII:36–37.

[36] *Istinjā*ʾ is cleaning oneself after fulfilling one's natural needs.

[37] Bukhārī Khan trans. I:108. By facing Syria is meant facing Jerusalem, which some also considered to be forbidden.

[38] In *Muṣaffā* 1, pp. 40–41, Walī Allāh notes that al-Shāfiʿī forbade both facing towards Mecca or Jerusalem while outside, not while indoors, while Abū Ḥanīfa considered both reprehensible in all circumstances. al-Shāfiʿī, *al-Risāla*, 217–218.

In summary, the opinions of the Companions of the Prophet, may the peace and blessings of God be upon him, varied and each one of the Successors learned whatever he was able from them, in like manner. Thus the Successor memorized whichever hadiths of the Prophet and opinions of the Companions he heard and thought them over, and he reconciled the variations in so far as he was able and preferred some opinions over others. Some of the sayings vanished from their consideration even though they had been reported from some of the most important Companions. For example the opinion reported from ʿUmar and Ibn Masʿūd concerning (the invalidity of) performing the ablution with sand on the part of a person who was ritually impure in the major sense (*junūb*),[39] faded away once the hadith reports from ʿAmmār and ʿUmrān ibn al-Ḥaṣīn and others became abundantly transmitted.[40] At this point each learned scholar among the Successors came to have his very own school, so that within every city there stood out a leading scholar like Saʿīd ibn al-Musayyab (712) and Sālim ibn ʿAbd Allāh ibn ʿUmar (725) in Medina, and after them al-Zuhrī (742) and al-Qāḍī Yaḥyā ibn Saʿīd (761) and Rabīʿa ibn ʿAbd al-Raḥmān (753), and ʿAṭā' ibn Abī Rabāḥ (732) in Mecca, and Ibrahīm al-Nakhā'ī (715) and al-Shaʿbī (c. 728) in Kufa, and Al-Ḥasan al-Baṣrī (728) in Basra, and Ṭāwūs ibn Kaisān (c. 720) in Yemen, and Makḥūl (730–736) in Syria. Thus God made people avid and desirous of their knowledge so that they learned from these scholars hadiths of the Prophet, legal opinions and sayings of the Companions; while they also learned from them their own legal opinions and verifications. Those who needed legal opinions would consult them, legal issues were discussed among them, and cases were put before them to judge. Saʿīd ibn al-Musayyab and Ibrahīm (al-Nakhā'ī) and their peers compiled together all of the categories of jurisprudence, and within each topic they possessed principles that they had learned from the pious ancestors.

Saʿīd and his associates believed that the people of the two Holy Cities were the most reliable in jurisprudence so they based their school on the legal opinions of ʿAbd Allāh ibn ʿUmar, ʿĀ'isha, and Ibn ʿAbbās, and the verdicts of the judges of Medina. They com-

---

[39] That is, the person requires a full bath following the emission of semen.

[40] Bukhārī I:208–209 Tayammum 8, has the versions. Also Muslim, Nasā'ī Ṭahāra 195, 199–201.

piled whatever they were able of these, then they examined them with respect to reliability and scrutinized them. Whatever they found to be agreed upon by the learned scholars of Medina they firmly held to, and whenever they found them in disagreement, they took the strongest opinion and the most preponderant, either due to the numbers of them who had held this, or due to it agreeing with a strong analogy or a clear inference from the Qur'ān and sunna, or for some other similar reasons. When they didn't find among what they had preserved from those scholars any response on the issue they derived it on the basis of their sayings, and traced its allusions or logical entailments.[41] Thus they came up with many cases in which each topic has sub-divisions.

Ibrahīm and his associates thought that 'Abd Allāh ibn Mas'ūd and his associates were the most reliable persons in jurisprudence as shown by what 'Alqama (680/81) said to Masrūq (682/83). "Is anyone of them more reliable than 'Abd Allāh (Ibn Mas'ūd)?"[42] And the saying of Abū Ḥanīfa, may God be pleased with him, to Al-Auzā'ī, "Ibrahīm is better at *fiqh* than Sālim, and if not for the virtue of ('Abd Allāh Ibn 'Umar's) having been a Companion, I would have said that 'Alqama had more legal acumen than 'Abd Allāh ibn 'Umar; and 'Abd Allāh (ibn Mas'ūd) is in a class by himself."[43] The basis of his (al-Nakhā'ī's) school consists of the *fatwas* of 'Abd Allāh ibn Mas'ūd, the judgments and *fatwas* of 'Alī, may God be pleased with both of them, and the judgments of Shuraiḥ and other Kufan judges, so that he combined whatever he was able of this. Then he did for their reports what the Medinan scholars had done for the reports of the Medinans. He derived as they had derived, then outlined for this the issues of jurisprudence according to sub-divisions under each topic. Sa'īd ibn al-Musayyab was the spokesman for the jurists of Medina and had memorized

---

[41] The allusions (*īmā'āt*) are meanings implied by a text while the logical entailments (*iqtidā'āt*) are meanings which are necessarily understood from a text although they are not explicitly stated. Both are forms of textual indication (*dalālat al-naṣṣ*).

[42] The occasion is explained in more detail in the following chapter on p. 429.

[43] The occasion of this remark was that al-Auzā'ī had asked Abū Ḥanīfa why he did not lift his hands during the ritual prayer, while he knew of a hadith transmitted by Zuhrī from Sālim from Ibn 'Umar that the Prophet used to raise his arms at the time of the *Takbīr* of respect, bowing, and returning to the upright position. Abū Ḥanīfa said that he had heard a hadith from Ḥammād from Ibrahīm from 'Alqama from Asad from 'Abd Allāh Ibn Mas'ūd that the Prophet raised his hands only at the beginning of the prayer. *Kashshāf fī Tarjuma al-Inṣāf* (Urdu translation by Muḥammad Aḥsan Siddīqī (Delhi: Mujtabā'ī, 1891), 18.

more judgments of ʿUmar and hadith of Abū Huraira than any of them, while Ibrahīm (al-Nakhāʾī) was the spokesman for the Kufan jurists. Thus, if they said something without attributing it to someone, this usually was attributed to one of the pious ancestors either explicitly, by allusion, or in some other way. The jurists of their two cities (Medina and Kufa) concurred on their accuracy, learned from them, reflected on what they learned, and drew further inferences on the basis of it, and God knows better.

CHAPTER 83

The Causes for Disagreement among the Schools of the Jurists[1]

Know that God, may He be exalted, brought into being a generation of scholars after the era of the Successors who conveyed knowledge in fulfillment of the promise of the Prophet, may the peace and blessings of God be upon him, when he said "A just person of every succeeding generation will convey this knowledge."[2] Thus these persons learned from those who had been with the Prophet the manner of performing the lesser and greater ablutions, prayer, pilgrimage, marriage, business transactions, and all other commonly occurring things. They transmitted the hadith reports of the Prophet, may the peace and blessings of God be upon him, heard the judgments of the *Qaḍī*s of the various cities and the *fatwa*s of their *muftī*s and they inquired about legal issues, and carried out *ijtihād* concerning all of these things. Then when they became leaders of the people and were consulted about all religious matters, they followed in the footsteps of their teachers and did not fail to study the allusions (*īmā'āt*) and logical implications (*iqtiḍā'āt*) (of revealed texts). Thus they judged, gave legal opinions, transmitted hadith, and taught, and the procedure of (all of the) ulema of this generation was similar.

The essence of the procedure of these scholars was to hold to both the hadith which went back uninterruptedly to the Prophet (*musnad*) and those related about him but not directly on a Companion's authority (*mursal*). They deduced knowledge using the sayings of the Companions and Successors which might be hadiths transmitted from the Messenger of God, may the peace and blessings of God be upon him, which they considered to be less authoritative, so they termed them (these hadith) interrupted (*mauqūf*) before reaching the Prophet. An example is what Ibrahīm (al-Nakha'ī) said while he was transmitting a hadith about the Prophet, may the

[1] This chapter appears in the same form in *al-Inṣāf*, 10–16.
[2] *Mishkāt*, 57.

peace and blessings of God be upon him, forbidding crop futures contracts and the sale of fresh dates still on the tree for dried dates[3] It was asked of him, "Have you learned from the Prophet of God(s) any other hadith than this?" He answered, "Yes, but I prefer to say "'Abd Allāh said," or "'Alqama[4] said." Likewise al-Sha'bī[5] said, when he was asked about a hadith which was said to go back to the Prophet, may the peace and blessings of God be upon him, "I prefer to say that it goes back to a great person at a lesser rank than the Prophet, may the peace and blessings of God be upon him, so that if there should be any addition to or deletion from it, it comes from someone lesser than the Prophet."[6] In other cases they deduced from sayings of the Companions using inferences (*istinbāṭ*) on their part from revealed sources or by reasoned conclusions (*ijtihād*) based on their own opinions. In all of these things their procedure was better, and they were more accurate, closer in time, and knew more religious sources by heart, than those who came after them.

For this reason implementing their rulings was prescribed unless they disagreed or a hadith of the Prophet of God, may the peace and blessings of God be upon him, manifestly conflicted with their opinion. It was also their method in cases when the hadith reports of the Prophet of God, may the peace and blessings of God be upon him, were at variance with one another about some issue, to refer to the opinions of the Companions. If the Companions held that some hadiths were abrogated or that their apparent meaning should be disregarded; or if they did not pronounce them abrogated but concurred on leaving them aside and not holding them obligatory—this would be tantamount to rejecting any legalistic force (*'illa*) in them, or to a ruling that they were abrogated or should be interpreted—and the jurists used to follow the Companions in all of these matters. An example is the doctrine of Mālik concerning the hadith about a dog's saliva.[7] This hadith has been reported but

[3] Muzābana (exchanging dry harvested dates for fresh dates in equal quantities) and Muhāqala (renting land in return for part of the wheat harvest) were forbidden—Khan trans. *al-Bukhārī*, III:214–215. In the Shāfi'i school this was due to its being interest and according to Mālik's school due to its being a form of gambling. *Muṣaffā* I:352. al-Shāfi'ī, *al-Risāla*, 235–6, 325–6.

[4] 'Alqama ibn Qais (680–692) was a Successor and Kufan legal expert, d. 721–83.

[5] al-Sha'bī, 'Amīr ibn Sharāḥil was a Successor who compiled hadith and was a legal expert.

[6] Dārimī I:84.

[7] This refers to the hadith "If a dog drinks from one of your utensils it must be washed seven times in order to be pure." Mālik considered dogs to be ritually clean,

I don't understand what it really means. In his *Mukhtaṣar al-Uṣūl* Ibn Ḥājib[8] related it, but I do not see the jurists acting upon it.

When the opinions of the Companions and Successors differed about an issue, then the preference of every scholar was for the opinion of the people of his city and his teachers since he was more able to distinguish their sound opinions from their faulty ones, was more cognizant of the principles connected to these opinions, and would be predisposed toward their superiority and erudition.

The way of 'Umar, 'Uthmān, Ibn 'Umar, 'Ā'isha, Ibn 'Abbās, Zaid ibn Thābit, and their associates such as Sa'īd ibn al-Musayyab who among them knew best 'Umar's judgments and the hadith of Abū Huraira; and those such as 'Urwa, Sālim, 'Aṭa' ibn Yasār, Qāsim, 'Ubayd Allāh ibn 'Abd Allāh, al-Zuhrī, Yaḥyā ibn Sa'īd, Zayd ibn Aslam, and Rabī'a was more worthy of being followed than others according to the people of Medina due to what the Prophet, may the peace and blessings of God be upon him, had explained about the virtues of Medina, and due to its being the abode of the jurists and the gathering place of the scholars in every age. Therefore you find Mālik following their methods of reasoning and it is well-known that Mālik upheld the consensus of the people of Medina and that al-Bukhārī consecrated a chapter to "Taking Up What the People of the Two Holy Cities Agree Upon."

The school of 'Abd Allāh ibn Ma'sūd and his associates, and the judgments of 'Alī, Shuraiḥ, and al-Sha'bī, and the *fatwa*s of Ibrahīm were more worthy of being followed than others according to the people of Kufa and this is represented by 'Alqama's saying when Masrūq[9] leaned toward the opinion of Zaid ibn Thābit concerning giving equal shares of inheritance. 'Alqama said, "Is anyone of you more reliable than 'Abd Allāh (ibn Mas'ūd)?" Then he replied, "No, but I saw Zaid ibn Thābit and the people of Medina doing *tashrīk*[10] and if the people of the region concur about something

---

and held that this ruling could be implemented as an act of piety but was not obligatory which seems to be at variance with the explicit meaning of the hadith. Bukhārī Wuḍū' 33, Muslim, Abū Dāwūd, Tirmidhī, Nasā'ī, Ibn Mājah, Dārimī, Ibn Ḥanbal.

[8] Ibn Ḥājib, 'Uthmān ibn 'Umar, 1175–1249 who wrote *Mukhtaṣar al-Muntahā al-Uṣūlī* (Beirut: Dār al-Kutub al-'Ilmiyya, 1983).

[9] Masrūq ibn al-Ajda' (d. 682). A Kufan Successor in the tradition of 'Abd Allāh ibn Mas'ūd.

[10] *Tashrīk* is that a woman's husband, mother, mother's brothers, and parent's brothers will inherit from her and that the parent's brothers will share in a third. 'Umar said that they should not inherit.

they firmly hold to it."[11] About this same type of issue Mālik said, "The established practice about which there is no disagreement among us is such and such, and if people differ, then act upon the stronger and preferable opinion—either due to the large number of those who hold it, or due to it agreeing with a strong analogy, or due to it being derived from the Qur'ān and the practice of the Prophet. This is the type of instance in which Mālik said, "This is the most correct (opinion) among those which I have heard."

If the scholars did not find among the sources that they had memorized a response to an issue they derived it on the basis of their (the Companions') sayings and sought out allusions and logical implications (within these sayings). In this generation they were inspired to record hadiths so that Mālik and Muḥammad ibn ʿAbd al-Raḥmān Ibn Abī Dhiʾb (775) recorded them in Medina, and Ibn Jurayj (767) and Ibn ʿUyayna (814) in Mecca, and al-Thaurī (778) in Kufā, and Rabīʿ ibn al-Ṣabīḥ (777) in Baṣra. All of them followed the procedure which we mentioned, and when (the Caliph) Manṣūr performed the pilgrimage he said to Mālik, "I have decided to order that the books which you have compiled (the *Muwaṭṭaʾ*) should be copied and then I will send copies to every garrison town of the Muslims and command them to follow what is in them, and not to go beyond them to any other source." Mālik replied, "O Commander of the Faithful, don't do this, for sayings (of the Companions and Successors) are already known to people, and they have heard hadith and have transmitted reports. Each group follows what they already know and differences have arisen, so leave people with what each locality has chosen for itself. A story has also been told about Harūn al-Rashīd, that he consulted Mālik regarding having the *Muwaṭṭaʾ* hung up in the Kaʿba, and urging the people to act according to it. Mālik replied, "Don't do that, for the Companions of the Messenger of God, may the peace and blessings of God be upon him, differed about the branches of the law (*furūʿ*)[12] and they dispersed to various localities and each practice (of the Prophet) is already carried out. Thereupon Harūn said, "May God grant you success, O Abū ʿAbd Allāh!" Al-Suyūṭī related this.

Mālik was the most reliable of them concerning the hadiths which the Medinans reported from the Messenger of God, may the peace

---

[11] Dārimī *Sunan* II:349–350 Farāʾiḍ. ʿAlī didn't and Zaid did.

[12] The systematic elaboration of the law, or positive law.

and blessings of God be upon him, and the most trustworthy of them concerning the chain of transmitters and the most knowledgeable of them concerning the judgments of 'Umar and the opinions of 'Abd Allāh ibn 'Umar and 'Ā'isha and their associates among the seven jurists.[13] Through him and other scholars like him the science of hadith transmission and juristic opinions was established. Once he became the established authority he taught hadith, gave legal opinions, benefited the people and distinguished himself. This saying of the Prophet, may the peace and blessings of God be upon him, "Soon people will travel great distances seeking knowledge, but they will find none more knowledgeable than the scholar of Medina,"[14] truly applies to him (Imām Mālik). This hadith was narrated by Ibn 'Uyayna and 'Abd al-Razzāq—and the word of these two should be sufficient. Mālik's associates collected reports and preferred opinions and summarized them, edited them, explained them, derived rulings on issues from them and discussed their principles and proofs. They (eventually) dispersed to the Western lands and all reaches of the world so that through them God brought great benefit to His people.

If you would like to verify what we have said concerning the basis of Mālik's school then consult the book *al-Muwaṭṭa'* and you will find it to be as we have reported.

Abū Ḥanīfa, may God be pleased with him, was the closest of them to the way of Ibrahīm (al-Nakha'ī) and his contemporaries and very rarely departed from his teachings. He was extremely talented in making legal derivations based on Ibrahīm's school and was a precise inquirer into the meanings of the derivations, and he gave the fullest attention to positive law (*al-furū'*). If you wish to verify the truth of what we have said then go over the statements of Ibrahīm and his contemporaries in the book *al-Āthar* of Muḥammad (Abū Yūsuf, 798), may God have mercy on him, and the *Jāmi'* of 'Abd al-Razzāq (827),[15] and the *Muṣannaf* of Abū Bakr ibn Abī Shayba;[16] then compare these with his school and you will find

[13] The seven early jurists of Medina who are particularly distinguished are Sa'īd ibn Musayyab, 'Urwah ibn Zubayr, Qāsim ibn Muḥammad ibn Abī Bakr al-Ṣiddiq, Abū Bakr ibn 'Abd al-Raḥmān Makrūmī, Kharijah Zaid ibn Thābit, 'Ubaid Allāh ibn 'Abd Allāh ibn 'Utba Ma'sūdī, Sulaimān ibn Yassār Hilālī.

[14] Tirmidhī V:48 'Ilm 18, Ibn Ḥanbal.

[15] Abd al-Razzāq has a work known as *Muṣannaf* (Beirut, 1970–72), not a *Jāmi'*.

[16] Ibn Abī Shayba, 'Abd Allāh ibn Muḥammad, 775/6–849.

that he doesn't diverge from this procedure except on insignificant occasions and that even on these minor occasions he did not go beyond what the jurists of Kufa held.

The most well-known of his students was Abū Yūsuf (731/2–798), may God be pleased with him, and he held the post of chief-judge during the reign of Harūn al-Rashīd and thus he was instrumental in the emergence of Abū Ḥanīfa's school and in judgments being based on it in the regions of Iraq, Khurasan, and Transoxiana. The best compiler and most assiduous student among them was Muḥammad ibn al-Ḥasan (al-Shaibānī, 805), and it is reported that he studied law with Abū Ḥanīfa and Abū Yūsuf and then went on to Medina where he studied the *Muwattā'* with Mālik. After that he went over it on his own and correlated the school of his associates with the *Muwaṭṭā'*, issue by issue, whenever they could be harmonized.

If they could not be so harmonized and he saw that a group of the Companions and the Successors held the same opinions as his associates, then he held this to be his doctrine; but if he found the jurists using a weak analogy or a feeble derivation which disagreed with a sound hadith or which was opposed by the practice of most of the scholars, he abandoned it in favor of one of the opinions of the pious ancestors which he found preferable. These two (Abū Yūsuf and Muḥammad ibn Ḥasan al-Shaibānī) followed the way of Ibrahīm al-Nakhā'ī and his contemporaries in so far as they were able, just as Abū Ḥanīfa, may God be pleased with him, had done.

These (three Hanafī jurists) only were are variance in either one of two cases. Either the two students disagreed with a derivation which Abū Ḥanīfa had made based on Ibrahīm's opinion, or the opinions held by Ibrahīm and his peers were at variance and the two students disagreed with Abū Ḥanīfa concerning which opinion they found to be preferable over the others. Muḥammad (ibn Ḥasan), may God have mercy on him, compiled and gathered the opinions of these three (Ibrahīm, Abū Ḥanīfa and Abū Yūsuf) and this benefited many people.

The followers of Abū Ḥanīfa, may God be pleased with him, devoted themselves to these compilations by abridging them, explicating them, commenting on them, making derivations, establishing fundamental principles, and making deductions. Later they dispersed to Khurasan and Transoxiana and this became known as the school of Abū Ḥanīfa.

Al-Shāfi'ī came on the scene during the early emergence of these two (Mālikī and Ḥanafī) schools and at a time when their legal theory and positive law had begun to be elaborated. He examined the procedure of the earliest figures and found in it certain matters which kept him from following their method and this he discusses at the beginning of his *Kitāb al-Umm*.[17]

1) Among these matters is that he found that they accepted hadiths not connected to the Prophet through a Companion (*mursal*) and hadiths which were otherwise interrupted (*munqaṭiʿ*), and that these two types of hadith in many cases were defective. When he collated the chains of transmission of the hadiths it became evident that many of the hadiths not transmitted directly by a Companion from the Prophet were baseless, and many conflicted with those hadiths which were uninterruptedly transmitted (*musnad*). He decided not to accept a *mursal* hadith unless certain conditions were fulfilled, and these are mentioned in the books of juristic theory.[18]

2) A second among the factors which dissuaded him from following their way was that the rules for collating the variants had not been rendered precise by the earlier figures, so that due to this defects had entered into their *ijtihād*s. Therefore al-Shāfiʿī established principles for doing this and recorded them in a book[19] and this was the first recording made of the theoretical bases of jurisprudence (*uṣūl al-fiqh*). An exemplification of this is what we have heard concerning al-Shāfiʿī, i.e. , that he went over to Muḥammad Ibn Ḥasan's while the latter was challenging the scholars of Medina about their giving a judgment concerning one witness (being sufficient) for giving an oath. Ibn al-Ḥasan held that this judgment was augmenting the Qur'ān.[20] Al-Shāfiʿī said, "Is it affirmed by you that it is not permitted to augment what the Qur'ān says on the basis of the report of a single individual?" He replied, "Yes." al-Shāfiʿī said, "Then why do you hold that the will in favor of an heir is not permitted based on the Prophet's, may the peace and

---

[17] While some recent available printed texts of *Kitāb al-Umm* begin with the "Book of Purity," manuscripts and earlier editions often began with al-Shāfiʿī's *al-Risāla* which contains this theoretical material.

[18] For example, in his *al-Risāla*, 279–284.

[19] al-Shāfiʿī's, *al-Risāla*.

[20] The Ḥanafī's do not allow a singly transmitted hadith to abrogate the Qur'ān nor to augment it since they consider this also to be a type of abrogation in the case of a single tradition.

blessings of God be upon him, saying, "Know that there is no will in favor of an heir,"[21] while God, may He be Exalted, said in the Qur'ānic verse, "It is prescribed when death is drawing near to one of you,"[22] (that if some property is to be left a will should be made in favor of parents and relatives).[23] He raised a number of objections of this sort to him so that Muḥammad ibn al-Ḥasan was silenced.

3) A third reason is that some of the sound hadiths were not known to the ulema among the Successors who were charged with delivering legal opinions, so that they performed independent reasoning based on their personal opinions, made generalizations, or followed one of the deceased Companions, delivering legal opinions according to his authority. Then when these hadith reports later became known in the third generation, they were not implemented out of the supposition on their part that these conflicted with the practice and custom of the people of their city about which they all agreed, and that this constituted a reason for rejecting these hadith and a case for not taking them into consideration. Or (in some cases) these hadith did not come to light in the third generation, but only after that at the period when the hadith scholars deeply investigated the chains of transmission and traveled to all corners of the earth seeking them out from the bearers of traditional knowledge, so that the body of those hadith which had only been transmitted by one or two persons among the Companions, and passed on from them by only one or two persons, proliferated, and in this way the matter continued. Thus these had not been known to the jurists (*ahl al-fiqh*) and only came to light in the time of the memorizers who collated the chains of many of the hadith, for example, those transmitted by the people of Basra, although those in other regions were ignorant of them. Al-Shāfiʿī explained that the knowledgeable people among the Companions and Successors never ceased seeking out the hadith reports relative to an issue, and if they didn't find any then they would seize on to some other means of deduction (*istidlāl*). Then if they became aware

---

[21] Bukhārī Waṣāyā 6, Abū Dāwūd, Tirmidhī, Nasā'ī, Ibn Mājah, Dārimī, Ibn Ḥanbal.

[22] Qur'ān 2:180. ". . . if he bequeath wealth, that he bequeath unto parents and near relatives in kindness. (This is) a duty for those who ward off (evil)."

[23] al-Shāfiʿī, *al-Risāla*, 142–144 considers that bequests to parents are invalid since their right to inheritance certain shares was confirmed elsewhere in the Qur'ān. Bequests to others valued up to one-third of the estate were allowed. In the case of this hadith, the Ḥanafī's hold that it is not singly but rather, multiply, transmitted.

of a hadith after that they would revoke their answer based on *ijtihād* in favor of the hadith. Therefore if this were the case, their (the Companions') failure to have (previously) adhered to the hadith did not constitute a reason for rejecting it, never indeed—unless they explained the reason behind this rejection.

An illustration of this is the hadith about the two large jars[24] for this is a sound hadith transmitted by many chains, the majority of them going back to Abū al-Walīd Ibn Kathīr from Muḥammad ibn Ja'far ibn al-Zubair from 'Abd Allāh—or from Muḥammad ibn 'Ibād ibn Ja'far from 'Ubayd Allāh ibn 'Abd Allāh—both of these from Ibn 'Umar. Then after that the chains of transmission branched out further. These two hadith transmitters (Muḥammad ibn Ja'far and Muḥammad ibn 'Ibād), although they were considered to be reliable, were not among those who were authorized to give legal opinions, nor did people depend on them.

For this reason this hadith did not come to light in the period of Sa'īd ibn Musayyab, nor in the time of al-Zuhrī, and neither the Mālikīs nor the Ḥanafīs proceeded according to it, so they did not implement it, while al-Shāfi'ī did. Similarly the hadith about contract options remaining open as long as the parties are in each others' company (*khiyār al-majlis*), for it is a sound hadith,[25] transmitted by many chains of reporters. Among the Companions Ibn 'Umar implemented it and so did Abū Ḥuraira but it was not known among the seven jurists and their contemporaries so they did not hold it. Therefore Mālik and Abū Ḥanīfa held this to be a reason for rejecting the hadith while al-Shāfi'ī implemented it.

4) The fourth of them is that the opinions of the Companions were collected at the time of al-Shāfi'ī, so that these came to proliferate, disagree with one another, and branch out, and he saw that many of them opposed sound hadiths since the Companions had not been aware of those hadiths. He saw that the pious ancestors had never ceased giving preference to the hadith in such cases so he abandoned rigid adherence to their opinions when these sayings did not agree saying, "They are (only) human beings and so are we."

5) The fifth reason is that he saw that a group of the jurists

---

[24] That if an amount of water reaches an amount which could be held in two large jars it remains ritually pure for the purposes of major ablution (*ghusl*). Abū Dāwūd, Sunan I:17 65 Ṭahāra 33, Tirmidhī, Nasā'ī, Ibn Mājah, Dārimī, Ibn Ḥanbal.

[25] Bukhārī Buyū' 19, 20, 42, 43, 44, 46, 47; Muslim, Abū Dāwūd, Tirmidhī, Nasā'ī, Ibn Mājah, Dārimī, Muwaṭṭa', Ibn Ḥanbal. al-Shāfi'ī, *al-Risāla*, 227–8.

mixed personal opinion, which the divine law did not sanction, with analogical reasoning, which it affirmed; so that they did not distinguish the one from the other and they sometimes termed this *istiḥsān*.[26] What I mean by personal opinion (*ra'y*) is that they ascribe the anticipated source of some hardship or benefit (*maṣlaḥa*) as being the reason for legislation (*ʿilla*) behind the ruling, while *qiyās* would be syllogistically extracting the reason for legislation from the ruling of the revealed sources and basing the ruling on this reason.

Al-Shāfiʿī completely nullified this type of personal opinion (*ra'y*), saying, "whoever does *istiḥsān* wants to become the lawgiver." Ibn al-Hājib related this in *Mukhtaṣar al-Uṣūl*.[27] An example of this is the ruling concerning reaching puberty or the maturity of an orphan which is a covert matter. Therefore they established the expected time of maturity as reaching twenty-five years in its place saying, "When an orphan attains this age his property should be remitted to him." They opined that this was *istiḥsān*, while the derivation based on analogy is that it should not be remitted to him.[28]

In sum, since Al-Shāfiʿī found things like this occurring among the procedure of the preceding figures he started jurisprudence over from the beginning and set out its theoretical foundations, drew out their practical ramifications, and compiled books distinguishing himself and benefiting humanity. The jurists concurred with him and devoted themselves to summarizing, commenting on, making deductions, and deriving rulings from his books. Then they dispersed to the various cities so that this became the Shāfiʿī school, and God knows better.

---

[26] *Istiḥsān*. That a less apparent analogical conclusion (*qiyās*) may be preferred to other more obvious ones if it fulfills the general purpose of the *sharī'a* better. Likewise in other cases where the result of the *qiyās* process is passed over in favor of other results from the sources of Islamic law (Qur'ān, sunna, or consensus) or due to necessity or the force of customary practice.

[27] *Mukhtaṣar al-Muntahā al-Uṣūlī* II (Beirut: Dār al-ʿIlmiyya, 1983), 288 with slightly different wording "*qāla al-Shāfiʿī man istaḥsana faqad sharaʿa*." "The one who performs *istiḥsān* has legislated." (based on his own opinion).

[28] Some of the Ḥanafīs, including Abū Ḥanīfa set the age of maturity at twenty-five years, at the attainment of which the orphaned person's property should be put at his disposal irrespective of his mental capacity or judgment. Abū Yūṣuf disagreed and preferred analogy on the basis that incompetence or stupidity would be a reason that property should not be remitted, despite the age of the person. ʿAbd al-Raḥmān al-Jazīrī, *Al-Fiqh ʿalā Madhāhib al-Arbāʿa*. Vol. II (Beirut: Dār al-Fikr, 1986), 350–352.

CHAPTER 84

The Difference Between the People of the Hadith and Those Who Exercise Personal Opinion[1]

You ought to know that there were among the ulema at the time of Sa'īd ibn al-Musayyab, Ibrahīm, and al-Zuhrī[2] and in the time of Mālik and Sufyān, and after that, a group who despised engaging in the use of personal opinion, and feared giving *fatwa*s and making deductions except in cases of unavoidable need, and their greatest concern was for transmitting the reports of the Prophet of God, may the peace and blessings of God be upon him.

'Abd Allāh ibn Mas'ūd was asked about a matter and he said, "I would hate to permit for you something which God had forbidden to you, or that I should forbid a thing which God had permitted you."[3] Mu'ādh ibn Jabal said, "O People, don't hasten to calamity before it has struck, for there will always remain those among the Muslims who, if asked, will respond with a prophetic tradition."[4] Similar statements were reported from 'Umar, 'Alī, Ibn 'Abbās, and Ibn Mas'ūd concerning the dislike of speculative discussion of matters which had not occurred. Ibn 'Umar said to Jābir ibn Zaid,[5] "You are one of the jurists of Basra so don't give legal opinions unless they are based on a conclusive Qur'ānic injunction or an established prophetic practice, for if you do otherwise you will perish and cause the ruin of others."[6] Abū al-Naṣr said, "When Abū Salama arrived at Basra, Ḥasan (al-Baṣrī) and I came to see him and he said to Ḥasan, 'Are you Ḥasan? There is no one in Basra whom I would rather meet than you, and this is because I heard that you give *fatwa*s on the basis of your personal opinion, so (in the future) don't give a legal opinion based on your personal opinion

---

[1] The earlier section of this chapter appears in the treatise *al-Inṣāf*, 16–24.
[2] al-Zuhrī, Muḥammad ibn Muslim, d. 742.
[3] Dārimī I:55.
[4] Dārimī I:49.
[5] Jābir ibn Zayd, 642-c. 721. A famous Basran jurist affiliated with the Ibāḍī sect.
[6] Dārimī I:59.

unless there is a sunna from the Prophet, may the peace and blessings of God be upon him, or a revealed Qur'ānic verse.'"[7] Ibn al-Munkadir said, "The scholar is a mediator between God and His servants—so he had better find a way out for himself."[8] Al-Sha'bī was asked, "What did you do when you were asked (about legal matters)?" He replied, "You have asked the expert. If when any person were asked about an issue, he would say to his associate, "give a *fatwa* on the question," then in the very same way this would go on from one to another until it wound up back up at the first person." Al-Sha'bī said, "Accept whatever these persons reported to you from the Messenger of God, may the peace and blessings of God be upon him, and what they said on the basis of their own opinion, throw in the toilet."[9] Al-Dārimī related all these reports.

Then the recording of the Prophet's hadith and reports from the Companions and the writing of the Qur'ān manuscripts spread in the Islamic regions until at last there remained very few hadith transmitters who did not have (made of their hadith collections) a recorded copy, a collation or a manuscript due to their need of this on some important occasion. Thus the great scholars of that time who had attained knowledge circulated among the regions of the Hijaz, Syria, Iraq, Egypt, Yemen, and Khurasan and collected the books, studied the manuscripts, and carefully scrutinized the less known and rare hadiths. Through the great endeavors of these people there were collected hadiths and sayings which no one had ever gathered before, and they could do what had never before been possible, and many chains of hadith transmission became known to them, so much so that some hadiths were known to them through over one hundred or more lines of transmission. Some of the chains brought to light what had been obscure about certain others and they recognized the status of each hadith in being transmitted by a single person or by a wide variety of transmitters. They were enabled to investigate the concurring (*mutābi'āt*) and supporting (*shawāhid*)[10] hadiths and many sound hadiths came to light for them

[7] Dārimī I:58–59.

[8] Dārimī I:53.

[9] Dārimī I:67.

[10] *Mutābi'āt* are hadiths about a certain matter all traceable to the same Companion. *Shawāhid* are hadiths on a matter traceable to two or more Companions. Ṣiddīqī, *Kashshāf*, 34.

which had not been known to the people previously giving legal opinions. Al-Shāfi'ī said to Aḥmad (ibn Ḥanbal), "You are more knowledgeable about the hadiths than I, so if there exists a sound report, please inform me so that I can follow it, whether it is Kufan, Basran, or Syrian." Ibn al-Humām related this.

This is because a few sound hadiths were only related by the people of a particular locale such as the Syrians or Iraqis or the people of a particular family such as the manuscript of Barīd transmitted from Abū Burda from Abū Mūsā,[11] and the manuscript of 'Amr ibn Shu'aib (736) transmitted from his father and from his grandfather.[12] In other cases a certain Companion might have been of minor influence and obscure so that only a small group of hadiths were passed on from him. Thus most of the people giving legal opinions were unaware of these types of hadiths. This generation of scholars had available to them the reports of the jurists of each city who were Companions and Successors, for before their time a person had only been able to collect the hadiths of his city or associates. Those before them had relied for knowing the names of the transmitters and the degree of their reliability on what was available to them based on situational and circumstantial evidence. This generation went deeply into this discipline (of biography) and made it a distinct field for recording and investigation. They debated the rulings of hadith soundness, etc., so that through this putting in writing and debate there were disclosed to them things which had previously been unknown in terms of the hadith going back uninterruptedly to the Prophet or being interrupted. Sufyān, Rakī' and ones like them had made the greatest efforts but had only been able to find less than one thousand uninterrupted hadith going back to the Prophet, as Abū Dāwūd al-Sijistānī mentioned in his letter to the people of Mecca, while the people of this generation transmitted about forty thousand hadiths.

It is true that al-Bukhārī condensed his *Ṣaḥīḥ* to six thousand hadiths and that Abū Dāwūd limited his *Sunan* to five thousand, and that Aḥmad made his *Musnad* a standard by which to recognize

---

[11] This manuscript had been passed down to Barīd from his grandfather, Abū Burda (d. c. 722) who was named as a judge in Kufa and noted as a hadith transmitter. Abū Burda was the son of Abū Mūsā al-Ash'arī (c. 614–663) who was a Companion of the Prophet, and governor of Basra. The hadith scholars differ about its reliability.

[12] This transmission was said to go back to 'Abd Allāh ibn 'Amr ibn al-'Aṣ (682/3 A.H.), great-grandfather of 'Amr ibn Shu'aib.

the hadith of the Messenger of God, may the peace and blessings of God be upon him. Thus, whichever hadith is found in the *Musnad*, even if reported by one chain could be valid, and if not, it would have no validity. The chief hadith scholars of this generation were Abd al-Raḥmān ibn Mahdī (813), Yaḥyā ibn Saʿīd al-Qaṭṭān (813), Yazīd ibn Harūn (736), ʿAbd al-Razzāq (827), Abū Bakr ibn Abī Shaiba (849), Musaddad (ibn Musarhad (843), Hannād (ibn al-Sarīd, 857), Aḥmad ibn Ḥanbal (855), Isḥāq ibn Rāhawayh (852/3), al-Faḍl ibn Dakain (748), ʿAli al-Madīnī (849) and their peers, and this generation was an excellent model for the subsequent generations of hadith scholars.

The researchers among them, after mastering the discipline of hadith transmission and recognizing the ranking of hadith, next turned to jurisprudence. They didn't hold the opinion that people should agree to perform *taqlīd*[13] of a person who had gone before due to the fact that they observed that each of these schools contained contradictory hadith and reports. Thus they took up evaluating the Prophet's hadith and the reports of the Companions, Successors and *Mujtahid*s according to rules which they themselves established—and I will explain to you in a few words what these principles are.

1) They held that if there were found a conclusive Qur'ānic verse pertaining to an issue, it was not permitted to turn from this to something else.

2) If the Qur'ān could support various interpretations then the sunna would be used to rule on the issue.

3) If they didn't find (the answer to an issue) in the Divine Book they used a sunna of the Messenger of God, may the peace and blessings of God be upon him, whether it was abundantly reported, current among the jurists, known only to the people of a certain region or family, or reported through a particular chain of transmission, and whether the Companions and jurists implemented it or they did not.

4) When there existed a hadith about the issue they wouldn't follow any report from the Companions or any *ijtihād* of a scholar which opposed it.

5) Once they had concluded their efforts in tracing the hadiths and had not found any hadith relevant to the matter, they would

[13] *Taqlīd* is to accept or follow a previous knowledge or judgment, as opposed to reinvestigating the sources through independent reasoning (*ijtihād*).

accept the opinions of a group of the Companions and Successors and not restrict themselves to one group to the exclusion of another or one region to the exclusion of another, as those before them had done.

6) If the majority of the Caliphs and Jurists had agreed on something, they accepted this, and

7) If they disagreed they would accept the saying of the one who was the most knowledgeable and pious, or the most accurate or the one who was most well-known among them.

8) If they found a matter in which two opinions held equal force this was considered an issue in which both could be held to be valid.

9) If they were unable to do even this then they would look attentively into what is generalizable from the Qur'ān and the sunna, their referents by way of allusion, and what they logically entail, and they would bring parallel cases to bear on the issue in order to respond when these two cases were obviously close to each other.

In this they did not rely on principles of legal theory but on what could be arrived at through pure human understanding and what would assure the heart, just as the standard of concurrent traditions (*tawātur*) is not the number of transmitters, nor their status, but rather the certainty in the hearts of people which follows hearing the report, as we have previously recounted concerning the status of the Companions of the Prophet.[14]

These principles were derived on the basis of the practice of the first generations and their pronouncements. It is reported from Maimūn ibn Mihrān (734) that he said, "Whenever a dispute was laid before Abū Bakr he used to consult the Book of God, and if he found something in it by which to adjudicate among them he judged by it, and if it wasn't in the Qur'ān and he knew of a sunna from the Messenger of God, may the peace and blessings of God be upon him, pertaining to the matter, he judged by it, and if he failed in this he would go out and ask the Muslims, saying, "Such and such a case has been referred to me, so do you know if the Messenger of God, may the peace and blessings of God be upon him, had made any judgment on this?" Thus sometimes all of the people would gather around him mentioning a judgment from the Messenger of God, may the peace and blessings of God be upon him,

[14] In Ch. 82.

about this, and then Abū Bakr would say, "Praise be to God, who put among us those who have preserved reports of our Prophet." If he failed to find a sunna of the Prophet, may the peace and blessings of God be upon him, about it he would gather the pious and reliable people and the best among them and he would consult them. Then if their opinion concurred on a matter he would judge according to this.[15]

It is reported from Shuraiḥ that 'Umar ibn al-Khaṭṭāb wrote to him, "If you find something in God's book judge according to it and don't let others divert you from this, and if something arises which is not in God's book, then look at the practice (sunna) of the Prophet, may the peace and blessings of God be upon him, on it and judge on the basis of this. If there arises something which is not in God's book and neither is there a sunna of the Prophet, may the peace and blessings of God be upon him, about it, then consider what people have concurred on, and act on this. If there arises something which is not in God's book, nor covered by any sunna of the Prophet, may the peace and blessings of God be upon him, nor has any one before you discussed it—then choose either of two courses of action. If you wish to do independent reasoning (*ijtihād*) based on your own opinion, and proceed thusly, then proceed. If you wish to leave it aside, then leave it aside and I consider leaving it aside as nothing but good for you."[16]

From 'Abd Allāh ibn Mas'ūd it is reported that he said, "A time has come for us in which we do not judge nor are we capable of judging. God has decreed that we should arrive at this (situation) which you see. Thus, whoever is presented with a case to judge after this time, should rule on it based on what is in the book of God, may He be Great and Exalted, and if something comes up which is not in the book of God then he should rule on it based on what the Prophet, may the peace and blessings of God be upon him, ruled. If something comes up which is neither in God's Book, nor did the Prophet, may the peace and blessings of God be upon him, rule on it, then he should judge according to what the righteous ones did (i.e., by *ijmā'*), and he should not say, 'I am afraid', or 'I hold the opinion', for, 'the forbidden is clear and the permitted is clear, and between them are ambiguous matters—so leave

[15] Dārimī I:58.
[16] Dārimī I:60. Also Nasā'ī.

aside what you are dubious about in favor of that in which you have no doubt.'"[17]

Ibn 'Abbās, when asked about a matter, informed about it if it was in the Qur'ān, and if it was not in the Qur'ān but was ruled on by the Prophet, may the peace and blessings of God be upon him, he related it, and if not, then he related what Abū Bakr and 'Umar had ruled, and if not, then he gave his own opinion about it.[18]

It is reported from Ibn 'Abbās, "Don't you fear that you will be punished or be made to sink into the ground for saying, 'The Prophet, may the peace and blessings of God be upon him, of God said such and such, and some person said . . .'"[19] Qutāda said, "Ibn Sīrīn recounted to a man a hadith from the Prophet, may the peace and blessings of God be upon him, then that man said, 'so and so said such and such a thing.'"[20] Then Ibn Sīrīn said, 'I tell you a hadith from the Prophet, may the peace and blessings of God be upon him, and you said, 'so and so said such and such a thing!'"[21] Al-Auzā'ī[22] said, "'Umar ibn 'Abd al-'Azīz gave an order that no one could give personal opinions about what was in the Qur'ān and the leaders of the legal schools could only give opinions concerning things which the Qur'ān had not revealed, nor had a sunna of the Prophet been transmitted about them, nor could anyone hold their own personal opinion about a matter for which there existed a sunna of the Prophet. Al-A'mash[23] said "Ibrahīm (al-Nakha'ī) used to say that the *muqtaḍī*[24] should stand on the left (of the prayer leader), then I related to him a hadith from Samī' al-Ziyāt from Ibn 'Abbās, that the Prophet, may the peace and blessings of God be upon him, set him (Ibn 'Abbās) on his right side,[25] so Ibrahīm adopted this."

Al-Sha'bī reported that a man had come to him asking about an

---

[17] This report in found in Dārimī I:59. The quote is a hadith of the Prophet found in al-Bukhārī Īmān 39, Buyū' 2, Muslim, Abū Dāwūd, Tirmidhī, Nasā'ī, Ibn Mājah, Dārimī, Ibn Ḥanbal.

[18] Dārimī I:59.

[19] For raising some other person's opinion on par with a statement of the Prophet. Dārimī I:114.

[20] Dārimī I:117.

[21] Dārimī I:117.

[22] al-Auzā'ī, 'Abd al-Raḥmān ibn 'Amr (d. 774), a Syrian jurist. Dārimī I:114.

[23] al-A'mash, Sulaimān ibn Mehrān, c. 680–765. Traditionalist and Qur'ān reader who studied with Mālik.

[24] *Muqtaḍī* is a single individual who is following the prayer leader (*imām*) when only the two of them are performing the prayer. Ibn 'Abbās's hadith is in Khan. trans. Bukhārī I:377.

[25] The hadith of Ibn 'Abbās is in Dārimī I:286.

issue so that he replied that Ibn Mas'ūd had said such and such a thing about it. The man then said, "Tell me your opinion about it." Al-Sha'bī said, "Aren't you amazed at this person, I told him what Ibn Mas'ūd said and he asked about my opinion. My religion is more important to me than that! By God, I would rather burst into song than inform you on the basis of my opinion."[26] Al-Dārimī gathered all of these reports.

Al-Tirmidhī reported from Abū al-Sā'ib who said, "We were at Wakī's and he said to one of those persons who had the habit of giving his own opinion, 'the Prophet of God, may the peace and blessings of God be upon him, used to practice *ish'ār*.[27] Did Abū Ḥanīfa hold that it (*ish'ār*) is *mathla*?'[28] The man said, 'It had been reported that Ibrahīm al-Nakha'ī said, "*Ish'ār* is the same as *mathla*."' Then (Abū Sā'ib) related, "I saw Wakī' get very angry and he said 'I tell you that, "the Prophet of God, may the peace and blessings of God be upon him, said" and you say "Ibrahīm said". It's better that you should be imprisoned and not set free until you repudiate what you have just said.'"[29]

It is reported that 'Abd Allāh ibn 'Abbās, 'Aṭā, Mujāhid and Mālik ibn Anas, may God be pleased with them, used to hold that except for the Prophet of God, may the peace and blessings of God be upon him, there was no one whose speech could not either be accepted or refuted.

In summary, once (the scholars) had laid out jurisprudence according to these principles there remained no issue among those that had been previously discussed nor among those that had come up in their era but that they had found a hadith pertaining to it, whether going back uninterruptedly to the Prophet (*marfū'*), having all transmitters mentioned (*muttaṣal*), being interrupted at the level of a Companion (*mursal*), or the statement of a Companion (*manqūl*), whether sound, good, or being worthy of being considered, or that they had found a statement of Abū Bakr or 'Umar, or the other caliphs or the judges of the early Islamic garrison cities and the

---

[26] Dārimī I:47.

[27] *Ish'ār* was a practice of branding a camel vowing it as a sacrifice for Allāh's sake. Abū Ḥanīfa disapproved of *Ish'ār* in the case that the animal was tortured or harmed by it. *Ṣaḥīḥ Muslim*, Hadith #2865 and note, p. 632.

[28] *Mathla* is mutilation which defaces the appearance. For the classification of Abū Ḥanīfa's opinion as *istiḥsān* or *ra'y* see Schacht, J. *The Origins of Muhammadan Jurisprudence*. (Oxford: Oxford University Press, 1950), 112.

[29] Tirmidhi, *Sunan* II:195 Ḥajj Ch. 66 #908.

legists of the (early) regions or an inference (*istinbāṭ*) through a generalization, allusion, or entailment. In this way, Allāh facilitated implementing the sunna for them. The highest of the scholars in dignity, the one who transmitted hadith most extensively, the most knowledgeable of the ranking of hadiths, and the most astute in jurisprudence was Aḥmad ibn Ḥanbal, then Ishāq ibn Rāhwayh.[30]

The organization of jurisprudence along these lines thus depended on collecting a great number of hadiths and accounts to the point that Aḥmad (Ibn Ḥanbal) was asked if (knowing) 100,000 hadith would suffice a person to be able to give a legal opinion. He replied, "No," until the number 500,000 hadith was suggested. Then he said, "I hope so." This is quoted thus in *Ghāya al-Muntahā*.[31] He meant that this basis (would suffice for) giving *fatwas*.

Then God brought forth a later generation who observed that their predecessors had spared them the trouble of gathering hadiths and laying out jurisprudence on their foundation, so they were free to turn their attention to other disciplines such as singling out those sound hadiths concurred on by the great masters of the hadith scholars such as Zaid ibn Harūn, Yaḥyā ibn Sa'īd al-Qaṭṭān, Aḥmad, Ishāq and ones like them; collecting the legislative hadith upon which the jurists of the garrison towns and the ulema of the early regions had built their legal schools; as well as ruling on each hadith according to its merits such as the anomalous (*shādhdha*) and singly transmitted (*fādhdha*) hadiths which the earlier reporters had not transmitted, or following up their lines of transmission that earlier scholars had not traced in which there might be found an uninterrupted connection to the Prophet or an elevated chain, or the transmission from one juristic expert to another jurist, or from one memorizer to another and so on with this type of technical topic. These ones are al-Bukhārī, Muslim, Abū Dāwūd, 'Abd ibn Ḥumayd (863), al-Dārimī, Ibn Mājah, Abū Ya'lā (1066),[32] al-Tirmidhī, al-Nāsā'ī, al-Dāraquṭnī, al-Ḥākim, al-Baihaqqī, al-Khāṭib, al-Dailamī, and Ibn 'Abd al-Barr (1070)[33] and their like. In my opinion, the

---

[30] A traditionalist contemporary of Aḥmad ibn Ḥanbal who was also hostile to the people of personal opinion (ra'y). He died in 852.

[31] Of Marī ibn Yūṣuf al-Karmī (1624). Some other versions of this report mention 300,000 as the number, for example Abū Ṭālib al-Makkī in *Qūt al-Qulūb* I (Cairo: Muṣṭafā al-Bābī al-Ḥalabī, 1961), 300.

[32] Ibn al-Farra, Abū Ya'la Muhammad ibn al-Husain, 990–1066

[33] Ibn 'Abd al Barr, 978–1070. A scholar distinguished in *fiqh* and genealogy.

ones among them who are the most famous, the most knowledgeable, and whose writings are the most useful, are four, approximately contemporary to one another.

The first of them is Abū 'Abd Allāh al-Bukhārī (870) whose goal was sorting out the sound, abundantly transmitted hadith which went directly back to the Prophet from the others, and inferring from them jurisprudence, prophetic biography and Qur'ān interpretation. Thus he compiled his collection, *al-Ṣaḥīḥ*, remaining faithful to his conditions. We heard that a pious man saw the Prophet of God, may the peace and blessings of God be upon him, in a dream and he said, "What's wrong with you that you have become preoccupied with the jurisprudence of Muḥammad ibn Idrīs (al-Shāfi'ī) and gotten away from my book." He asked "O Prophet of God, what then is your book?" He replied, "*Ṣaḥīḥ al-Bukhārī.*" By my life it has achieved fame and acceptance to a degree beyond which none could possibly aspire.[34]

The second of them is Muslim al-Nīsāpūrī (875), who aimed to isolate those sound hadith which hadith scholars had agreed upon, which were uninterruptedly transmitted from the Prophet, and from which the Prophetic sunna could be inferred. He wished to popularize them and facilitate the inference of jurisprudence from them. Thus he did an excellent job of organizing them, assembling the chains of transmission of each hadith in one place so as to clarify as fully as possible textual variants and the branches of the lines of transmission, and he correlated the variants so that there remains no excuse for the person who is cognizant of the Arabic language in turning away from the sunna to something else.

The third of them is Abū Dāwūd al-Sijistānī (889) whose concern was with collecting the hadiths in which jurists found the indicants (*istadalla*) for rulings and which were current among them and on which were founded the rulings of the ulema of the early cities. To this end he compiled his *Sunan* collecting in it the sound, good, without defect (*līn*), and proper to be implemented (*sāliḥ l-il 'amal*) hadith. Abū Dāwūd said, "I did not cite in my book any hadith which people had agreed to leave aside." He exposed the weakness

[34] Shāh Walī Allāh wrote on al-Bukhārī's methodology in a separate treatise, *Sharḥ Tarājim Abwāb Ṣaḥīḥ al-Bukhārī* (Hyderabad, India: Dā'ira al-Ma'ārif al-'Uthmāniyya, 1949), 1–6. One of his letters in the collection *Kalimāt al-Tayyibāt* (Delhi: Mujtabā'ī, 1309 A.H.) also addresses the role of al-Bukhārī. Nadvī, 110.

of any weak hadith among them, and whichever of them contained a deficiency, he explained this in a way that the expert in hadith studies would understand, and he explained in the case of each hadith whatever (ruling) a scholar had deduced from it, or whichever opinion a knowledgeable person had based on it, and therefore al-Ghazzālī and others have stated that his book would suffice for the legal scholar doing independent reasoning (*mujtahid*).

The fourth of them is Abū ʻIsā al-Tirmidhī (892), and it's as if he perfected the method of the two shaikhs (Bukhārī and Muslim) insofar as they clarified and did not obscure, and the method of Abū Dāwūd insofar as he collected everything on which an opinion had been given. Thus he combined each of the two methods and added to them the explanation of the methods of the Companions, Successors, and jurists of the early garrison towns. He compiled a comprehensive book and elegantly abridged the hadith chains. Thus he would cite one chain while pointing out what he had omitted and he explained the status of each hadith in its being sound, good, weak or undetermined, giving the reason for defectiveness so that the student of hadith would be informed concerning its status and recognize those hadith which could properly be taken into consideration from those which could not. He also indicated whether a hadith was transmitted by a wide variety if persons or by a single narrator, and he mentioned the schools of the Companions and Jurists of the early Islamic cities, giving the first names when necessary and supplying the *kunyas*[35] if necessary. He left nothing hidden from the knowledgeable person and therefore it is said that his book suffices the *mujtahid* and is more than enough for the *muqallid*.[36]

In contrast to these persons there was in Mālik's and Sufyān's time and after them a group of people who were not reluctant to delve into the issues, nor did they fear giving legal opinions and they held that jurisprudence was the foundation of religion so that it must become widespread. They rather feared the transmission of Prophetic hadith which were being made to reach back to him, such that al-Shaʻbī said, "We prefer (a hadith) going back to someone

---

[35] *Kunya* is the patronymic element of a name, i.e. Abū (father of) or Umm (mother of) plus a name.

[36] *Muqallid* is one who does imitation or follows (*taqlīd*) the previous rulings of a legal scholar.

other than the Prophet, may the peace and blessings of God be upon him, for if there is any addition or deletion from it involves someone other than the Prophet, may the peace and blessings of God be upon him." Ibrahīm (al-Nakhaʿī) said, "I say, that "ʿAbd Allāh said,' and "ʿAlqama said,' is preferable according to us." When Ibn Masʿūd related hadith which he had heard from the Prophet his face streamed with tears and he said, "(He said) exactly this or something along these lines, and so on."[37] ʿUmar said when he sent a group of the Anṣār to Kufa, "You are going to Kufa to a people who weep when they recite the Qurʾān so they will come to you saying 'the Companions of Muhammad have arrived, the Companions of Muḥammad have arrived.' Then they will come to you and ask you about hadith so try to be sparing in giving reports from God's Messenger, may the peace and blessings of God be upon him."[38] Ibn ʿAun (933) said, "When al-Shaʿbī was presented with an issue he was cautious, and Ibrahīm used to expound on it at great length."[39] Al-Dārimī reported these accounts.

In short, the writing down of hadith, jurisprudence, and specific legal issues occurred due to their need for another approach and this was because the ulema did not have enough hadith reports and accounts from the Companions to suffice in inferring (*istinbāṭ*) jurisprudence according to the principles which the People of the Hadith had chosen. They did not take pleasure in studying the pronouncements of the religious scholars of the (various) regions, collecting and investigating them, for they considered this to be a dubious method. They believed, however, that their leaders (Imāms) were at the highest level of inquiry and they were very much biased toward their colleagues and ʿAlqama said, "Was anyone among them (the Companions) more reliable than ʿAbd Allāh (ibn Masʿūd)?" and Abū Hanīfa said, "Ibrahīm has more legal acumen than Sālim, and if not for the virtue of being a Companion I would have said, "ʿAlqama has more legal acumen than Ibn ʿUmar."[40] They were astute, intuitive, and quick in shifting the intellect from one thing to another which enabled them to deduce the answer to issues based on the pronouncements of their teachers. "For everyone will find

[37] Dārimī I:84.
[38] Dārimī I:85.
[39] Dārimī I:52.
[40] Recounted in detail in Chapter 82 on page 425.

it easy to do that for which he was created."[41] "Each sect rejoicing in its own tenets."[42]

Thus they laid out jurisprudence on the principle of derivation (*takhrīj*) which is as follows.

Each jurist memorizes the book of the one who was the spokesman for his associates and the most knowledgeable of the group's pronouncements and the most correct in examining its preference of opinions (*tarjīḥ*) so that in each case he takes into consideration the interpretation of the ruling.

Whenever he is asked about a matter or needs some information he will look into the pronouncements of his associates which he had memorized in case he finds the answer there, and if not:

1) He will examine the generalization of their sayings so as to make the matter conform to this form.

2) He will take into account an indication implicit in the statement so that he can infer the response on the basis of this.

3) Sometimes there may be an allusion (*īma*) or *iqtidā* (logical entailment) of certain statements from which the intent can be understood.

4) Sometimes the stated issue may have a parallel instance to which it can be referred.

5) Sometimes he can look into the reason for legislation (*ʿilla*) of the ruling which has been stated through derivation (*takhrīj*), simplification, or ellipsis so that its ruling can be applied to a case other than what had been originally been pronounced upon.

6) Sometimes there would be two statements about a case which if combined according to the format of a conjunctive syllogism[43] (*qiyās iqtirānī*) or hypothetical syllogism[44] (*sharṭī*) will produce the answer to the issue.

7) Sometimes there would be in their statements a thing known through pattern and category but not through a comprehensive

---

[41] As in al-Bukhārī Khan trans. IX: 480–481 Tauḥīd, 54, where he translates it, "Everyone will find it easy to do such deeds as will lead him to his destined place for which he was created." Muslim, Abū Dāwūd, Tirmidhi, Ibn Mājah, Ibn Ḥanbal.

[42] Qur'ān 23:53, 30:32.

[43] A conjunctive syllogism is of the type that if the two initial premises are true then a third may be deduced. "If all A's are B and all B's are C then all A's are C." In Islamic legal reasoning an example would be, "All intoxicants are forbidden, wine is intoxicating. Therefore wine is forbidden."

[44] A hypothetical or disjunctive syllogism is of the form, "If A then B." "A is true, therefore B is true."

exclusive definition, so that they would have recourse to the linguistic experts and take pains to establish its essential properties, in order to determine its comprehensive exclusive definition, settle its ambi-guities, and distinguish its problematic aspects.

8) Sometimes their sayings might have two possible interpretations so they would attend to establishing preference for one of the possibilities.

9) Sometimes the mode of argumentation of the proofs (*taqrīb al-dalāʾil*) for the issues would be obscure so that they would elucidate this.

10) Sometimes certain of those using the methods of derivation would make (legal) deductions based on the action of the founders of their school, or upon their remaining silent, and so on.

It may be said of all of these forms of derivation that "the opinion derived from such and such a person is thus," or "it is said according to the school of so and so," or "according to the principle of so and so," or "according to the opinion of so and so"—"that the response to the question is such and such." Those ones (who practice legal derivations in this manner) are termed "those exercising legal reasoning (*mujtahids*) within a legal school." A person who holds that whoever memorizes the *Mabṣūt*[45] is a *mujtahid*, even if he has no knowledge at all about hadith transmission, nor even knowledge of one hadith is referring to this type of *ijtihād* according to this principle. Thus the process of derivation took place in every school, and proliferated. The school which had famous members who became judges and givers of legal opinions, whose writings became well-known among people and who taught openly, spread to all regions of the world, and still continues to spread all the time. The school which had undistinguished members who were not entrusted to judge and give *fatwa*s, and who were not liked by people, died out after a time.

---

[45] *Kitāb al-Mabsūṭ* of Muḥammad ibn Aḥmad al-Sarakhsī, 11th C. A basic book in the Ḥanafī school

CHAPTER 85

An Account of the Condition of People Before the Fourth Century and After[1]

You ought to know that during the first and second centuries people did not unanimously follow any particular *madhhab.* Abū Ṭālib al Makkī (d. 996) said in his *Qūt al-Qulūb*, "Books and compilations are all later developments, as is holding to the statements which people have made, giving legal opinions based on the school of a single individual, holding to his opinion, emulating him in every thing, and conducting jurisprudence according to his school. This was not the way of the people who preceded us in the first and second centuries."[2]

I hold that some amount of making legal derivations (*takhrīj*) had arisen after the first two centuries, although the people of the fourth century were not agreed on the absolute imitation (*taqlīd*) of the school of a single person and conducting jurisprudence according to it and emulating his opinion, as will be clear from the following exposition.

Rather among them were the religious scholars and the common people. In cases involving consensual issues about which there was no disagreement among the Muslims and among the majority of the *mujtahids*, the common people only performed *taqlīd* of the master of legislation (the Prophet). They used to learn the manner of ablution, full bath, prayer, *zakāt* and so on from their forefathers or the teachers of their cities—and they acted according to this. If some uncommon situation arose they would ask for a legal opinion about it from whichever *muftī* they found without specifying a legal school.

---

[1] Some sections of this chapter are duplicated and rearranged in the treatise by Shāh Walī Allāh, *al-Insāf fī Bayān*, while others appear in *'Iqd al-Jīd*, as will be indicated in the text. Some sections of the latter work were translated by Daud Rahbar, "Shah Waliullah and *Ijtihād*." *The Muslim World* 45 (December 1955): 346–358.

[2] al-Makkī, *Qūt al-Qulūb* I, 324.

It was a trait of the specialists that the People of the Hadith among them were deeply involved with hadith scholarship so that they possessed hadiths of the Prophet, may the peace and blessings of God be upon him, and traditions of the Companions besides which nothing else would be required for (deciding) the issue, and abundantly transmitted hadith and sound hadith which had been implemented by some of the jurists, due to which there is no excuse for not acting upon them. Or (they had available) publicly declared opinions of the majority of the Companions and Successors which may not properly be opposed. If a person still didn't find any answer to the issue which would satisfy him due to conflicting transmissions and lack of clear preference (for any one over the other), and so on, he could refer to the discussion of one of the past jurists. Then if he found two opinions he could choose the more reliable of them, whether it came from the people of Medina or Kufa.

Those among them who were People of Legal Derivations (*ahl al-takhrīj*) carried this out in cases where they found no clear pronouncement (*maṣraḥ*) (about an issue) and they used independent reasoning within a school and became affiliated with the school of a certain person so that it was said, "So and so is a Shāfiʿī," and, "So and so is a Ḥanafī." Even one of the People of the Hadith might become associated with a legal school due to his usually concurring with it, for example al-Nasāʾī and al-Baihaqqī who were referred to as Shāfiʿīs. Thus the positions of giving judgments and *fatwa*s came to be entrusted only to *mujtahid*s and only *mujtahid*s were called jurists.[3]

Then following these centuries other people veered off to the right and left and new developments took place.

1) Among them were dispute and disagreement in jurisprudence. The elaboration of this, based on what al-Ghazzālī said, is that when the era of the Rightly Guided Caliphs came to an end the Caliphate passed to people who held it illegitimately. They were not self-reliant in the science of giving *fatwa*s and the (sharīʿa) rulings, thus they were forced to ask for assistance from jurists and to associate with them in all circumstances.

A remnant of the scholars were faithful to the original mode and held to the purity of the religion. Thus if they were sought

[3] This section ends on *al-Inṣāf* p. 28. The next paragraph resumes on p. 38 of *al-Inṣāf*.

after they fled and shunned (the sultans). Thus the people of those times saw the greatness of the scholars and the interest of the leaders in them despite their avoidance of them (the sultans). Later the scholars abandoned (this) refusal and pursued knowledge in order to gain access to achieving honors and attaining high rank. Thus the jurists went from being sought after to becoming the seekers, and they went from having been dignified by their avoiding the sultans to being despicable in their running after them, except those ones whom God made successful (in their resolve). Even before them, persons had compiled works in theology and multiplied the (scholastic) questioning and answering, objecting and responding, and laying the groundwork of argumentation. This had already made an impression on them before the time when some of the officials and kings became disposed toward debates about jurisprudence and determining the primacy between the schools of al-Shafi'ī and Abū Ḥanīfa, may God have mercy on him. After that people abandoned theology and the disciplines of knowledge of the religious sources (*'ilm*) and became interested instead in contentious issues, in particular those between al-Shāfi'ī and Abū Ḥanīfa, may God be pleased with him, while they were tolerant of the disagreements among Mālik, Sufyān, Aḥmad ibn Ḥanbal and others. They claimed that their goal was deducing the finer points of the religious law and determining the reasons for legislation according to the legal school and laying out the principles of legal opinions. Thus they multiplied the compilations and deductions concerning this and they schematized the types of disputations and classifications and they persist in this until today. We don't know what God the Exalted will decree in later times. (End of the gist of what al-Ghazzālī said.)[4]

2) Among them are that people came to depend on *taqlīd*, and this *taqlīd* slowly crept into their hearts while they remained unaware of it.

One reason for this was competition among the jurists and their disputing among themselves, so that when competition in giving legal opinions occurred among them, whoever gave a *fatwa* about something was contradicted about that fatwa. He then replied to this so that the discussion was not brought to a conclusion except

---

[4] al-Ghazzālī, *Iḥyā' 'Ulūm al-Dīn*, I:41-42. Ends on *al-Inṣāf*, 38 the next paragraph begins on *Inṣāf*, 41.

through recourse to the pronouncement of someone who had given a verdict on the issue in an earlier era.

An additional reason for *taqlīd* was the injustice of the judges, for once most of the judges had become unjust and were no longer reliable, only that on which the common people did not cast doubt was accepted from them, i.e., something which had been ruled on previously.

An additional reason was the ignorance of the leaders of the people, and people's asking opinions from those with neither knowledge of the hadith nor of the method of deductive inference, as you may observe apparent in most of the recent ones. Ibn al-Humām and others warned about this. At that time non-*mujtahid*s began to be called jurists.[5]

3) Most of them began to be over-specialized in each discipline so that some claimed to have laid the foundation of the discipline of knowing the hadith transmitters (*ʿilm al-rijāl*) and recognizing their ranks in being reliable or unreliable (*jarḥ wa taʿdīl*). Then they would go on from this to ancient and recent times. Among them were ones who sought out the unusual and rare reports even if they lay within the scope of fabrication. Among them were ones who increased the argumentation concerning the roots of jurisprudence and each deduced in support of his peers principles of argumentation, so that he posed an issue, then exhausted it, responded, sought its conclusion, defined, classified, and edited, sometimes lengthening the discussion and at other times condensing it.

Some of them began to concoct remote instances which were not worthy of the attention of a reasonable person and they liked the generalizations and allusions in the discussions of the legal interpreters and those of a lower rank, to whom neither the knowledgeable person nor the ignorant one would care to listen.

The harmfulness of this disputation, disagreement, and hair-splitting was close to that of the first crisis (of the Muslim community) when people quarreled over rulership and took up sides. Just as the former resulted in a tyrannical rulership and events of severity and folly—similarly these latter (disputes) led to ignorance, interpolations, doubts, and conjecture from which there is no hope of deliverance. Subsequent to them generations arose who relied purely on *taqlīd*, neither distinguishing the true from the false nor the

---

[5] Ends on *al-Inṣāf*, 41 the next paragraph begins on *al-Inṣāf*, 43.

argument from the inference (*istinbāṭ*). The *faqīh* of this time was a prattler and wind-bag who indiscriminately memorized the opinions of the jurists whether these opinions were strong or weak, and related them in a loud-mouthed harangue; and the hadith scholar (*muḥaddith*) became a person who counted up the hadiths whether sound, faulty, or nonsensical, and recited them quickly like an entertainer, flapping his jaw full-force.

I don't say that this is so in all cases, for God has a group of His worshippers unharmed by their failure, who are God's proof on His earth even if they have become few. No time has come after that except that the crisis has increased and *taqlīd* has become more prevalent, and integrity has become more and more absent from people's hearts until they have become content to leave off examining religious matters and so that they say, "We found our fathers following a community and we follow in their footsteps."[6] The complaint may be raised to God and He is the one to turn to for help. He is reliable and our trust is in Him.

Subsection

At this point people should be alerted to issues in whose deserts the intellects went astray, the feet stumbled and the pens blotted.

1) Among them are that (in the case of) these four schools which have been recorded and formulated—the community has agreed, or those whose opinions are worth considering among them have agreed—on the permissibility of performing *taqlīd* of them up until our time. In these are benefits which are not concealed—especially in these days in which people's endeavors fall very short, their hearts have become of self-seeking and everyone delights in his own opinion.[7]

Ibn Ḥazm said,

> *Taqlīd* is forbidden, it is not permitted for anyone to follow the opinion of someone other than the Prophet of God, may the peace and blessings of God be upon him, without proof, due to God's, may He be Exalted, saying, "Follow what was revealed to you from

[6] Qur'ān 43:22.

[7] The treatise *al-Inṣāf* concludes with this passage (p. 43). The following section of the text is duplicated in the treatise *ʿIqd al-Jīd*, pp. 14–16 (Arabic).

your Lord and do not follow guardians besides Him," (7:3) and "If it were said to them obey what God has revealed to you, they say rather we obey what we found our ancestors doing." (2:170). (God) said in praise of the ones who don't perform *taqlīd*, "Give good news to my worshippers who hear advice and follow the best of it. Such are those whom Allāh guides, and such are those possessed of understanding." (39:17–18) And He said, may He be exalted, "If you disagree among yourselves about something refer it to God and the Prophet if you believe in God and the Last Day." (4:59)

Thus God, the Exalted, in time of dispute did not allow reference to anyone besides the Qur'ān and sunna. In this (Qur'ānic verse) He forbade referring in time of dispute to any person's opinion because it is not the Qur'ān or sunna. The consensus of all of the Companions, from the first of them to the last, and the consensus of the Successors from first to last, confirmed the refusal and interdiction of any one of them from imitating the opinion of any contemporary or preceding person, so that he accepts it totally.

Therefore it should be known that whoever follows the totality of Abū Ḥanīfa's, Mālik's, al-Shāfi'ī's or Aḥmad's opinions, may God be pleased with them, and does not leave aside any opinion of a follower of theirs, or of anyone else in favor of that of someone else, and does not rely on what is in the Qur'ān and the sunna without submitting it to the opinion of a particular person—this person has surely and indubitably opposed the consensus of the whole Muslim community from its beginning to its end and he will not find any pious elder or person among all of the three praiseworthy first generations (in agreement with him). Therefore he has chosen a path other than that of the believers. We take refuge with God from this position.

In addition, all of these jurists forbade *taqlīd* other than the imitation of the pious ancestors, thus whoever follows these jurists contravenes their own prohibition. Also, what is it that could make a person among them (the founders of the legal schools) or anyone else, more worthy of being imitated, than say, 'Umar ibn al-Khaṭṭāb, 'Alī ibn Abī Ṭālib, Ibn Mas'ūd, Ibn 'Umar, Ibn 'Abbās or 'Ā'isha, mother of the believers, may God the Exalted be pleased with them—for if *taqlīd* were permitted then each one of these people would be more worthy of being imitated than anyone else.

This statement (of Ibn Ḥazm) applies to any person who has some inkling of *ijtihād* even if only in one issue, and to whomever it is clearly apparent that the Prophet, may the peace and blessings of God be upon him, commanded one thing and forbade another, and that it is neither abrogated by tracing the hadiths and the opposing and concurring opinions about the issue nor by finding anything abrogating them—nor by seeing a large group of those scholars steeped in learning acting upon it, for he sees that the one who opposes it has no proof other than analogical reasoning, deduction, or something like this.

In this case there is no reason for opposing a hadith of the Prophet, may the peace and blessings of God be upon him, except concealed hypocrisy or overt stupidity. This is what Shaikh 'Izz al-Dīn ibn 'Abd al-Salām[8] meant when he said,

> It is one of the most amazing wonders that one of the jurists who practices *taqlīd* agrees on the weakness of something taken from his Imām because there is found no defense for its weakness, while in spite of this he imitates his (the Imām's) decisions about it and ignores the one whose opinion is attested to by the Book, sunna, and sound analogies—rigid in his adherence to practicing *taqlīd* of his Imām. Indeed he concocts things which oppose the manifest meaning of the Book and the sunna, and exegetes them by remote esoteric interpretations in defense of the person he imitates.

He (further) said,

> People always used to ask (opinions from) whichever scholar they happened to run across without being restricted to a legal school, and without rebuke to any questioner, until these legal schools appeared and those who were prejudiced in their favor among the ones who practiced *taqlīd*. Thus one of them would follow his Imām despite the remoteness of his opinion from textual justification (*adalla*), imitating him in what he held as if he were a messenger sent from God—and this is far removed from the truth, far from what is correct, and unacceptable to any reasonable person.

The Imām Abū Shāma said,

> It is incumbent upon one who engaged in jurisprudence not to confine himself to the school of one Imām, and that he should hold in every issue the soundness of what is closer to the indication (*dalāla*) of the Qur'ān and the established practice of the Prophet (sunna *maḥkama*). This will be easy for him if he is well-versed in most of the traditional disciplines. Let him avoid partisanship (to a school) and studying the recent modes of disagreement for these are a waste

---

[8] 1181–1262.

> of time and will disturb his serenity. It is confirmed that al-Shāfi'ī forbade performing *taqlīd* of himself or anyone else.

His associate al-Muzanī said at the beginning of his *Mukhtaṣar*,

> I summarized this book of al-Shāfi'ī's teaching and the meaning of his opinions in order to make it available to whomever wishes, while I apprise him of his (al-Shāfi'ī's) forbidding performing *taqlīd* of himself or of any other, so that this person should study it for the sake of his religion and should take care—i.e., I admonish whoever wishes to study al-Shāfi'ī's teaching, that he himself forbade *taqlīd* of himself or anyone else.[9]

(Ibn Ḥazm's saying applies to) the person who is not learned and follows a particular one of the jurists believing that no one like him could err and that what he said must definitely be correct, and who has secreted in his heart not to leave off following him even if a proof opposing him would come to light. (On this point) there is what al-Tirmidhī reported from 'Adī ibn Ḥatim—that he said, "I heard him, i.e., the Prophet of God, may the peace and blessings of God be upon him, reciting this verse, "They took their rabbis and monks as Lords besides Allāh."[10] The Prophet said, "They didn't used to worship them, rather if these ones permitted something for them, they considered it to be permitted; and if they forbade a thing they forbade it."[11]

As (it applies) to the one who does not allow a Ḥanafī, for example, to ask for a legal opinion from a Shāfi'ī jurist and vice versa, and does not allow a Ḥanafī to follow Imām Shāfi'ī for example, this person has opposed the consensus of the early generations and contradicted the Companions and Successors. This statement (of Ibn Ḥazm) does not apply to the one who obeys only the sayings of the Prophet, may the peace and blessings of God be upon him, and only considers permitted what Allāh and his Prophet made permissible, and only considers forbidden what God and his prophet have forbidden. However if he doesn't have information about what the Prophet, may the peace and blessings of God be upon him, said, neither by way of correlating conflicting statements about what he said, nor by means of deduction from his saying, he may follow a rightly-guided learned person pro-

---

[9] al-Muzānī, *al-Mukhtaṣar* (Beirut: Dār al-Ma'rifa, n.d.), 1.
[10] Qur'ān 9:31.
[11] Tirmidhī IV:342 #5093 Tafsīr Sūra 9, where it is classified as *gharīb*.

vided that he is correct in what he says, and that he gives a clear legal opinion clearly based on the sunna of the Prophet, may the peace and blessings of God be upon him. Then if this person should oppose what he thinks (to be correct) he should part company with him immediately without dispute and insistence. For how can anyone condemn this, when asking for legal opinions and giving them has gone on among Muslims since the time of the Prophet, may the peace and blessings of God be upon him, and there is no difference between always asking the same person for legal opinions and asking that person on some occasions and another person at other times, once what we have agreed on what was mentioned above. How can this be gainsaid when we don't believe that a jurist, whoever he may be, received jurisprudence through Divine revelation, and that God made obeying him obligatory upon us, and that he is infallible. Thus if we follow a jurist, this is due to our knowing that he is knowledgeable concerning God's book and the sunna of His Prophet, and that his opinion must either be based on a pronouncement of the Qur'ān or the sunna or be deduced from them through some variety of deductive apparatus or that he knows from the context that the ruling (*ḥukm*) in a certain case is contingent on a particular cause for legislation (*ʿilla*) and his heart is confident in this recognition. Thus he draws analogies from something which is textually revealed to what is not stated in the revealed texts and it is as if he were saying, "I believe that the Prophet of God, may the peace and blessings of God be upon him, would say, 'Whenever I find this reason for legislation (*ʿilla*) present then the ruling (*ḥukm*) in the case will be thus'"—and the analogized thing is gradually obtained through these generalizations, so that this also is ascribed to the Prophet, may the peace and blessings of God be upon him, but in this method there are conjectures (*ẓunūn*).

If this were not so then no believer would follow a *mujtahid*, since if a hadith from the infallible messenger whose obedience God made obligatory upon us reached us by a correct chain of transmission, indicating something which conflicted with his (the *mujtahid*'s) opinion and we were then to ignore the hadith in favor of obeying that guesswork—who would be more evil than us, and what would be our excuse on the day when people will stand before the Lord of the Worlds?[12]

[12] Concludes section on p. 16 of *ʿIqd al-Jīd*.

2) Among these (difficult issues) is making derivations (*takhrīj*) according to the statements of the jurists and following the literal meaning of the hadith. Each has a fundamental basis in the religion, and in each era researchers among the ulema have employed each of them. Among them there have been those who minimized one of them and emphasized the other, and vice versa. Thus it is not suitable to neglect one of them entirely as did the majority of the factions. Rather the pure truth is to correlate one with the other and to compensate for the defects of each through the other. This is the opinion of Ḥasan al-Baṣrī, "Your practice, by God, besides Whom there is no other God, should lie between the two—between the excessive and the deficient." Thus, he who is one of the People of the Hadith must subject what he selects to critical examination and uphold it against the opinion of the *mujtahid*s among the Successors, while whoever is one of those using deductive reasoning (*takhrīj*) must make something part of his methods (*sunan*) only while taking care that it cannot oppose the sound obvious hadith, and while guarding himself against speaking from personal opinion in a case about which there exists a hadith or report from a Companion, insofar as he is able.

The hadith scholar (*muḥaddith*) does not have to be over-scrupulous about observing the principles which his associates established for which there are no textual stipulations of the law-giver, so that through this he would reject a sound hadith or analogy, such as the rejection of whatever has the least flaw in reaching back to the Prophet or being uninterrupted. Ibn Ḥazm did this when he rejected the hadith forbidding musical instruments due to a suspicion of a break in the transmission of al-Bukhārī, despite the fact that on its own the hadith was soundly connected to the Prophet.[13] Rather, one should have recourse to something like this only in the case of another conflicting report. Another case is the hadith scholars' saying, "So and so preserved more hadiths of a certain person than someone else, so we prefer his version to the hadith of the other for this reason"—even if there were one thousand reasons for preferring the other's version.

The concern of the majority of hadith transmitters when trans-

---

[13] Bukhārī VII:345. Ibn Ḥazm claimed that the chain of reporters was broken between al-Bukhārī and Hishām. This hadith is referred to in *Ḥujjat Allāh al-Bāligha* II, Chapter "Clothes, Adornments, Utensils, etc.," pp. 189–196.

mitting the meaning of the hadith was with expressing the essentials of the meanings, not the (precise) expressions which are recognized by those experts in the Arabic language. Thus they drew inferences from things like the "*fa*" or the "*waw*"[14] and one word preceding or coming after another and other sorts of hair-splitting. Often another transmitter will express this same narration, replacing one word instead of another. The truth is that whatever the transmitter reports should be literally taken as the speech of the Prophet, may the peace and blessings of God be upon him, then if another hadith or evidence comes to light it must also be taken into account. The person using deductive methods should not deduce a meaning which his peers would not find conveyed by the same expression and which neither native speakers nor scholars of the language would understand from it. Nor should he derive an opinion based on identifying the reason (*takhrīj al-manāṭ*) for legislation in a case where judgment was pronounced for no apparent reason, or applying a parallel case to it about which the interpreters disagree and opinions contradict each other, for if his associates had been asked about this issue perhaps they would have drawn a parallel to a parallel instance which would exclude it, or perhaps they would have cited a reason for legislation (*'illa*) other than that which he himself derived. In fact, derivation is only permitted because it is a form of following (*taqlīd*) of a *mujtahid* and it is only effected based on what may be understood from his statement. He must not reject a hadith or report of a Companion on which the Muslims have agreed in favor of a principle which he himself or his peers derived, such as in the case of the hadith of the milk-giving camels,[15] or like the annulment of the share of those with a relationship.[16] Indeed, taking account of the hadith is

[14] Particles in the Arabic language.

[15] This refers to the practice of leaving camels or cattle unmilked or tying up their udders some days before they are sold to make them appear more productive. In this case the hadith says that the buyer should have a purchase option of three days and then if he gives the animal back he should give a *ṣā'* of dates. The debate concerns the approving of the purchaser's option and remittance of this set amount of food which is allowed according to al-Shāfi'ī and not allowed according to Abū Ḥanīfa because the amount of dates is fixed while the amount and type of milk may vary, thus conflicting with the answer one would arrive at by analogy. Ṣiddīqī, *Kashshāf*, p. 52. Shāh Walī Allāh returns to this issue in *Hujjat Allāh* II "Forbidden Sales". *Muṣaffā* I:367. Some hadiths on this topic are Bukhārī Buyū' 23, 26, 28, Shurūt 11, Muslim, Abū Dāwūd, Tirmidhī, Nasā'ī, Ibn Mājah, Dārimī.

[16] This refers to a share of 1/25 of the spoils of war being distributed to members

more necessary than caring about this derived principle and this is what al-Shāfiʿī meant when he said, "In the case when I have held something or established it as a principle, if there should later come to your attention some saying of the Prophet, may the peace and blessings of God be upon him, conflicting with what I said, then what he, may the peace and blessings of God be upon him, said, must be upheld."

3) Among the (difficult issues) is that the investigative study of the Qur'ān and the sunna for the purpose of recognizing the sharīʿa rulings is at various degrees.

A) The highest of them is achieved by a person through actually knowing the rulings or virtually knowing them, which enables him to usually give an answer to the ones asking for legal opinions about certain circumstances insofar as his answer usually concerns some matter about which there is agreement. This is what is specified by the designation "*ijtihād*."

B) This ability (to perform *ijtihād*) is sometimes achieved through scrutinizing all of the reports and studying all of the anomalous and exceptional ones among them as Aḥmad ibn Ḥanbal indicated; together with the recognition of the referents of the speech in such a way that the rational person who knew the language would concur, as well as mastery of the knowledge of the reports of the pious ancestors through collating the discrepancies and organizing the inferences, and so on.

C) Sometimes the ability to perform *ijtihād* is acquired by becoming expert in the method of derivation (*takhrīj*) according to the legal school of one of the authorities in jurisprudence, together with knowing a sufficient body of prophetic sunnas and reports from the Companions, so that he can know that his opinion does not oppose the consensus, and this is the method of those who use derivation.

D) The middle level of study draws on both methods in that he acquires a knowledge of the Qur'ān and sunna which will enable him to recognize the preeminent issues of jurisprudence that have been agreed on together with their detailed proof texts (*adilla tafṣīliyya*). He should have achieved as well the highest degree of

---

of the Banū Hāshim clan of Banū Muṭṭalib by the Prophet and early caliphs. Later the juristic schools disagreed about whether this practice should be continued generally or limited. Ṣiddīqī, *Kashshāf* p. 52. Abū Dāwūd III pp. 145–147.

knowledge of certain issues of *ijtihād* through knowing about their proof texts (*adilla*), the preference of certain opinions over others, the criticism of derivations, and recognizing the correct from the false. Even if he has not perfected the critical apparatus to the same extent as the absolute *mujtahid* (*al-mujtahid al-muṭlaq*) still someone like him can select the better among two schools of opinion if he knows their proofs (*dalīl*), while realizing that his opinion is not operative in the same sphere as the *ijtihād* of the *mujtahid*, and is not admissible in the adjudication process of the judge, nor is it valid for the *muftī* in giving legal opinions. He is permitted to abandon certain derivations which people previously used if he learns that they lack validity and therefore the scholars who do not claim to be doing absolute *ijtihād* continue to make compilations, classify, make derivations, and give preference (in legal studies). Since *ijtihād* has become subdivided in the view of the majority and derivation has as well, and since the goal is only to obtain conjectural opinion (*ẓann*) and to base legal obligation on it, then nothing is disqualified by this.

E) As for people below this level, their course in those commonly arising questions is usually what they have acquired from their associates, ancestors, and compatriots among the legal schools that are followed. In those issues which rarely occur they follow the *fatwa*s of their *mufti*s, and in judgments they follow what their judges rule. We have found the reliable ulema from every legal school proceeding in this manner, formerly and recently, and this is what the founders of the schools bequeathed to their associates.[17]

In *al-Yawāqīt wa'l-Jawāhir* (Sapphires and Jewels)[18] it is reported that Abū Ḥanīfa, may God be pleased with him, used to say, "One who does not know my indicating factor (*dalīl*) must not give a *fatwa* based on my opinion," and he, may God be pleased

[17] In his treatise *'Iqd al-jīd fī aḥkām al-ijtihād wa-l-taqlīd* (Cairo: Maktaba al-salafiyya, 1965), Shāh Walī Allāh cites the classifications of the scholars al-Rafi'ī and al-Nawawī that there are within the category of absolute *ijtihād* (*ijtihād muṭlaq*) two levels; independent (*mustaqill*) and affiliated (*muntasib*) *ijtihād* (p. 5). Below this are further rankings; the *mujtahid* within the boundaries of a legal school (*al-mujtahid fi-l-madhhab*); the "*mujtahid al-fatayā*" or *mujtahid* who can give legal opinions since he is well acquainted with the literature of a school (*mutabahhir fi-madhhab Imāmihi*), and finally the common person who must follow the legal opinion which he is given. In his treatise *al-Inṣāf fī Bayān Sabab al-Ikhtilāf* he shifts his categories somewhat so that independent (*mustaqill*) *mujtahid* is the highest ranking.

[18] By al-Sha'rānī, 'Abd al-Wahhāb ibn Aḥmad, 1493–1565/6.

with him, whenever he gave a *fatwa*, used to say, "Al-Nu'mān ibn Thābit, i.e., his own name, has this view and this is the best we were able to do, so if someone comes up with something better, this is more correct." Imām Mālik, may God be pleased with him, used to say, "Anyone's opinion may either be accepted or rejected except that of the Prophet of God, may the peace and blessings of God be upon him."

Al-Ḥākim and al-Baihaqqī reported that al-Shāfi'ī used to say, "If there is a sound hadith, that becomes my opinion," and in another report, "If you see that my opinion opposes this hadith, then act according to the hadith, and throw my opinion out the window." One day he said to Muzanī, "O Ibrahīm, don't emulate me in everything I say, but look into it on your own, for this is the religion." He used to say, may God be pleased with him, "There is no final word (*ḥujja*) in anyone's saying except that of the Prophet of God, may the peace and blessings of God be upon him, even if there are many who hold such an opinion; nor in an analogy, nor in anything else, and moreover there is nothing at this level except that obeying God and his Prophet with full acceptance is mandatory." Imām Aḥmad (ibn Ḥanbal), may God be pleased with him, used to say, "No one is allowed to argue with God and his Prophet," and he also said to a man, "Neither perform *taqlīd* of me, nor Mālik, nor al-Auzā'ī, nor Nakha'ī, nor others, and follow the rulings insofar as they took them from the Book and the sunna. No one should give a legal opinion unless he knows the opinions given by the ulema in making *sharī'* rulings on issues and knows their legal schools. Thus when he is asked about an issue he will know that the scholars whose legal school he follows agreed upon it, so there is no harm for him in saying that this is permitted and that is not permitted, for his opinion is by way of reporting. If there have been divergent opinions about the issue, there is nothing wrong in his saying, "This is permitted according to so-and-so's opinion and not permitted according to so-and-so's opinion." It is not up to him to choose, for in that case he would be responding with the opinion of one of them whose proof (*ḥujja*) he did not know.

It is reported that Abū Yūsuf, Zufar,[19] and others, may God be pleased with them, said, "It is not permissible for anyone to give legal responses based on our opinions without knowing from where

[19] An early Ḥanafī jurist (d. 775).

we got them." It was said to 'Uṣām ibn Yūsuf, may God be pleased with him, "You usually disagree with Abū Ḥanīfa may God be pleased with him." He replied, "Abū Ḥanīfa, may God be pleased with him, was given a level of comprehension which we were not, and he discerned through his comprehension things which we don't understand, thus it is not permissible for us to give legal opinions on the basis of his statements when we don't understand them."

Muḥammad ibn al-Ḥasan was asked, "When is it permitted for someone to give legal opinions?" Muḥammad replied, "When he is right more often than he is wrong." Abū Bakr al-Askāf al-Balkhī was asked, "if there is a scholar in a city who is more knowledgeable than anyone else, is it permitted for him not to give *fatwas*?" He replied, "If he is one who is capable of doing *ijtihād*, it is not permitted for him (not to respond)." He was asked, "What makes a person capable of doing *ijtihād*?" He replied, "That he knows the reasons (indications) of the issue and is able to debate it with his contemporaries if they disagree." It was said, "The minimal condition for *ijtihād* is having memorized *al-Mabsūṭ*." (End of quotes from *Sapphires and Jewels*.)

In the *Al-Baḥr al-Rā'iq* (The Pure Sea) it is reported from Abū al-Laith[20] that he said,

> Abū Naṣr was asked about an issue which had been put to him previously, "What would you say, May God have mercy on you, if you had four books before you—the book of Ibrahīm ibn Rustam,[21] the *Adab al-Qāḍī* in the recension of al-Khaṣṣāf,[22] *Kitāb al-Mujarrad*,[23] and *Kitāb al-Nawādir* in the recension of Hishām.[24] First of all, would you permit us to give legal opinions based on them, and secondly, are these books commendable, in your opinion?" He replied, "What has been correctly reported from our associates is a body of knowledge which is approved, appreciated, and worthy of acceptance, but as for giving *fatwas*—while I do not think that anyone should give

---

[20] Abū al-Laith al-Samarqandī. A Ḥanafī theologian and jurist of the 10th C. The *Baḥr al-Rā'iq* is a work on Ḥanafī *fiqh* by Zain al-'Ābidīn ibn Nujaim al-Misrī (d. 1562/3).

[21] Ibn Rustam d. 211 A.H. A Ḥanafī pupil of Abū Yūṣuf.

[22] *Adab al-Qāḍī* (Cairo, 1978) by al-Khaṣṣāf, Aḥmad ibn 'Umar (al-Shaybānī), a Ḥanafī lawyer d. 874/5.

[23] Perhaps *al-Mujarrad* of Ḥasan ibn Ziyād.

[24] "Kitāb al-Nawādir." A number of *Nawādir's* were composed in the Ḥanafī school by students of Abū Yūṣuf including ones by Hishām (291 A.H.) and Ibn Rustam (211 A.H.). They are not considered among the major authoritative collections of the school because of their derivative nature. Nadvī, *Arā'*, 147.

> legal opinions based on what he doesn't understand, nor should he try to take up people's burdens—however if it concerns well-known issues which have become apparent and been made clear by my associates, I would hope that it would be possible for me to rely on them."

It is also cited in the *Baḥr al-Rā'iq*, "If someone has blood drawn or backbites and he considers himself to have broken the fast, so that he eats—then if this person did not ask a jurist for a legal opinion nor did the (correct) hadith reach him—must he make recompense (*kaffāra*) because this is merely ignorance, and there is no excuse for this within the domain of Islam? Then if he had asked a jurist for a legal opinion and he had given him one, there would be no penalty against him, because the ordinary person must perform *taqlīd* of the knowledgeable scholar (*'ālim*) if he has confidence in his fatwa, so he should be excused for what he did, even though the *muftī* was in error in the opinion he delivered. If the man did not (personally) request the opinion but he knew about the hadith, i.e., the saying of the Prophet, may the peace and blessings of God be upon him, "The cupper (one who has his blood drawn) and the cupped have broken their fast"[25] and his pronouncement, may peace be upon him, "Backbiting breaks the fast,"[26] and he didn't know that this had been abrogated nor its interpretation, then there should be no penalty assessed against him according to the two of them[27] since it is obligatory to act according to the manifest sense of the hadith. Abū Yūsuf, however, disagrees with this since he holds that it is not up to the ordinary person to act upon a hadith since he doesn't know about what abrogates or is abrogated. If a person has touched or kissed a woman out of lust or applied kohl, "so that he presumes that he has broken his fast, then he eats, he must pay a penalty, unless he had asked a jurist for a legal opinion and the jurist told him to break his fast, or he had heard a hadith report about this." If a person had made the intention to fast before noon, and then broke his fast, he would

---

[25] Hadiths for and against permitting cupping for the fasting person are found in al-Bukhārī, III:90–92. Shāh Walī Allāh held that this is not repugnant (*makrūh*) unless the person fears that he will be weakened by it. Therefore one who has blood drawn should continue fasting. *Muṣaffā* I:241–242. Bukhārī Ṣaum 32, Abū Dāwūd, Tirmidhi, Ibn Mājah, Dārimi, Ibn Ḥanbal.

[26] Not stated formally as a hadith.

[27] The Ḥanafi jurists Imām Muḥammad and Zufar.

not have to make the compensatory payment according to Abū Ḥanīfa, may God be pleased with him, thus he contradicts both of the other two in this way, as reported in the *Muḥīṭ*[28]

From this exposition it has become evident that the school of the common person is the *fatwa* of his *muftī*. Also found in the *Muḥīṭ* in the chapter on "Making up for the Missed Prayers" is that if a common person doesn't have a specified legal school then his course of action should be based on the legal opinion of his *muftī* as the ulema have declared. Thus if a Ḥanafī gives him an opinion he should make up for the afternoon and sunset prayers, and if a Shāfiʿī gives him an opinion then he cannot make them up, and there is no consideration of his personal opinion. If he doesn't ask anyone for a *fatwa*, or finds out what is sound according to the school of a *mujtahid* then this is permitted for him, and he has no need to revise this.

Ibn al-Ṣalāḥ (1245)[29] said, "If a Shāfiʿī finds a hadith which contradicts his school, then he should investigate further. If he has full competence in the apparatus of absolute *ijtihād* or competence concerning that topic or issue, he can choose independently in acting upon the hadith. However if he is not fully competent to do *ijtihād* and after he investigates he finds that which opposes the hadith to be problematic, and finds no satisfactory answer to this objection, then he may act according to this hadith on the condition that any non-Shāfiʿī independent Imām (founder of a school) did so, and he is excused in this instance for abandoning the school of his Imām. Al-Nawawī approved of this and affirmed it.[30]

4) Among them (difficult issues) are that most instances of disagreement among jurists, especially in cases where there appear sayings of the Companions which fall on two sides, such as the *Takbīrs* (pronouncing *Allāhu Akbar*) of the Days of *Tashrīq*,[31] the number of *Takbīrs* of the two *ʿĪd* prayers,[32] the marriage of one in

[28] *Al-Muḥīt bi-l-Taklīf* (Beirut: Maṭbaʿa Kathūlīkiyya, 1965), of Asadābādī, ʿAbd al-Jabbār ibn Aḥmad, d. 1025/6.

[29] 1181–1245. A Shāfiʿī expert in hadith. Author of the *Kitāb Maʿrifat Anwāʿ ʿIlm al-Ḥadīth*, commented on by al-Nawawī in *al-Taqrīb*.

[30] The following portion of the text duplicates *al-Inṣāf*, 41 ff.

[31] The three days after the *ʿĪd* of Sacrifice, when "*Allāhu Akbar*" is said aloud in various litanies. The Ḥanbalīs and Shāfiʿīs consider this practice sunna (established as recommended), the Ḥanafīs consider it obligatory, and the Mālikīs recommended (*mandūb*). al-Jazīrī, *al-Fiqh ʿalā al-Madhāhib al-arbaʿa* I:355–357.

[32] The Kufans used to increase the *Takbīr* (saying *Allāhū Akbar*) to three times in

Iḥrām,[33] (the special sanctified state of one on the Hajj pilgrimage), the manner of doing *Tashahhud* of Ibn 'Abbās and Ibn Ma'sūd,[34] silently (or loudly) pronouncing the *Bismillāh* and the *Āmīn*,[35] pronouncing the formula of the call to prayers twice or once during the *Iqāma*,[36] and so on—have to do with giving preponderance to one of the two opinions. The pious ancestors did not disagree on the essential legality on all of these opinions, but rather their disagreement concerned which was the more correct of the two things, and a parallel to this is the differing of Qur'ān reciters on the (acceptability of) variant modes of reading of the Qur'ān.[37]

They usually explained this matter by saying that the Companions differed although they were all correctly guided and therefore the ulema continue to endorse the legal opinions of the muftis in issues of independent reasoning, and to accept the judgment of the judges, and on some occasions they act so as to differ from their legal schools. In these situations you will see the leaders of the legal schools holding each opinion to be valid and dealing with disagreement about the opinion of one of them in such a way that he will say, "This is the more prudent," "this is preferable," and, "I like this better." Or he may say, "We only know about this opinion," and this occurs often in the *Mabsūṭ*, the *Āthār* of Muḥammad (Abū Yūsuf), may God be pleased with him, and the discussions of al-Shāfi'ī, may God be pleased with him. Then there

---

each prayer cycle (*raka'*) and the Medinans to seven *takbīrs* in the first *raka'* and five in the second. *Muṣaffā* I:178. This is examined in *Ḥujjat Allāh al-Bāligha* II, Chapter "The Two *'Īds*." Also, al-Shāfi'ī says that the *Takbīr* should be pronounced loudly on *Īd al-Fiṭr*, Abū Ḥanīfa does not.

[33] This is invalid according to al-Shāfi'ī and most others but Abū Hanifa allows it due to this belief that Prophet Muḥammad married Maimūna at this time Musaffā 1:289–. Shāh Walī Allah prefers that one shout not marry. *Ḥujjat Allah-Bāligha II*, "Characteristics of Ḥajj Observances," 59.

[34] *Tashahhud* (bearing witness) is a litany recited silently during the ritual prayer and each school varies slightly in the exact wording of the litany, position of the feet, etc. Walī Allāh prefers al-Shāfi'ī who follows the wording of Ibn 'Abbās. See *Muṣaffā* 1:115–116. al-Shāfi'ī, *al-Risāla*, 206–208.

[35] Discussed in *Ḥujjat Allāh al-Bāligha* II, Chapter on "Recitations During the Prayer."

[36] *Iqāma* is the second call to prayer given inside the mosque. The Ḥanafīs repeat the formula twice as in the call to prayer (*adhān*) while other schools say them only once except for the phrase "*qad qāmat as-Ṣalāt*." (Prayer has begun). *Muṣaffā* I:83.

[37] That is, each one of the variant readings of the Qur'ān is sound and acceptable and no preference is stipulated al-Shāfi'ī, *al-Risāla*, 208–210.

succeeded them a body of people who abbreviated the discussions of the jurists in such a way that they emphasized the disagreement and maintained the preferences of their leaders, and whatever was reported from the pious ancestors which reinforced remaining within the school of their associates and not going outside of it in any circumstance. This is either due to human nature, for every person likes what his peers and nation have chosen even in dress and cuisine; or due to some arbitrary leap arising in considering the proof, or due to other reasons of this sort. Some took this to be fanaticism in religion, but they were completely free from this. Among the Companions and Successors there were those who recited the *Basmala* and those who did not, and those who pronounced it aloud and those who did not,[38] and those who did the *Qunūt*[39] prayers at the time of the dawn prayer and those who did not; and those who performed the ablution after having blood drawn, nosebleeds, and vomiting, and those who did not; and those who believed in doing ablution after touching a woman out of lust or touching the male member, and those who did not; and among them were those who did ablution after eating things cooked in fire and those who did not; and those who did ablution after eating camel's meat and those who did not.[40]

In spite of these differences, they used to pray behind one another, as Abū Ḥanīfa or his associates and al-Shāfiʿī and others, may God be pleased with them, used to pray behind the Imāms from Medina who were Mālikīs and others even if they neither recited the *Basmala* silently nor aloud; and Harūn al-Rashīd[41] led the prayer as Imām after having blood drawn and Imām Abū Yūsuf prayed behind him and didn't repeat the prayer. Imām Aḥmad ibn Ḥanbal

---

[38] That is the recital of, "In the name of Allāh, the Merciful, the Compassionate," during the prayers.

[39] *Qunūt* refers to an extra supplication made during the *witr* or other prayers, especially when Muslims are struck by a calamity. There are various types: a) *Qunūt nāzila*. This is made at a time of calamity and may be offered with any prayer but preferably at morning, sunset, or night. b) *Qunūt Witr*. The Ḥanafīs offer this during the final *rakʿa* of the *witr* prayer after the night prayer and do not do it during the *fajr* (morning) prayer. The Shāfiʿis only pray the *Qunūt* prayer as part of the *witr* prayer during Ramaḍān, but do pray it at morning prayer. Shāh Walī Allāh refers to this in *Ḥujjat Allāh al-Bāligha* II, Chapter "Recitations During the Prayers," *Muṣaffā* I:112.

[40] These matters are dealt with in *Ḥujjat Allāh al-Bāligha* II, Chapter "Things Which Make Ablution Necessary." *Muṣaffā* II:36.

[41] Harūn al-Rashīd (d. 809), the Abbasid caliph.

held that ablution was necessary after a nosebleed and being leeched so someone once asked him, "If the Imām had experienced a flow of blood and had not done ablution, would you pray behind him?" He said, "How could I not pray behind Imām Mālik and Sa'īd ibn al-Musayyab?"

It is reported that Abū Yūsuf and Imām Muḥammad used to do the two *'Īd* prayers performing the two *Takbīr*s according to Ibn 'Abbās[42] because Harūn al-Rashīd preferred the way of performing the *Takbīr* of his ancestor. Once al-Shāfi'ī, may God be pleased with him, prayed (in the morning) near the grave of Abū Ḥanīfa, may God be pleased with him, and did not perform *Qunūt* out of respect for him.[43] He also said, "sometimes we incline toward the Irāqī (Hanafī) school (of law)." We have previously cited the answers given by Mālik, may God be pleased with him, to al-Manṣūr and Harūn al-Rashīd.[44] In the *Fatāwā al-Bazzāziyya*[45] it is reported that the second Imām—i.e., Abū Yūsuf, may God be pleased with him, prayed the Friday prayer having performed the full ablution at a public bath. He led the prayer and then the congregation dispersed. After that he was informed that a dead mouse had been found in the well of the bath-house. He then said, "In this case, we will use the response of our brothers from the Medinan (Mālikī) school that if the water reaches the amount held by two large jars it won't become ritually impure."[46]

Imām al-Khujandī[47] was asked, may God be pleased with him, about the case of a man from the Shāfi'ī school who had not prayed for a year or two, then he transferred to the school of Abū Ḥanīfa, may God be pleased with him. How should he make up for these missed prayers? Should he make them up according to the Shāfi'ī school or the Ḥanafī school? He replied, "He should make them up according to either of the schools provided that he believes in

---

[42] That is, seven *Takbīr*s in the first *ra'ka* and five in the second.

[43] This was respectful since the followers of Abū Ḥanīfa do not perform the *Qunūt* supplication in the morning prayer.

[44] See Chapter 83, p. 430.

[45] A treatise of Ḥanafī law by Ibn al-Bazzāz, Muḥammad ibn Muḥammad. Printed on the margin of *Fatāwā 'Alāmgīrī* (Beirut, 1973).

[46] Otherwise as a Ḥanafī, he would consider the water unusable for ablution and the ritual bath. Abū Dāwūd I:17 #65 Ṭahāra 33, Tirmidhī, Nasā'ī, Ibn Mājah, Dārimī, Ibn Ḥanbal. Ends section from *al-Inṣāf*, 43.

[47] Abū 'Imrān Mūsā ibn 'Abd Allāh from Samarqand. Died before 360 A.H. *Al-Ansāb*, V, al-Tamīmī al-Sam'ānī (Hyderabad, India: 1966), 53–54.

its validity." In the *Jāmi' al-Fatāwā*[48] it is stated, "If a Ḥanafī said that he had married a certain woman who had been divorced by the triple formula;[49] then he asked a Shāfi'ī for a legal opinion and he answered that she was not divorced and that his vow was invalid—there was no harm in his following the Shāfi'ī in this issue because many of the Companions were on his side."

Muḥammad (Abū Yūsuf), may God have mercy on him, said in his *Amālī*, "Even if a jurist says to his wife, you are definitely divorced, and he considers that this is equivalent to a triple divorce, then if a judge gives a judgment that the divorce is revocable, he is permitted to live with her."[50]

Likewise is every department about which the jurists disagree, whether in forbidding, permitting, freeing slaves, taking property, and so on. The jurist who has received a verdict counter to his own view must act according to the judgment of the Qāḍī and forgo his own opinion, and he must hold himself to do what the judge requires and act according to what he told him. Muḥammad (Abū Yūsuf), may God have mercy on him, said, "Likewise is the case of a man who does not have knowledge and is confronted by some problematic situation, so that he asks the jurists about it and accordingly they give him a legal opinion concerning what is permitted or forbidden. However, the judge of the Muslims hands down a judgment against him which contradicts the fatwa, for in fact this is a matter about which the jurists disagree. In this instance, the person must accept the judgment of the judge and leave aside what the jurists had responded to him."

5) Among (new developments) are that I have found some of them claiming that everything that is found in these voluminous commentaries and thick tomes of legal opinions are the opinions of Abū Ḥanīfa and his associates; so that they do not make a distinction between an actual original statement and the derived statement. They do not understand the meaning of the jurists' statement that according to the derivation of al-Karkhī the ruling on a issue is thus, and according to al-Ṭahawī it is thus, nor do they make a distinction between their saying, "Abū Ḥanīfa said thus," and their

[48] Of al-Ḥamīd al-Ḥanafī d. 1475. Brockleman II, p. 226.

[49] If the three divorces are pronounced all at one time, some consider this to be equivalent to one time only, so that she is still married.

[50] This is called "*Ṭalāq al-batta*." Some schools consider saying "definitely," equivalent to a triple divorce formula, while the Shāfi'is do not. *Muṣaffā* II:51.

saying, "the response to this case according to the school of Abū Ḥanīfa is thus," or "based on the principle of Abū Ḥanīfa is thus," nor do they heed what the Ḥanafī scholars like Ibn al-Ḥumām and Ibn Nujaim said about the case of the ten by ten (water)[51] and similarly the case of the condition of having to be a mile distant from water in order to do the ablution with sand and other cases like these—i.e., that these are derivations done by members (of a legal school) and not, in fact, part of the school. Some people claim that the legal school is founded upon these controversial disputes mentioned in the *Mabsūt* of al-Sarakhsī[52], the *Hidāya*,[53] the *Tabyīn*[54] and works like these. They don't realize that the first ones among whom these disputations appeared were the Mu'tazila, and that their legal school is not founded upon these. Then the ones who came later liked using these (disputes) for expanding and honing the minds of the students, and whether it was for some other reason than that, God knows better. Many of these ambiguities and doubts may be resolved through what we have set out in this chapter.[55]

6) Among them (new developments) are that I have found some of them claiming that the basis of the disagreement between Abū Ḥanīfa and al-Shāfi'ī, may God be pleased with both of them, is founded on those principles which are mentioned in the book of al-Pazdawī[56] and other similar ones. Rather the truth is that most of them are principles derived on the basis of their opinions.

According to my view, the statements that:

A) the specific pronouncement (*khāṣṣ*) is clear and it needs no explanation,
B) that some additional phrase (in a hadith) can abrogate,
C) that a general statement (*'āmm*) is as certain as a particular one (*khāṣṣ*),
D) that there is no preference given (to a hadith) due to a greater number of transmitters,

---

[51] The size of a pond whose water may be considered pure for the ritual bath. See *Ḥujjat Allāh al-Bāligha* II:183–185, "Rules About Water."

[52] The *Mabsūṭ* is a commentary on the *al-Kāfī* of al-Ḥākim al-Shahīd by Shams al-Aimma al-Sarakhsī d. 1090, a Ḥanafī scholar.

[53] Of al-Marginānī, 'Alī ibn Abī Bakr d. 1196/7, translated into English by Charles Hamilton and used by the British in India as a model of Islamic Law.

[54] *Tabyīn al-Ḥaqā'iq* of al-Zayla'ī d. 1342. A work on Ḥanafī law.

[55] The following section corresponds to p. 38ff. of *al-Inṣāf*.

[56] D. 1089. Ḥanafī author of *Kanz al-Wuṣūl ilā Ma'ifat al-Uṣūl*.

E) that implementing the hadith of a non-legal expert is not obligatory when it would block the option of using personal opinion,
F) that particularizing a general statement by the import (*mafhūm*)[57] of a condition and a quality[58] is absolutely out of the question
G) that what a command requires is absolutely incumbent;

and other issues like these are principles derived from the statements of the founders (of the legal schools). These are not soundly transmitted from Abū Ḥanīfa and his associates, and holding to them and taking trouble to refute what contravenes them among the practices of the earlier ones in their inferences as Al-Pazdawī and others did, is not more correct than holding what opposes these and responding with what refutes them.

Examples of Type A

An example is that they (the Ḥanafīs) made a principle that the specifying expression (*khāṣṣ*) is clear, and that no explanation should be appended to it and they derived this based on the work of the earlier ones concerning His, may He be Exalted, saying, "Bow down and prostrate yourselves,"[59] and the Prophet's, may the peace and blessings of God be upon him, saying, "The prayer of a person is not rewarded unless he straightens his back in the bow and the prostration," insofar as they did not hold that coming to rest (during the bow and prostration) to be obligatory,[60] nor did they consider that the hadith was in explanation of the Qur'ānic verse. So there was raised as an objection to them what the earlier ones had made of His, may He be Exalted, saying, "Rub your hands over

[57] The *mafhūm* is what is understood from the language of a statement without being explicitly stated.

[58] Quality (*waṣf*) is one of the type of allusions (*imā'āt*) in which the meaning must be inferred from a statement. The quality (*waṣf*) is not relevant in the statement's legal force unless it functions as the reason for the ruling. Thus it may become clear from a sequence of cases which element is critical to the ruling. For example, the Prophet refused to enter a house where there was a dog but entered it when there was a cat, explaining that the cat was not ritually impure. Robert Brunschwig "Raisonnement par analogie," 65–67. al-Ghazzālī, *Mustaṣfā* II:74.

[59] Qur'ān 22:77.

[60] The Shāfi'īs consider this obligatory, most Ḥanafī's do not since they hold that the Qur'ānic wording "bow down:" and "prostrate" is already specific and cannot be further specified by a hadith. *Muṣaffā* I:114.

your heads"[61] and the Prophet's, may the peace and blessings of God be upon him, rubbing it up to his forelock, insofar as they (the earlier ones) had made it an explanation;[62] and His, may he be Exalted, saying, "The male fornicator and the female fornicator, scourge each of them"[63] and His, may He be Exalted, saying "Cut off the hand of the thief, male or female,"[64] and His, may He be Exalted, saying, "Until she marries a husband other than him,"[65] and those things which had been appended to these (specific injunctions) as explanations after that, so that they had to take great pains in responding as is mentioned in their books.

Examples of Type B

They formulated the principle that the general statement (*'āmm*) is as (legally) definitive as the particular one (*khāṣṣ*) and they derived it on the basis of what the preceding ones had done with His saying, may He be Exalted, "Recite of the Qur'ān what is easy for you,"[66] and the Prophet's, may the peace and blessings of God be upon him, saying, "There is no prayer without the opening chapter (al-Fātiḥa) of the Book"[67] insofar as they had not considered it (the Prophet's report) a specification,[68] and in his, may the peace and

[61] Qur'ān 5:6.

[62] That is, during the ablution (*wuḍū'*) preceding the ritual prayer some Ḥanafīs state that the Qur'ānic injunction cannot be further specified. while some do take the hadith into account. Some of these cases are cited in the article by Robert Brunschwig, "Raisonnement par analogie," 65–66.

[63] Qur'ān 24:2. This is understood in jurisprudence as specified in applying to free, married, adulterers only, since female slaves were only to be scourged, on the basis of hadith and unmarried fornicators were not stoned. al-Shāfi'ī, *al-Risāla*, 105–6.

[64] Qur'ān 5:38. This is considered to be specified by further conditions, including that the stolen goods are worth more than 4 dinars, the theft is intentional, etc. al-Shāfi'ī, *al-Risāla*, 105.

[65] Qur'ān 2:230. She cannot remarry a man who has divorced her until she marries another man and the marriage is consummated. al-Shāfi'ī, *al-Risāla*, 149–151. All of the above Qur'ānic verses are considered by legal scholars to contain specific (*khaṣṣ*) injunctions. They are, however, considered by some, if not all schools to be further explained by hadith or early practice. This contravenes the principle that a specific injunction does not require further explanation.

[66] Qur'ān 73:20.

[67] This is the wording of a chapter title for hadiths on this topic rather than a hadith itself. It is close to the hadith "*Lā ṣalāt illā bi (qirā'a) fātiḥa al-kitāb*," Tirmidhī Mawāqīt 69, Abū Dāwūd II:822 Ṣalāt 36, "*Lā ṣalāt li man lā yaqra' bi . . .*"

[68] The Arabic word "*mā*" (*what* is easy) in the Qur'ānic verse is considered to be general (*'āmm*) and thus should not be specified by the hadith. Ṣiddīqī, *Kashshāf*, 74.

blessings of God be upon him, saying, "There is a one-tenth *zakāt* (*'ushr*) on spring-watered land"[69] and his, may the peace and blessings of God be upon him, saying "There is no *ṣadaqa* (alms tax) on what is below five *Awāq* (of silver),"[70] insofar as they didn't consider the first[71] hadith to be specified by the second, and so on with other subjects. Then this objection was presented to them that His, may He be Exalted, saying, "Such a sacrifice as can be afforded,"[72] which is general (*'āmm*) is (specified as) the female sheep and what is worth more according to the explanation of the Prophet, may the peace and blessings of God be upon him,[73] so they were reluctant to answer this.

Examples of Type F

They made a principle that there should be no consideration (in a command) of the object (*mafhūm*) of a condition and a description, and they derived this based on what the earlier ones had done with His saying, may He be Exalted, "And whoever among you does not have the capacity."[74] Then many objections were raised to them based on the other rulings such as his, may the peace and blessings

---

[69] In Arabic this phrase also contains the general word "*mā*"—(*what* is watered by springs). Bukhārī Zakāt 55, Muslim, Abū Dāwūd, Tirmidhī, Nasā'ī, Ibn Mājah, Dārimī, Muwaṭṭa', Ibn Ḥanbal.

[70] Bukhārī Zakāt 32, 42, 56. Muslim, Abū Dāwūd, Tirmidhī, Nasā'ī, Ibn Mājah, Dārimī. Muwaṭṭa', Ibn Ḥanbal. *al-Bukhārī*, III:328. One *ūqiyya* equals forty Ḥijāzī dirhems.

[71] Bukhārī III:328. The issue here is whether a general and specific injunction conflict. Most jurists combine the two hadiths in this case and hold that there is one-tenth *zakāt* (*'ushr*) payable if the land production exceeds a minimum. Abū Ḥanīfa disagrees and holds that 1/10 is due on any quantity of agricultural produce since the measures stated in the hadiths are used in the grain trade (*wasq*) or in terms of money (*awāq*), and not in terms of other produce. Shāh Walī Allāh discusses this issue in *Ḥujjat Allāh al-Bāligha* II, "Amounts Stipulated for *Zakāt*."

[72] Qur'ān 2:196.

[73] That is during the *Ḥajj al-Tamattu'* (when the state of consecration [*iḥrām*] is broken between the Hajj and the *'umra*) the sacrifice should be a at least a sheep. *Muṣaffā* I:280–281. Bukhārī Ḥajj 102, Khan trans. II:436. The general Qur'ānic injunction is thus specified by a hadith.

[74] Qur'ān 4:25. In this Qur'ānic verse, the context refers to having the resources to marry a Muslim lady of respectable family. If one does not fulfill the condition, of having enough resources, he is encouraged to marry a woman taken as captive in war. If one held the condition to be intrinsic to the force of the order it would imply that a person able to marry a free woman should not marry a slave or captive.

of God be upon him, saying, "There is *zakāt* on the camel which is a pasture animal"[75] so that they had to make a lot of efforts to respond to these.

Examples of Type E

They made a principle that the hadith of a non-jurist does not need to be acted on if recourse to personal opinion[76] is blocked by it, and they derived this from what they did in rejecting the hadith about the female animals which are sold without being milked for some time.[77] Then there was raised to them an objection to the hadith about laughing aloud (during prayer),[78] and the hadith about the fast not being invalidated by eating out of forgetfulness, so they were reluctant to respond.

Cases of what we mentioned are many, not hidden from the one who pursues the investigation, whereas extensive expositions beyond having this pointed out will not suffice in the case of the person who does not investigate. The opinion of the researchers should suffice you as proof about this issue, i.e., that it is not necessary to act on the basis of a tradition of a person who is known for accuracy and justice but not for legal acumen since the option of using personal opinion is blocked, such as in the case of the hadith about the female animals which are sold without having been

---

[75] This is a case of "*mafhūm ṣifa*", i.e., the intended object of a descriptive characteristic, in this case "pasturing". From it can be understood that there is no *zakāt* on "stall-fed" (*maʿlūfa*) animals. Ibn Hājib, *Mukhtaṣar al-Muntahā*, II:173–174. The Shâfi'ī's consider the condition and description to be taken as specifying a particular case while the Ḥanafī's do not.

[76] Here (*ra'y*) personal opinion, refers to the results of analogical reasoning (*qiyās*).

[77] *Maṣarrāt*. Walī Allāh prefers the judgment of al-Shāfiʿī, *al-Risāla*, 330–1, who decides that the milk is the property of the buyer in analogy to the case of young born to animals after they are purchased. He follows a hadith which states that a fixed compensation will be paid to the seller for such milk. The buyer may also return the animal if the milk given does not meet expectations. This was rejected by the Ḥanafīs who considered the fact that the value of the milk would vary to be problematic and thus appear to reject the hadith in favor of analogy. *Muṣaffā* I:367.

[78] Some of the Ḥanafīs accept a hadith holding that laughing aloud invalidates prayer, while other schools hold this to be weak. al-Ghazzālī, *al-Mankhūl fī Taʿlīqāt al-Uṣūl*, ed. Muḥannad Ḥasan Hitu, (Damascus, 1970), 376. Al-Sarakhsī, *Kitāb al-Mabsūt* (Cairo: Maṭbaʿa al-Saʿāda, 1324 A.Ḥ [1907/8]), 172–73. Shāh Walī Allāh comments on this in *Ḥujjat Allāh al-Bāligha* II in the chapter on "Things Which Make Ablution Necessary." These hadiths which are accepted despite the fact that they conflict with analogical cases and were transmitted by non-jurists.

milked for some time. This is the opinion of ʿĪsā ibn Ibān[79] and many of the later ones preferred this, while al-Karkhī[80] and many of the ulema followed him in holding that the condition that the transmitter have legal acumen does not hold due to the precedence of a report over an analogy. They said, "This opinion was not transmitted from our leaders, rather what was transmitted from them is that a single hadith report has precedence over any analogy." Don't you see that they implemented the report of Abū Ḥuraira concerning the fasting person who eats or drinks out of forgetfulness, even if this opposes analogy, so that Abū Ḥanīfa, may God be pleased with him, said, "If not for the report I would have held the analogy (to be correct)." You should also be guided to what is correct by their disagreement over many of the derivations taken from their practice and the fact that some of them refute others.

7) Among (new developments) is that I have found that certain people claim that there are only two groups with no third—"the Literalists" (*ẓāhiriyya*) and "the People who Exercise Personal Opinion" (*ahl al-ra'y*)—and that whoever uses analogy or deduction is one of the people of personal opinion—no, by God! Rather isn't what is meant by personal opinion the same as using understanding and reason? This is not absent from any scholar, nor is this the personal opinion which is absolutely not based on the sunna, for absolutely no Muslim would claim to be doing this, nor is what is meant the ability to make deductions and use analogy (*qiyās*), for Aḥmad, Ishāq, and even al-Shāfiʿī too, unanimously were not "People of Personal Opinion," while they used inferential methods and analogical reasoning. Rather what is meant by "People of Personal Opinion" is a group who reopen for derivation issues agreed on among Muslims or among the majority of them on the basis of one of the early persons. Thus what they do in most cases is to relate parallel cases to one another, and to refer to one of the theoretical principles without consulting the hadiths and reports. The Ẓāhiri (literalist) is one such as Dāwūd Ibn Ḥazm,[81] who neither

---

[79] A Basran Ḥanafī d. 221 A.H. *al-Jawāhir al-Mudiya fī Ṭabaqāt al-Ḥanafiyya* II (Cairo: Maṭbaʿa ʿĪsā al-Bābī al-Ḥalabī, 1978), 678–680.

[80] Abū al-Ḥasan ʿUbayd Allāh al-Tamīmī al-Samʿānī, *Ansāb* XI (Hyderabad: Maktaba Dā'ira al-Maʿārif, 1970), 75.

[81] 994–1064. Andalusian jurist, theologian, poet, and historian. Codifier of the literalist (Ẓāhirī) doctrine which approaches the sources of the religion through their literal meaning.

accepts using analogies nor accepts using the reports of the Companions and Successors. Between the two groups are researchers among the People of the Sunna such as Aḥmad ibn Ḥanbal and Ishāq (ibn Rāhwayh).

We have gone on about this here at very great length so that we have ranged beyond the discipline which was our subject in this book, although this is not our habit. This is due to two reasons. One of them is that God, may He be Exalted, put into my heart at one time a measure by which to recognize the cause of every difference arising in the religious community of Muḥammad, may peace and blessing be on its master, and what is correct according to God and His Messenger, and He also enabled me to confirm this by rational and textual proofs so that there should remain no ambiguity or doubt. Thus, I intended to write a book called "The Summit of Fairness in Explaining the Causes for Juristic Disagreement" and to unequivocally clarify in it these subjects and to copiously cite evidence, examples, and ramifications while sticking to the middle course between excess and negligence at each stage, comprehending all sides of the debate and the principles of what is intended and sought. Up until now I have not been free to do this so that when the discussion here reached the source of the disagreement, I was led by my inner motivation to explain whatever portion I easily could.

The second reason for going on at some length is the factionalism of the people of today and their disagreement and confusion concerning some of the things that we mentioned, to the point that they almost assault those who recite to them God's verses, and "our Lord is the Merciful, the one to ask for help against the blasphemies you utter."[82]

However, this is the end of what we wished to mention in the first section of the book, "The Conclusive Argument from God: Concerning the Inner Meanings of the Hadith." All Praise be to God, at first and last, outwardly and inwardly, and, God willing, the second volume will follow it which will explain in detail the meanings of the hadith reports emanating from the Prophet, may the peace and blessings of God be upon him.

---

[82] Qur'ān 21:112.

BIBLIOGRAPHY

Works of Shāh Walī Allāh

For annotated bibliographies attempting to make a chronology of his works see J.M.S Baljon. *Religion and Thought*, pp. 9–14, S.A.A. Rizvi. *Shāh Walī Allāh and His Times*, pp. 220–228, and in G.M. Qasimi's introduction (in Arabic) to *al-Tafhīmat al-Ilāhiyya*. vol. 1, pp. 15–37.

Alṭāf al-Quds. (Original Persian with Urdu translation by 'Abd al-Ḥamīd Swātī). Gujranwala: Madrasa Nuṣrat al-'Ulūm. 1964. English translation. *The Sacred Knowledge*. G.H. Jalbani and D. Pendelberry. London: Octagon Press. 1984.

Anfas al-'Arifīn. (Urdu translation of the Persian original). Sayyid Muḥammad Farūqī al-Qādirī. Lahore: Al-Ma'ārif, 1974.

al-Budūr al-Bāzigha. (Arabic text). Hyderabad. Sindh: Shāh Walī Allāh Academy, 1970. English translations by J M S. Baljon. Lahore: Ashraf, 1988 and G.H. Jalbani. Islamabad: Hijra Council, 1985.

al-Dhikr al-Maymūn. Urdu translation of *Surūr al-Makhzūn*. Dakka: Ashrafiya, 1965.

Fatḥ al-Raḥmān bi Tarjumat al-Qur'ān. (Persian). Karachi: Nūr Muḥammad, n.d.

al-Fauz al-Kabīr fi-Uṣūl al-Tafsīr. Urdu translation by Maulānā Rashīd Aḥmad Anṣarī Lahore: Maktaba Burhān. 1963. English translation by G.N. Jalbani, *The Principles of Qur'ān Commentary*, Islamabad: National Hijra Council, 1985.

Fuyūḍ al-Ḥaramain. (Arabic with Urdu translation). Karachi: Muḥammad Sa'īd, n.d.

Hama'at. (Persian original). Hyderabad. Sindh: Shāh Walī Allāh Academy, 1964.

Hawāmi'. (Persian). Delhi: Maṭba' Aḥmadī, 1308 A.H.

Ḥujjat Allāh al-Bāligha. (Arabic). Vols. 1 & 2. Al-Qāhira: Multazim al-Ṭab' wa-l-Nashr Dār al-Kutub al-Ḥadīth, 1952–1953. There is also an edition published by Kitābkhāna Rashīdiyya, Delhi: 1953, and Cairo: Bulāq, 1877.

Ḥujjat Allāh al-Bāligha. Urdu translation *Ni'mat translation al-Sābigha* by Abū Muḥammad 'Abd al-Ḥaqq Ḥaqqānī. Karachi: Asaḥḥ al-Mutābi', n.d.

al-Inṣāf fi-Bayān Sabab al-Ikhtilāf. (Arabic). edited by Rashīd Aḥmad Jelānduri. Lahore: Hi'at al-Auqāf bi-Ḥukūmat al-Banjāb, 1971. *Inṣāf al-Bayān fī Sabab al-Ikhtilāf*. Cairo: Muḥibb al-Dīn al-Khaṭīb, 1965.

Inṣāf. edited by 'Abd al-Faṭṭāh Abū Ghuddah. Beirut: Dār al-Nafā'is, 1978.

As'āf min Tarjumat il-Inṣāf. (Urdu translation by Muḥammad 'Abd Allāh Balyavī). Lucknow: 1304 (1886).

Kashshāf fī Tarjuma al-inṣāf. (Urdu translation by Muḥammad Aḥsan Siddīqī. Delhi: Mujtabā'ī, 1891.

'Iqd al-Jīd fi-Aḥkām al-Ijtihād wa-l-Taqlid. (Arabic). Cairo: Maktaba al-Salafiyya, 1965. See Daud Rahbar article for partial English translation.

Irshād 'ila Muhimmat 'ilm al-isnād. Lahore: Sajjād Publishers, 1960.

Ittiḥāf al-Nabīh. Lahore: al-Maktaba al-Salafiyya, 1969.

Izālat al-Khafā' 'an Khilāfat al-Khulafā'. Lahore: Suhail Akaidimī, 1976.

Al-Juz' al-Laṭīf fi-Tarjamat al-'Abd al-Ḍa'if. (Persian original). In *Journal of the Asiatic Society of Bengal* 14 (1912): 161–175 with English translation by M. Hidayat Ḥusain.

Kalimāt-i-Ṭayyibāt. (Persian letters). Delhi: Maṭba' Mujtabā'ī, 1309 A.H.

al-Khair al-Kathīr. (Arabic). Maktaba al-Qāhira. 1974. English translation by G.H. Jalbani. Hyderabad, Sindh: Shah Waliullah Academy, 1974.

Lamaḥāt. (Arabic). Hyderabad, Sindh: Shāh Walī Allāh Academy. undated. English translation by G.H. Jalbani. Hyderabad. Sindh: 1970. Reissued and re-edited by D.B. Fry as *Sufism and the Islamic Tradtion. Lamahat and Sata'at of Shah Waliullah of Delhi*. London: Octagon Press, 1986.

Musawwa Muṣaffā. (Arabic/Persian). 2 vols. Karachi: Muḥammad 'Alī Karkhāna-i-Islāmī Kutub, 1980.

Maktūbāt-i-Dīnī wa 'Ilmī. Unpublished Persian letters in the collection of the Indian Institute of Islamic Studies. Tughlaqabad. India.

Maktūbāt Shāh Walī Allāh Dihlavī Farsī. ed. Muḥammad 'Aṭā Allāh Ḥanīf. Lahore: Maktaba al-Salafiyya, 1983.

al-Qaul al-Jamīl. (Arabic with Urdu translation). Translated into Urdu with title *Shifā' al-'Alīl*. Bombay: 'Alī Bha'i Sharf 'Alī and Company, n.d.

Qurrat al-'Aynain fi-Tafḍīl al-Shaikhain. (Arabic). Lahore: Maktaba al-Salafiyya, 1976.

Risāla Dānishmandī. Gujrānwālā: Madrasa Nuṣrat al-'Ulūm, n.d.

Saṭa'āt. (Arabic). Hyderabad, Sindh. Shāh Walī Allāh Academy, 1964. English translation by G.H. Jalbani Hyderabad. Sindh: 1970. Translation reissued and re-edited (see *Lamaḥāt*. above), Urdu Translation by Muḥammad Matīn Hāshimī Lahore: Idāra Thaqāfiyya Islāmiyya, 1986.

Sharḥ Tarājīm Abwāb Saḥīḥ al-Bukhārī. Hyderabad, India: Osmania, 1949.

al-Tafhīmāt al-Ilāhiyya. (Arabic and Persian). 2 vols. Hyderabad, Sindh: Shāh Walī Allāh Academy, 1973.

Ta'wīl al-Aḥādīth. (Arabic). Ed. G.M. al-Qāsimī. Hyderabad, Sindh: 1966. English translation by G.H. Jalbani, Hyderabad, Sindh: 1972, and J.M.S. Baljon (abridged). see under Baljon below.

Works about Shāh Walī Allāh and His Family

'Abdel 'Aal, Khalīl 'Abdel Ḥamīd. "God. the Universe. and Man in Islamic Thought: The Contribution of Shāh Walīullah of Delhi (1702–1762)." Ph. D. Dissertation. University of London, 1970.

Aḥmad, 'Azīz. "Political and Religious Ideas as Shāh Walī Allāh of Delhi." *The Muslim World* 52 (January 1962): 22–30.

———. "The Waliullahi Movement." In *Islamic Culture in the Indian Environment*. Oxford: Oxford University Press, 1964.

———. *The Intellectual History of Islam in India*, Edinburgh: Edinburgh University Press, 1969.

———. *Studies in Islamic Culture in the Indian Environment*. London: Oxford University Press, 1964.

'Āshiq, Muḥammad. *al-Qaul al-Jalī*. Persian Biography of Shāh Walī Allāh. Delhi: Shāh Abū'l-Khair Akāḍmī, 1989. Urdu translation by Maulānā Ḥāfiẓ Taqī Anwār 'Alavī Kakorvī. Kakorī: Maktaba Anwārī, 1988.

Baljon, J.M.S. "Psychology as Apprehended and Applied by Shāh Walī Allāh Dihlavī" In *Acta Orientalia Neederlandica*. Leiden: E.J. Brill. 1977, pp. 53–60.

———. "Ethics of Shāh Walī Allāh of Delhi." In *Acts of VII Congress of Arabists and Orientalists*. Gottingen: Vandenhoeck & Ruprecht. 1976, pp. 63–73.

———. *A Mystical Interpretation of Prophetic Tales by an Indian Muslim: Shāh Walī Allāh of Delhi's Ta'wīl al-Aḥādīth*. Leiden: E.J. Brill, 1973.

———. "Shāh Walī Allāh's Terminology of Creation." *Actes du 8ème Congress de l'Union Europeene des Arabisants et Islamisants. Aix-en-Provence*, 1976, pp. 17–22.

———. *Religion and Thought of Shāh Walī Allāh Dihlavi*. Leiden: E.J. Brill, 1986.

Baqā, Maẓhar. *Uṣūl-i-Fiqh aur Shāh Walī Allāh*. Islamabad: Idāra Taḥqīqat Islāmī, 1979.

Berque, Jacques. "Un contemporain islamo-indien de Jean-Jacques Rousseau." *L'Islam au temps du monde*. Paris: Sindbad, 1984, pp. 113–146.

Halepota, A. *Philosophy of Shah Waliullah*. Lahore: Ashraf, 1970.
Hermansen, M.K. "Tension between the Universal and the Particular in an Eighteenth Century Theory of Religious Revelation: Shāh Walī Allāh of Delhi's *Ḥujjat Allāh al Bāligha*. *Studia Islamica* 63 (1986): 143–157.
———. "Shah Walī Allāh's Theory of the Subtle Spiritual Centers (Laṭā'if): A Sufi Theory of Personhood and Self-Transformation." *Journal of Near Eastern Studies* (January 1988):1–25.
———. "The Current State of Shāh Walī Allāh Studies." *Hamdard Islamicus* XI (3, 1988): 17–30.
Husain, M. Hidayat. "The Persian Autobiography of Shāh Walīullāh bin 'Abd al-Raḥim al-Dihlavī." *Journal of the Asiatic Society of Bengal* (1912): 161–175.
Jalbani, Ghulam Hussain. *Life of Shah Waliyullah*. Lahore: Ashraf, 1978.
———. *Teachings of Shah Waliullah of Delhi*. Lahore: Ashraf, 1967.
Metcalf, Barbara Daly. *Islamic Revival in British India: Deoband, 1860–1900*. Princeton: Princeton University Press, 1982.
Nizami, Khaliq. *Shāh Walī Allāh Ke Siyāsī Maktūbāt*. Delhi: Nadwat al-Muṣannifīn, 1969.
———. "Shah Waliullah of Delhi: His Thought and Contribution" *Islamic Culture* 54 (1980): 141–152.
Rahbar, Da'ud. "Shah Waliullah and Ijtihād." *The Muslim World* 45 (December 1955): 346–358.
Rahman, Fazlur. "The Thinker of Crisis—Shāh Waliy-Ullah." *The Pakistan Quarterly* (Summer 1956): 44–48.
Rizvi, Sayyid Athar 'Abbas. *Muslim Revivalist Movements in Northern India in the Sixteenth and Seventeenth Centuries*. Lucknow: Balkrishna Book Co., 1965.
———. *Shāh Walī Allāh and His Times*. Canberra: Ma'arifat, 1980.
———. "The Political Thought of Shāh Walī Allāh." *Abr Nahrain* 16 (1975–76): 91–107.
——— *Shah 'Abd al-'Aziz*. Canberra: Ma'arifat, 1982.
Sindhi, Maulana 'Ubaid Allāh. *Shāh Walī Allāh aur unka Falsafa*. Lahore: Sindh Sagar Akaidami, 1947.
———. *Shāh Walī Allāh unkī siyāsī taḥrīk*. Lahore: 1970.
———. *Urdu Sharḥ Ḥujjat Allāh al-Bāligha*. Lahore: Maktabiyyat Hikmat, n.d.
Valiuddin, Mir. *Contemplative Disciplines in Sufism*. London: East-West Publications, 1980.
Voll, John O. "Hadith Scholars and Tarīqahs: An 'Ulema' Group." *Journal of Asian and African Studies* 15 (July–October 1980): 262–273.

Reference Works and Works and Articles on Islamic Thought

Hadith Collections

al-Bukhārī. *Ṣaḥīḥ al-Bukhārī*. Trans. Muhammad Muhsin Khan. Chicago: Kazi Publications, 1979.
al-Dārimī. *Sunan*. Beirut: Dār Iḥyā' al-Sunna al-Nabawiyya, 1974.
Ibn Hanbal. *Musnad*. Reprint of Cairo 1985 edition, Beirut: Maktaba al-Kutub al-Islāmī, 1969.
Mālik. *al-Muwaṭṭa'*. Trans. Muhammad Rahimuddin. Lahore: Ashraf, 1980.
Muslim. *Ṣaḥīḥ Muslim*. Trans. Abdul Hamid Siddiqi. Lahore: M. Ashraf, 1971–75.
al-Muttaqī al-Hindī 'Alī. *Kanz al-'Ummāl fī Sunan al-Aqwāl wa'l-af'āl*. (on margins of *Musnad Ibn Hanbal*).
al-Nasā'ī. *Sunan*. Cairo: Muṣṭafā al-Bābī al-Ḥalabī, 1964.
Robson, James. Trans. *Mishkāt al-Maṣābiḥ* Lahore: Ashraf, 1963.

al-Tirmidhī. *Sunan. al-Jāmi' al-Ṣaḥīḥ*. ed. 'Abd al-Raḥmān Muḥammad 'Uthmān Dār Beirut: Dār al-Fikr, 1983.

Other Works

al-Āmidī. *al-Iḥkām fī Uṣūl al-Dīn*. Cairo: Dār al-Ḥadīth, 1984.

Brunschwig, Robert. "Raissonnement Juridique par analogie d'apres al-Ghazālī". in *Studia Islamica* 34 (1971): 57–88.

al-Ghazzālī. *Iḥyā' 'Ulūm al-Dīn*. Beirut: Dār al-Ma'rifa, 1982.

———, *al-Mankhūl fī Ta'līqāt al-Uṣūl*. ed. Muḥannad Ḥasan Hitu, Damascus: 1970.

———, *al-Mustaṣfā*, 2 vol., Cairo: Maṭba'a al-Amīriyya, 1334 A.H.

Hallaq, Wael. "Development of Logical Structure in Sunni Legal Theory", *Der Islam* 64, (1987): 42–67.

——— "On the Origins of the Controversy About the Existence of Mujtahids and the Gate of Ijtihād", *Studia Islamica* 63 (1986): 129–142.

——— "Was the Gate of Ijtihād Closed?" in *International Journal of Middle East Studies*. 16 (1984): 3–41.

———. "Notes on the Trem Qarīna in Islamic Legal Discourse", *Journal of the American Oriental Society* (1989): 475–480.

Ibn Hājib. *Mukhtaṣar al-Muntahā al-Uṣūlī*, Beirut: Dār al-Kutub al-'Ilmiyya, 1984.

al-Jazīri. *al-Fiqh 'alā Madhāhib al-Arba'a*, Beirut: Dār al-Fikr, 1392.

Juynboll, G.H.A. *Muslim Tradition*. Cambridge: Cambridge U. Press, 1983.

Makdisi, George. "The Juridical Theory of al-Shāfi'ī", *Studia Islamica* 59 (1984): 5–47.

Peters, Rudolph. "Ijtihad and Taqlīd in 18th and 19th Century Islam", *Die Welt des Islams*, XX, 3–4 (1980): 131–145.

Rahman, Fazlur. *Islam*. Chicago: University of Chicago Press, 1971.

———. "Dreams, Visions and the 'ālam al-mithāl" in Gustave von Grunebaum ed. *The Dream in Human Societies*. Berkeley: University of California, 1966, pp. 409–419.

al-Sarakhsī. *Kitāb al-Mabsūṭ*. Cairo: Maṭba'a al-Sa'āda, 1907/8.

al-Shāfi'ī. *al-Risāla*. trans. Majid Khaddouri. Baltimore: Johns Hopkins, 1962.

———. *Kitāb al-Umm*. Beirut: Dār al-Ma'rifa, n.d.

al-Suyūṭi, Jalāl al-Dīn. *Al-Radd 'alā man akhlada 'alā Arḍ wa jahila anna al-Ijtihād fī kulli 'Aṣr Farḍ*. Td. Shaikh Khalīl al-Mais. Beirut: Dār al-Kutub al-'Ilmiyya, 1983.

INDEX OF QUR'ĀNIC CITATIONS

1:4 Thee alone do we worship 186
2:7 God has sealed their hearts and hearing 240
2:18 Blind, deaf and dumb, they will not repent 152
2:20n God is able to do anything 40
2:45 Except for the humble who think they will meet their Lord 327
2:81 Whoever acquires evil will be surrounded by his sin 87, 233
2:87n note on being aided by the Holy Spirit 251
2:106 We do not abrogate any verse 357
2:115 Unto God belong the East and the West 313
2:117n He says to a thing, "Be!" and it is 67
2:129n Teach them the book 253
2:142–145n note on changing the *Qibla* 357
2:158 Al-Ṣafā and al-Marwah are among the emblems of Allah 12, 312
2:159 Those who keep hidden the proofs and guidance 239
2:161–162 Punishment will not be reduced for them 88
2:170 We follow the way upon which we found our forefathers 351
2:174 Those who conceal what God revealed in the scripture 347
2:179 And there is life for you in retaliation 12, 403
2:180 It is prescribed when death is drawing near 434
2:183 Fasting is commanded for you 291
2:183 That you might purify yourselves 12
2:185 God desires for you ease 300, 322
2:187 God knew that you were deceiving yourselves 14, 403
2:189 They ask you about the new moons 256
2:193 Fight against them until there is no more persecution 138
2:196 Such a sacrifice as can be afforded 475
2:217 Fighting in it is a great sin 415
2:222 They will ask you about menstruation 415
2:222 He loves those who keep pure 217
2:230 Until she marries a husband other than he 474
2:246 Send us a king 137
2:260 I ask in order that my heart might be filled with confidence 21
2:268n note on *al-khair al-kathīr* 4
2:275 Usury is just like trade 259
2:279 Wrong not and you shall not be wronged 309
2:282 So that if one makes a mistake the other will remind her 403
2:284 Whether you show what is in your hearts or conceal it 93
3:5n Those firmly grounded in knowledge 336
3:14 A tremendous sum 309
3:49n Jesus creating a bird from clay 177
3:93 All food was lawful to the Children of Israel 262
3:110 The best community brought out for humanity 47, 252, 341
3:156 If they had remained here with us they would not have died 51
3:159 They would have dispersed from around you 322
3:169–70 Don't think that those who are killed in the path of God have died 103
3:180 That which they hoard will be their collar 12
4:22n note on the interpretation of this verse 316
4:23 Your mothers are forbidden to you 408
4:25 And whoever among you does not have the capacity 475

4:34 Men support the women 124
4:42n note on angels aiding Muslims during the battle of Badr 48
4:123 Whoever does evil will be requited for it 93
5:1 Decrees as He wills 161
5:3 Carrion is forbidden you 407
5:3 This day I have perfected your religion 192
5:6 Then perform the ablutions 408
5:6 Rub your hands over your heads 474
5:15n note on the Clear Book 83
5:19n note on "a great gap of time" 356
5:38 The religion of your father Abraham 361
5:38 Cut off the hands of the male and female thieves 277, 474
5:48 For each of you We have appointed a divine law 257
5:66 If they had acted according to the Torah and the Gospel 93
5:73 If you don't do it there will be a crisis on earth 403
5:95 That he may taste the evil consequences of his act 12
5:101 Don't ask about things which if they were known to would trouble you 269
5:103 A Baḥīra or a Sā'iba 187, 361n
6:40–41 No; upon Him you will call 186, 363
6:77 I love not things that set 313
6:90 With their guidance, be guided 398
6:91 Say, who revealed the book which Moses brought? 366
6:101n denial of God having a consort 178
6:137–140n note on the pagan Arabs 187
6:146n note on the forbidding of camel's meat to the Jews 283
6:149 The conclusive argument from God vi, 9, 68, 149, 211
7:12 You created me (Iblīs) from fire 350
7:93–97 If the people of the towns had believed and been pious 95
7:172 Your Lord brought forth from the children of Adam 67n, 199n, 238
7:189–190 He it is who created you from a single soul 188
7:195 Have they feet wherewith they walk? 177
8:39 Then fight them until there is no sedition 13
8:42 Let he who perished, perish by a clear proof 72, 275, 353n
8:66 Now God has lightened your burden 403
8:73 If you don't do it there will be sedition on the earth 358
9:31 They have taken their rabbis and monks as Lords 186 (twice), 352, 458
9:32 Even though the unbelievers detest it 254
9:34 Those who hoard gold and silver 222
9:45 So that in their doubt they wavered 362
9:102 Those "who mixed a righteous act with another evil one" 240
9:118 The world becomes straitened for him 231
10:49 Every people has an appointed hour 89
11:88 My success is only through God 27
11:107–108 As long as the heaven and earth endure 81
11:114 Good deeds annul bad deeds 82, 220
11:116 If only there had been in the generations before you people of excellence 347
12:2 An Arabic Qur'ān in order that you might understand 355
12:38 Most of the people are not grateful 77
12:53 The soul persistently inclines to evil 7
13:7 Every people has a guide 247

13:11 God does not change the condition of a people 70, 264
13:39 Allah effaces what He wills 200
14:1n To bring people from the darkness to the light 159
14:4 We never send any messenger except with the language of his people 355
14:18 Ashes blown away by a strong wind 166
16:22 Those who don't believe in the Hereafter 233
16:43–44 Ask the People of Remembrance if you do not know 310
16:60 God has the most sublime similitude 204
16:68n on the inspiration to the bee 107
17:13–14 On the Day of Judgment We will bring out a book 80
17:20 These and those are aided by the gifts of your Lord 59
17:23 Don't even say "uff" to them 397
17:67 All upon whom you call lose their way except Allah 363
17:84 Everyone acts according to his own manner 73
17:85 They ask you about the spirit 53
18:65n note on *'ilm ladunī* 6
18:66–83n note on the story of Moses and Khiḍr 36
18:75n note on Khiḍr 152
19:10 We sent our spirit unto her (Mary) 39
19:59 A generation which neglected the prayers 181
19:59n note on "those who are lost and follow their lusts" 346
20:5 The Merciful, who sits upon the Throne 326
20:14 Undertake prayers in order to remember me 11
20:41 I made you (Moses) for Myself 68
20:107 Any crookedness or distortion 353
21:22 If there had been in them Gods other than Allah 399
21:69–70n on the story of Abraham and the fire 34, 36
21:92 This, your community, is one community 72
21:112 Our Lord is the Merciful 478
22:18 punishment is decreed against many (for not prostrating) 63
22:18 God "decrees as He wills" 161
22:32 Whoever honors the emblems (*sha'ā'ir*) of Allah 210
22:37 Their flesh and their blood will not reach Allah 11
22:39 Permission is given to those who fought because they were wronged 360
22:40 Had God not driven some people back because of others 138
22:67 To each community We have given a devout ritual 257
22:78 The religion (*milla*) of your father, Abraham 267, 361
23:53 Each sect rejoicing in its own tenets 257, 264, 449
24:2 Scourge the male fornicator and the female fornicator 277, 474
24:24n note on the hands and feet "speaking" 82
24:35 Whose oil would almost glow forth (by itself) 152
24:37 Men whom neither merchandise nor selling tempt 290
25:7 What kind of Prophet is this who eats food and walks in the markets 366
26:4 Their necks bowed before it in humility 219
27:59ff Is Allah better? 185
27:87n note on the blowing of the Trumpet 55
30:17–18 Glory be to God in your evening hour 289
30:30 The original nature according to which God created human beings 84, 144
31:25 If you asked them who created the heavens and earth 363
32:3 That you might warn a people to whom a warner has not previously come 363

33:4, 58:3n	note on *ẓihār*, a form of Pre-Islamic divorce 269
33:56n	note on asking for blessings on the Prophet 212
33:62	You will never find any alteration in the Sunna of God 49
33:72–73	We offered the Trust to the heavens and the earth 57
34:3	Not an atom's weight escapes His notice 161
35:32	Those who wrong themselves and those who are in the middle 334
36:6	Warn a nation whose fathers were not warned 355
36:12n	note on "the Clear Register" 83
36:81–82	"Be!" And it is 178, 180
37:107	With a great sacrifice 291
37:171–173	Our army would certainly overcome 253
37:8n	note on the Highest Council 43
37:83	Abraham was from his (Noah's) following 267
38:25	He fell to the ground kneeling and was sincere 398
38:41,42n	note on the story of Job 34, 36
38:69n	note on the Highest Council disputing 44
39:3	Approach closer to God in proximity 177
39:56	I was remiss regarding God 211
39:56	For I was among the mockers 101
39:65	If you associate others with God then all your acts will be futile 82
39:69	The earth would be illuminated with the light of its Lord 7
40:7	They extol the praises of their Lord 46
40:7	They ask forgiveness for those who believed 46
40:7	Those who bear the throne 43, 67
41:37	Prostrate not to the sun nor to the moon 185, 313
41:44	If We had made it a Qur'ān in a foreign language 355
42:11	There is nothing like unto Him 66, 192
42:13	Establish the religion and be not divided therein 257
42:30	Those misfortunes which strike you are those which your own hands earned 93
42:51	Except through revelation or from behind a veil 195
43:22	We found our fathers following a community and we follow in their footsteps 455
44:2–4	We made it descend on a blessed night 67, 288
44:4	on which every wise command is decided 46
44:10	Today the heaven will bring forth clear smoke 40
46:9	I am no novelty among the Prophets 366
49:2n	note on not raising one's voice before the Prophet 212
50:29	The statement which comes from Me cannot be altered 410
53:42	The ultimate limit stops at your Lord 191, 255
55:13ff n	note on the benefits of Allah 206
55:15	He created the jinn from smokeless fire 33
55:29	Every day He is in a mode 50, 239
55:31	O you two dependents 95, 244
56:7–11	You will be of three types: those of the right hand 334
56:27n	note on people of the right side 155
57:27	But monasticism they invented 235
58:7	There are no three having a secret conversation 161
59:5n	note on burning the date palms 360
59:7	What the Prophet gives you, take 373
62:2	To the unlettered ones a messenger of their own 252, 355
65:3	God has made a measure for everything 267, 395

66:6n	note on angels 258, 364
69:17n	note on angels bearing the Throne of God 365
71:7n	note on Noah's people
72:18	Call upon no one except God 186
73:20	Recite of the Qur'ān what is easy for you 474
74:28	Does not let anything remain nor spare it 204
74:43	We were not among those who prayed 220
81:10n	note on the reading of the scrolls 82
85:22n	note on the Preserved Tablet 83, 106
88:7n	Does not nourish or satisfy hunger 6
89:27–30	O contented soul return to your Lord 45
92:5–10	We shall make these things easy for him 59
111:3n	note on Abū Lahab's punishment 198
112:3	He does not beget nor was He begotten 191

INDEX OF HADITH CITATIONS

This index includes both sayings of the Prophet and other reports found in Islamic Hadith Collections which are cited in *Ḥujjat Allāh al-Bāligha*

a cure for diarrhea 34
A just person of every succeeding generation will convey this knowledge 427
A man among the people of Paradise will ask his Lord for permission to plant crops 110
a man who died without associating anything else with God 175
a person who built a house, and made a feast in it 17
A soft sandy plain will be spread out for them 222
Abraham made Mecca a sacred place and asked for blessing on it 263
Abraham, Your Prophet and servant prayed for Mecca and I pray for Medina 273
actions are only judged through their intentions 11, 408
Acts will come forward on the Day of Judgment. First prayer will come 37
Adam and Moses met in the presence of their Lord 108
Affliction descends and prayer counteracts it 39
All blood revenge is abolished 308
Allah desires for you ease 322
An unbeliever will be afflicted by ninety-nine vipers in the grave 39
As far as (the range of someone's) eyesight 296
Asking permission (to enter a home) came because of seeing 13

Be easy (with the people) and don't make things difficult 332
Be keen to acquire (as best for Jihād) a black horse 374
Be on guard against the two causes for being cursed 403
Begin with what Allah begins with 313
By God, God will never debase you, since you honor the bonds of kinship 367
By God, if you take up the use of analogies, you will forbid the permitted 350

dead person being tormented by the weeping of his family over him 422
Death will be brought like a ram and sacrificed between Paradise and Hell 39
The devils eavesdrop 51
Do not saddle your camel except (to travel) to three mosques 188
Do you see what I see? I see civil strife falling in the midst of your homes 38
Does each one of you have two garments 14
Don't be too severe with yourselves or God will be severe with you 235
Don't bother the Abyssinians as long as they don't bother you 343
Don't you see that she, when she has her monthly period, does not pray or fast? 265

Error and forgetfulness have been removed from my community 408
Eve called her son 'Abd al-Ḥārith 189
Every child is created according to the original nature (*fiṭra*) 106, 107n, 238
Every intoxicating thing is forbidden 401

Every thing has a polish and the polish for the heart is the recitation of the Qur'ān 230
everyone will find it easy to do that for which he was created 408–9

Fasting is done for My sake and I reward it 225 (twice)
Fasting suppresses sexual desire 224
Forty virtues the highest of which is giving someone a female goat 294

Give to the charitable one a good replacement 45
God created human reason (*al-'aql*) and said to it, 'Come forward 39
God created man from a handful (of dirt) 49
God does not take away knowledge, per se, by removing it from the people 347
God has granted you the true opinion, O Ibn al-Khaṭṭāb 14, 402
God is single and loves the single 292
God looked at the people of the earth, and loathed the Arabs and the non-Arabs 359
God the Exalted will resurrect the days on the Day of Judgment just as they are 37–8
God will appear in it 287
God and His angels bless those in the right flanks of the row 333
Good and evil are creatures which will be set up for people on the Day of Judgment 37
Good deeds annul bad deeds 220

hadith of Khurāfa 374
hadith of the animals which are sold without having been milked for some time 22
hadith of the bridling 8
hadith of Umm Zara' 374
having many faces and tongues 46
He does not know where his hand has spent the night 13, 403
he (the devil) laughs when people yawn 286
he may make a mistake, and pray facing a different one 313
He measured the destinies 198
He recorded the destinies of all created things 198
he stretched forth his hand to pluck a bunch of grapes from the Garden 38
he was protected from the devil until evening 329
He will descend on it to the lowest heaven 287
Help me to achieve this for you through much prostration 220
His poetry believes but his heart does not 366
Human actions are presented before God on Mondays and Thursdays, 287

I am as My servant thinks Me to be 235
I am only a man, and when I order you to do something regarding your religion, accept it 374
I arose during the night and performed the ablution and prayed my share of prayer 44
I entered Paradise and saw a young girl, tanned and with dark red lips 110
I feared that it would become prescribed for you 262, 331, 381
I had forbidden you to visit graves, but now you may visit them 409
I have been sent with the tolerant, clear, Ḥanīfī religion 3, 304
I never saw any group better than the Companions of the Messenger of God 364

I only made a conjecture, so don't blame me for that opinion 374
I saw Ja'far ibn Abī Ṭālib as an angel with wings flying in paradise with the angels 45, 103
I was sent to bring the noble virtues to fulfillment 303
I was sent to eradicate musical instruments 303
I was sent with the tolerant, monotheistic, (Ḥanīfī) religion 304
Iblīs used an analogy and he was the first to use an analogy 349
If a community persists in fasting their devils will be chained 225
If God decrees that a person will die somewhere on the earth, a need will arise for him to go there 50
If God makes you enter the garden 110
If he asks to take refuge in Me, I will give him refuge 329
If I were to say yes, it would become a necessity 263
If not for recent unbelief among your people, I would have torn down the Ka'ba 324
If the fluid of the male precedes the fluid of the female the child takes after him 49
If the Prophet had seen what the women are doing these days 403
If the water reaches an amount to fill two jars 22–23, 325
if you graze her (a she camel) in a fertile place, you graze her by God's predestination 201
If you hear that some mountain had moved from its place, believe it 73, 78
In Hell he saw a thief who had robbed pilgrims to Mecca 38
In His hand is the scale, (with which) He raises the share and lowers it 50
in Nigella seed there is a cure for every illness except death 34
In one of you having intercourse (with his wife) there is a virtuous act (ṣadaqa) 322
in the Afterlife will take on the shape of a hairless snake 222
in the Garden he saw a prostitute who had given water to a thirsty dog 38
In the sexual act (with the spouse) of one of you there is a reward 15
In what state did you leave my servants? 70
Indeed an allegory for me and the message with which God has sent me is that of a man who came to a people 17, 72
Indeed during the night there is an hour 289
Indeed God will support this religion even by a profligate man 359
Indeed these are but your acts thrown back upon you 17
Indeed these are your acts which I will count for or against you 80
Indeed they (the Meccans) knew that the two of them never had recourse to divination 368
Indeed you will see your Lord (in the next life) 220
Indeed, in the body there is a piece of flesh 18
Inform them that God, may He be Exalted, has made charity a duty for them 12
It (the sun) rises between the horns of the devil 14
It is not unclean, rather it is an animal which comes and goes in your homes 13–4
It is to establish the remembrance of God 13
It was for fear of harm coming to the child 14
It won't be long before a man with a full stomach talks to you about this Qur'ān 347

Khusrau would be destroyed and there would be no Khusrau after him 308, 343

Leave me (and don't ask me) as I have left you, for those before you were destroyed through the excessiveness of their questionings 269–270
Leave the Turks as long as they leave you (alone) 342
Leave them (my leather socks), for I put them on when they (my feet) were clean 401
Let the one who is present inform the absent one 192
like an iron chain clattering upon smooth rocks 288
Lo, the children of Adam were created at various ranks 73

mountain goats (the Bearers of the Throne as) 365
My cistern is (the size of) what is between the Kaʿba and Jerusalem 292
My cistern is longer than the distance between Ailat and Aden 292
My Lord concurred with my judgment in three things 402
My Lord, may He be Blessed and Exalted, is angered today 268
My servant sinned and he knew that he had a Lord who forgives sins 161
My speech does not abrogate the speech of God 358

Neither (the hand of) the betrayer, nor the plunderer, nor the pilferer will be cut off 317
No Muslim is afflicted by an injury, be it sickness or some such thing 95
No Muslim standing to pray asks Allah for something 406
No one makes religion too strict but that it overwhelms him 349
No person believes until he believes in predestination, its good and its evil 197
Not one of you will be brought into paradise by his act 51
Nothing turns back destiny except petitionary prayer 222

O Muʿādh, do you know what is God's right over his servants, 203
On whatever (land) the sky pours down (rain), the *zakāt* of one-tenth applies 407
Or an understanding (of the Qur'ān) which was bestowed upon a Muslim 336
Our Lord descends every night to the lowest heaven 287

Paradise is surrounded by undesirable things and Hell by craved things 38–9
People are mines, like mines of gold and silver 73
Petitionary prayer is the core of worship 229
Prayer is the path of ascent for the believer 220
punished by a bridle of fire 8–9, 331
Purify yourself after urination 217
Reflect on the creation of God and do not speculate about God 255
Respect seniority 284

She said (ʿĀ'isha), "He sprang up" and she did not say, "he stood up" 412
Should I not apprise you of the best of your actions? 229
Sin is something which has a strong impact on your heart 235
So that the Jews will know that in our religion there is room for pleasure 322–3
Some fell in the fire, and some fell down and were saved, 240
Speculate about the creation, not about (the nature of) the Creator 191

talbīna gives rest to the heart of the sick person 34
Temptations strike the heart in the way that a reed mat is woven stick by stick 81

That the devil had spent the night inside his nose 13
That which they hoard will be their collar on the day of Resurrection 12
The acts of my community were presented to me, the good and the bad 294
the aged man permitted to kiss (his wife) while fasting, but not the young man 78
The angels of the day ascend to Him before the angels of the night 289
The angels pray for one of you as long as he remains seated at the end of his ritual prayer 44
The bad breath of the one who fasts is sweeter to God than the fragrance of musk 225
The believer is like the supple stalk of the plant 95
The believers' share of punishment is the hardship of this world 231
The best prayer is the midnight prayer, yet how few perform it 289
The chapters of the Qur'ān "The Cow" and "The Family of 'Imrān" will come 37
The congregational prayer is better than the individual prayer by twenty-seven degrees 295
the Dajjāl wants to kill the believing Muslim a second time 37
the debt of God is even more worthy of being paid off 328
the deeds of the day are brought to Him before the deeds of the night 70
The devil eats with his left hand 284
The devil... on the day of 'Arafāt 227
The Devil makes sport with peoples' posteriors 285–6
The devil spends the night in his nose 13, 403
the devils will not be able to do it 329
The drawstring on action is the two eyes 403
The evil eye is a reality (ḥaqq) 408
The fast has the effect of curbing lust 12
The first one to apply analogies was Iblīs 350
The first one to divide the day and the night into hours was Noah 296
The first thing that was revealed thereof was a chapter from al-Mufaṣṣal 324
The fruit used to be struck by blight, splitting, blistering from the heat, and so on 403
The funeral bier of a Jew was passing by 423
The Garden and the Fire were depicted for me 38
the governing angels get information on that day 199, about what will be his sustenance, his term, his life span, and whether his future life will be difficult or happy 244
The grave (of the believer) will be extended by seventy cubits 298
the grave will be extended seventy cubits by seventy 39
The Hajj is the day on which you make the pilgrimage 326
The hand of God is full 191
The hearts of men are between two of God's fingers which He turns as He wishes 202
the human heart, which when it is sound, everything is sound, and when it is corrupted, all is corrupt 175
The Imāms should be from the Quraish 343
The judgment in establishing lineage 409
The lapsing of the learned man, and the arguing of the hypocrite 350
the last man to come out of the fires of Hell 110
The lower soul wishes and desires and the sexual organ either confirms that or denies it 80
The lower soul wishes and desires while the sexual organ confirms that or denies it 84

the man who does not contribute the alms tax will find his wealth represented to him 400
The master is Allah 181
The Muslim who harms Muslims most is the one who asked about a thing so that it was forbidden because of his questioning, 16, 263, 270, 379
the old man and the sick man 82
The one (prayer) in the middle of the night 289
the one who does not recite the Umm al-Kitāb 401
(The one who gets up during the night) should not plunge his hand into a vessel 315
The one who gets up during the night (to pray) 290
The one who is present sees something that the absent one does not 375
The one who says "there is no God but God." 328
the one who seeks knowledge of a branch of Astrology 51
The pen was raised from three 302
the people who are pulled by chains to Paradise 214
The polytheists used to utter the following formula of "*talbiya*" 184
The prayer of a man in the congregation is greater than his prayer in his house 15
The prayer of a person is not rewarded unless he straightens his back in the bow and the prostration 473
The prayer of one of you is not accepted unless he performs the ablution 401
the Prophet changed the names of his companions ʿAbd al-ʿUzzā and ʿAbd Shams 189
the prophet forbade enfolding the smooth rocks 216
The Prophet is true (*ḥaqq*) 408
the Prophet will go in to his Lord while He is on His Throne 40
The prophets are all step-brothers, sons of the same father by different mothers 72
The Qibla is everywhere between the East and the West 326
The Qur'ān will be opened to the people 350
The rewards of my community were presented before me 294
the rule about the rider on a mount who prays the supererogatory prayer outside of the city 313
the saying of a man, "I slept with my wife during (the day in) Ramaḍān," 401
The situation of the Children of Israel was balanced until there arose among them those of mixed descent 352
The treaty which is between us and them is prayer, so whoever abandons it has become an infidel 220
The world shall be brought forth on the Day of Judgment in the form of an old woman 38
Their religious scholars forbade them these things, but they did not abstain from them 348
Then his parents make him a Jew or a Christian or a Magian 107
Then if it is obscured from you, complete the number (of days) of Shaʿbān at thirty 325
Then you can also make the pilgrimage on his behalf 399
There is a one-tenth *zakāt* (*ʿushr*) on spring-watered land 475
there is "heat" in a purgative grass (*shubrum*) 34
There is no alms tax on whatever is below five *ausuq* 407
there is no marriage without a guardian 408
There is no prayer except in a state of purity 408

There is no prayer without the opening chapter (*al-Fātiḥa*) of the Book 474
There is no *ṣadaqa* (alms tax) on what is below five *Awāq* (of silver) 475
there is *ribā'* in chick peas 399
there should be no speculation about the Lord 191, 255
There was God and there was nothing before Him 33
There was no prophet sent by God who did not have disciples and companions 353
There was no reason for us to do *ramal* 375
There were four rivers, two concealed and two apparent 38
There will be no cutting off (the hand in the case of) stealing fruit which is hanging (on a tree), nor in what is protected by the mountain 317
These are two books from the Lord of the worlds 39
These constitute God's punishment of His servant 80
These gardens are inhabited by *jinns*, and devils 285
They are (also) part of God's predestination 201
They are a community among the communities 311
They are weeping over her while she is being tormented in her grave 422
they should worship God, may He be exalted, at solar and lunar eclipses 313
They used to say that things were permitted for them, so that they regarded them as lawful 186
They were more hard on themselves, therefore God made it harder for them 271
They were only forbidden lest the people take them as proper 321
This is an hour when the gates of heaven open 13
This is the prayer of the hypocrite and the one who does this is not one of us 333
This is the time when the gates of heaven open 289
This is your time and the time observed by the prophets before you 291
This, your community, is one community (umma) 72
Three (types of people) who will have two rewards 294
Three to whom God, may He be Exalted, will not speak 294
Through this people have been destroyed before you 14
tore down the figured curtain and removed the shirt with designs on it 170
two cups of wine and milk were offered to the Prophet 108

'Umar ibn al-Khaṭṭāb cursed someone who asked about something hypothetical 415
Until I feared that it would be prescribed for you 16

We are an unlettered people who do not write or calculate 320
We have been given rain when such and such (a star) sets 51
Were it not that I would be hard on my community I would have imposed the duty of cleaning their teeth with a *siwāk* at every prayer 283, 322, 381
When God created ties of relationship 37
When God decrees an order in the heavens, the angels beat their wings in submission 43
When God loves a person he calls Gabriel and says, "I love so-and-so, so love him" 44
When He decides on a decree, the bearers of the throne glorify Him 43
When he lays down to sleep his joints relax 13
When one of you offers prayers, then God is facing him 213
When the corpse (of the believer) is put in the grave the setting sun is made to appear to him 39

When the star rises the scourge will be lifted 51
When two Muslims raise swords against each other 15
Where is God?" and she pointed to the sky 326, 338
Who among you can control the domination of passion as the Prophet of God used to 321
Who ever kills someone (in the *Jihād*) can keep his arms and clothing 375
Who is your Lord, and what is your religion and what do you say about the Prophet 102
Whoever does not believe in predestination, its good and its evil, I am free of him 197
whoever does not do it, it does not matter 400
Whoever eats during the day in Ramadan must make up for it 397
Whoever makes an oath by other than God commits *shirk* 188
Whoever meets Me (on the Day of Judgment) with enough sins to fill the earth 175
Why don't you get in rows as the angels get in rows 286

You are a person who can give kind treatment (to the ill) but God is the physician 52, 181
You didn't know because you didn't recite the Qur'ān 230
You have been sent to make things easy (for people) and not to make things difficult 252, 322
You should follow my sunna and the sunna of the rightly-guided caliphs after me 386
you will see Him as you see the full moon 12, 196
You will see your Lord 383

INDEX OF SUBJECTS AND TERMS

abrogation, of laws, 264, 357–360; (of Qur'ān), 410n; (of hadith), 406, 409, 423, 428; (of religions), 340–345
acquired traits, (*malakāt*), 85
ahl al-hadith. *See* People of the Hadith.
ahl al-sunna. *See* People of the Sunna.
aḥwāl and maqāmāt, states and stages, 64, 65, 77, 229
'ālam al-quds, the World of the Holy, 93, 94
amāna, the Trust, 57–58
amārāt, textual signs, 417
'āmm, generalized, 396, 407, 472, 474, 475
amr, command, 45, 46
angelic aspect of the person, v, 56, 58, 59, 63, 69, 73, 75, 87, 88, 93, 94, 102, 104, 148, 149, 155, 174, 199, 224, 225, 231, 232, 238, 250, 273, 287, 290, 328, 334, 337
Angels, 25, 43–48, 58, 67, 70, 71, 80, 87, 88, 93, 94, 96, 100, 101, 102, 103, 104, 108, 149, 150, 157, 159, 162, 180, 196, 199, 208, 214, 217, 226, 229, 244, 258, 265, 295, 296, 329, 332, 336, 337, 363, 364, 365, 400. *See also* Highest Council, Holy Enclave, Lower Council.
animalistic aspect of the person, v, 56, 58, 59, 63, 66, 74, 75, 88, 93, 94, 102, 108, 148, 149, 153, 155, 156, 158, 174, 199, 210, 215, 224, 225, 231, 232, 238, 286, 290, 328, 334, 335, 337
'Arafāt 227
aṣḥāb al-'arāf, People of the Heights, 338
aṣḥāb al-yamīn, People of the Right Side, 154, 155, 334, 336, 337, 338
aṣḥāb al-shimāl, People of the Left Side, 334
Ash'arites, 25n, 40, 67, 195; psychic speech (*kalām nafsī*) 67; as *ahl al-ḥaqq*, 40n
astrology, 51, 122, 244, 352, 368
Attributes of God, 25, 33, 66, 67, 73, 83, 162, 169, 170, 186, 190–6, 238, 255, 293
'azīma, 406, 409
barzakh, 56, 66, 101, 102–105
Bearers of the Throne, 43, 159, 364, 365

Caliph(s), 188, 137, 251, 286, 306, 342, 343, 375, 441; (Rightly-Guided), 386, 388, 452
Caliphate, vi, 76, 118, 137–139, 307–8, 336, 343, 359, 360, 369
categories of actions (in Islamic law), 65, 276, 312, 378, 381–2, 384, 400, 406, 414n; (*mubāh*), 405; (*mustaḥabb*), 405
celestial aspects, 67, 68, 71
change, God's modalities, 35, 49–51, 239; (in religions), 340–345, 353–372; (in reasons for requital), 72; (in codes-*manāhij*), 257, 258, 262–276, 304–5; (in shar'ias), 159, 162. *See* abrogation.
charity, 45, 283
child-rearing, 119, 124–125
Children of Israel, 137, 253, 262, 270, 306, 348, 352; (reports of-Isrā'īliyyāt) 394
choice, 85, 202, 205
Christianity, 28
Christians, 22, 176, 177–8, 182, 257, 330n, 351, 366, 402
circumcision, 367
coloration, 71, 331, 336
cistern (in Paradise), 109, 296
Companions, (of the Prophet), 23, 24, 25, 327, 336, 380, 384, 386, 388, 392, 394, 400, 402, 404, 405, 407, 410, 411, 412, 413, 414–421, 427, 428, 429, 430, 432, 434–435 439, 440, 444, 448, 452, 461, 462, 467, 468, 469, 478
conditions (of acts of worship), 259, 280 (defined), 281, 322, 324, 369, 378, 379, 384, 401, 414
consensus (*ijmā'*), (of the angels), xxii, 47, 194, 203, 204, 239, 253, 272, 354; (of Muslim community), 6, 11, 16, 27, 350–1; (legal), 350–1, 442
continuity in religion, 304–306
conventions, 142–144. *See* veils.
creation, 185, 198, 199; (of the world), 25, 33–36; (of humans) 49

dahrī, (atheist), 22, 206, 238
Dajjāl, (anti-Christ figure), 35
dalīl, pl. *dalā'il*, (proof, indicant) 25, 27, 414, 450, 462, 463; (*dalāla*-indication), 384, 387, 457; (*adilla tafṣīliyya*—detailed proof texts), 462, 463
Day of Assembly (*ḥashr*), 40, 105, 106–111
Days of God, (reminder of), 206
days of week, (special qualities), 287, 288; (Fridays), 37–38, 264, 288, 406
death, 93, 97–100, 101–104, 331
determinants, (of species—*aḥkām*), 61, 105, 106, 107
devil, 14, 95, 284, 329, 349, 350; devils, 48, 51, 79, 80, 104, 157, 217, 225, 265, 285, 295, 329, 332, 400
dhauq, (mystical insight), 82, 163, 206, 248, 285, 288
dhikr, (tablet), 198, 233; remembrance of God, 168, 219, 228, 229–231, 258, 301, 312, 328, 329, 332, 335, 337, 367, 380; people of Remembrance, 310
dīn (religion), xx–xi, 72, 257, 354
distortion in religion, 346–353, 361
Divine Management (*tadbīr*), 34, 50, 154, 176, 182, 183, 185, 196, 197, 252, 266
divorce, 124, 419, 470; (*ʿidda*) 124; (*ẓihār*), 269
dreams, 41–2, 47, 79, 102, 108, 157, 180, 196, 208, 209, 265, 331, 368
duʿā' (petitionary prayer), 204, 222, 229, 258, 367

eschatology, Islamic, 22, 82, 102, 109, 240; (bridge), 22, 24, 240
essence (*dhāt*) of God, 84, 184, 225, 255

faḥwā, (the import of an itterance), 397
fanā' (annihilation), 76, 106, 184
fāsiqūn (profligates, depraved ones), 339
fatwā (legal opinion, recommendation), 299, 408, 415, 418, 424, 427, 434, 437, 450, 452, 453, 458, 463, 464, 465, 466, 471; muftī(s), 427, 451, 463, 466, 468, 471
felicity (*saʿāda*), xx, 8, 28, 29, 146–170, 173, 175, 199
fiqh (jurisprudence), xvi, xxx, xxxi, 20, 21, 386, 448, 462; (of Companions), 404, 407; *fuqahā'* (jurists), 22, 327, 379, 382n, 410, 434, 435, 441, 453, 455, 458, 466, 469, 471
fiṭra (sound original nature), xxi, xxxiii, 72, 84, 106, 107, 108, 125, 144, 150, 156, 159, 173, 174, 199, 206, 236, 238, 239, 354
fortune tellers, 52
Four Virtues, xx, 64, 69, 156–160, 329, 336, 376; virtues, 307, 376; *see* purity, humility, magnanimity, justice.
al-furūʿ (applied jurisprudence), 414–426, 430; (Ḥanafīs and), 431

genus, genera (jins), 33, 34
ghusl (full bath), 164, 214–217, 366, 380, 402, 420, 472; (with sand), 419–20, 424
Gospel, 93, 351
grave, punishment in, 24, 39, 40–42, 102, 109, 217, 229–30, 236, 296

ḥadd, pl. *ḥudūd*, (offences whose punishments God defined in the Qur'ān), 12, 29, 236, 268, 369, 377
al-ḥadd al-jāmiʿ al-māniʿ (an exclusive exhaustive definition), 316, 396, 449–50
hadith sciences, xvi, xxiii, 4, 5, 6, 13, 17, 20, 22, 25, 27, 348; (disciplines), 454; (problems), 385; (weak), 394–5; (conflicting), 348, 405; (memorizing of), 386, 405–413; (interpretive methods), 373–375, 386n, 387–395; (books), 387–397
hadith terminology, *ḥasan* (a good hadith), 383, 388; *ṣaḥīḥ* (a sound hadith), 383, 388, 394; *maqlūb* (the word order is transposed), 388; *mauḍūʿ* (an invented hadith), 388; *mauqūf*, 427; *munqaṭiʿ*, (interrupted), 3389, 91, 433; *manqūl* (the statement of a Companion); *musnad*, 427, 433; *mustafīd* (transmitted by 3 or 4 Companions), 383, 387, 389; *mutābiʿāt* (concurring hadiths), 438; *mutawātir*, 383, 387, 389, 441; non-*mutawātir*, 383; *muttaṣal* (all transmitters are mentioned), 444; *shawāhid* (suppoting hadiths), 438, 444; *fadhdh* (singly transmitted), 445; *shadhdh* (anomolous), 388, 445

ḥajar-i-baht (perplexing stone), 208n
ḥajj, 12, 29, 184, 188, 212, 326, 332, 367, 375, 420, 421; (types of), 421; (inner dimensions of), 227-228, 258, 468; (pilgrimage in honor of other than Allah), 188, (to other sites), 227; ('Id after), 326, 467; (*sa'y*), 313; (Tashrīq days), 467; (*iḥrām*), 421, 422, 468, (*'umra*), 332
Ḥanīfī religion (monotheism), 3, 304, 330, 361, 369
ḥaqīqa farḍāniyya (individual essence), 54
ḥaqq (concept of the "right" of), 203–209; (God's), 258, 266
heirs of the Prophets, 76, 203
Hell, the Fire, 38, 43, 236, 288, 332
Highest Council (*al-mala al-a'lā*), xxii, 43 (defined)-48, 49, 68, 69, 71, 73, 79, 85, 86, 87, 100, 144, 149, 156, 157, 159n, 163, 173, 194n, 196, 204, 208, 221, 223, 228, 233, 234, 239, 244, 250, 251, 253, 254, 268, 272, 288, 289, 295, 303, 310, 312, 330, 331, 335, 354, 364, 378, 380
himma (concentrated resolve, zeal), 63, 77, 107, 149, 218; (of the angels), 49, 77, 87, 268, 312; (of the Prophet), 71, 268, 273
Hindus, 474
Holy Enclave (*ḥakīra al-quds*), xxii, 47 (defined); 70, 71, 87, 99, 106, 107, 108, 144, 194n, 196, 199, 200, 214; (light in the middle of), 71, 142, 204; (angelic light), 173, 214, 217, 218, 234, 239, 252, 272, 335, 378
al-ḥujjat al-bāligha (referred to), 9, 10, 60, 68, 154, 185
ḥukm pl. *aḥkām*, (legal ruling) 5, 275–280, 310, 354, 373, 418, 458
humility (*ikhbāt*) as a virtue, 157, 164, 175, 329, 330, 336
hypocrites (*munāfiqūn*), 338, 339

'ibād (worshippers), 336
'ibādāt (acts of worship), 66, 149, 315, 355, 357, 369, 373
ibdā' (absolute origination), 33
idols, 182, 361; (idolotry), 187, 361
iḥsān (righteousness), 29, 65, 99, 122
ijtihād (exercising independent reasoning), xxiii, xxix–xxxiii, xxxix, 15, 27n, 351, 373, 413, 428, 433, 435, 440, 442, 450, 451, 456, 462 (defined), 463; (of Prophet), 234, 351, 357–8; (of Companions), 413, 418, 428; (of Successors), 434
ijmā'. *See* consensus.
ikhtilāf (juristic disagreement), xxviii–xxxiii, xxxix, 367, 427–478; (of jurists), 471
ikhtiyār. *See* choice.
'illa (rationale, reason for legislation), 5, 18, 275, 276, 280 (defined), 312, 348, 354, 379, 399n, 401, 407, 410, 418, 428, 436, 449, 453, 458, 461; (disagreements over), 422–3, 461
'ilm ḍarūrī, necessary knowledge, 195, 249
'ilm dhātī (essential knowledge), 180
'ilm ḥuḍūrī (direct knowledge), 180n
'ilm ladunī, (divinely inspired knowledge), 6
'īmā', indirect allusion, 397 (defined), 401, 403, 427, 449
imāra (decisive sign), 187
inheritance, 296, 308, 358, 359, 416, 429 (*tashrīk*), 433–4, 436
iqtiḍā', (logical implication), 397 (defined), 427, 449
irtifāqāt, stages of civilizational development, xviii, xix, 28, 29, 63, 64, 65, 66, 113–141, 142, 147, 149n, 154, 165, 173, 230, 236, 263, 278, 286, 303–309, 310, 312, 315, 318, 338, 342, 353, 355, 357, 368, 369, 373, 376, 377. First *Irtifāq*, 117, 119–121, 130, 141; Second *Irtifāq*, 117, 121–128, 217, 230, 303, 304; Third *Irtifāq*, 117, 129, 230, 303, 304, 368; Fourth *Irtifāq*, 118, 137–139, 307–308, 355–6, 368
ishāra (textual indication), 5, 401, 403
ishrāk (associationism), 166, 169, 181, 232, 352. *See shirk*.
Islam/Islamic religion, 3, 236, 257, 338, 352, 358, 359
'iṣma (immunity from sin), 255, 351–2, 373
istidlālāt (deductions), 5, 256, 398 (defined), 411, 434, 446; (types), 398–399
istiḥsānāt (applying legal discretion), 256; (criticized), 349, 436
iṣṭilāh (inner integration) 74, 74, 76, 104, 250, 334, 337; (muṣṭaliḥūn-people of inner integration), 154
istinbāṭ (legal inference), 28, 256, 386,

428, 445, 448, 454; (types), 445
istiṭā'a (capacity), 49
i'tikāf (retreat in the mosque), 225–226

Jabarūt, 41n, 64, 70, 73, 77, 149, 153, 167, 203, 207, 208, 304, 329
Jews, 22, 174, 176, 182, 253, 257, 262, 264, 266, 322, 330n, 349, 351, 354, 360n, 366, 367, 402, 423; (Sabbath), 264, 354
jihād, (struggle in the path of God), 13, 36, 230, 355, 359, 360, 369, 374, 375; (booty in), 359, 375
jinn, 33, 51, 244n, 265, 285
Judgment Day, 8, 22, 37, 38, 102, 109, 233, 251, 258, 332
Judaism, 28
justice (*'adāla*), as a virtue, 158, 164, 330, 336

Ka'ba, 15, 211, 227, 282, 296, 324, 368, 421, 430
kaffārāt,(expiatory actions), 12, 369, 466
kalām, (theology), 255, 325, 352, 365; *mutakallimūn*, (theologians), 23
kalām manẓūm (composed speech, Asharite theory of the Qur'ān), 195
khalq, (creation of the material world), 33, 50; *'ālam al-khalq* (world of material creation), 46n
khaṣṣ (specification), 472, 473, 474
kings, 118, 132–139, 142, 347
kufr (unbelief), 182, 183n, 362; *kuffār*, 339

lailat al-qadar, 45–46, 71, 226
language, xvii, 119, 265, 316, 317, 328, 355, 461; (metaphorical and God's attributes), 190, 332; (understandings and interpretation), 328, 329, 332, 355, 396, 408
laṭīfa (subtle spiritual center), 207, 208, 209
literal interpretation, 23, 24, 28, 40–42; (of hadith), 382, 461, 477
Lower angels (*al-malā'ik al-sufla*) 45, 47 (defined), 70, 71, 74, 87, 102, 227, 288, 337
luxury, 128, 143, 304, 306, 307, 308, 317, 318, 342, 380

madhhabs. (legal schools), xxiii, xxix–xxxxiii, Ch. 82–85 passim.
al-madhhab al-ṭabī'ī (the natural religion), xxi, 281, 285, 290, 341
mafhūm (what is understood from an utterance), 473 (defined), 475, 476n
Magians, Magianism, 28, 107, 174, 289, 367, 417
magnanimity 158, 164
Malakūt, 2, 41, 64, 69, 71, 149, 186, 311, 329, 373
manāṭ (anchoring point), 402 (defined), 404, 407; (*takhrīj al-manāṭ*-identification of the anchoring point), 402 (defined), 404, 461
maqāmāt. See *aḥwāl*
ma'rifa (gnosis/realization), 21, 66, 77, 208
martyrs (*shuhadā'*), 335
maṣlaḥa, (beneficial purpose), xvi, xvii, xix, 6, 11, 16–17, 18, 20, 23, 34, 35, 64, 196, 264, 280–288, 310, 359; *maṣaliḥ* (beneficial purposes), xxii, 11, 18n, 346, 354, 374, 376–382, 387
maṣlaḥa kulliyya (universal beneficial purpose), xvi, xvii, 143, 149, 208, 303; (*maṣlaḥa fauqāniyya*-heavenly purpose), 204
maṭiyya (substratum, "riding animal"), 55, 97
Maturidites, 26
maẓinna, pl. *maẓānn* (anticipated source), xxii–xxiii, 182, 184, 206, 232, 234, 271, 278, 288, 318, 321, 328, 329, 330, 376, 410, 354, 357, 358, 359, 377, 378, 379
Mecca, 263, 273
Medina, 263, 273, 358
minhāj, pl. *manāhij* (codes), 257, 259, 265, 271–274
miracles, 20 (of Qur'ān), 24 (of saints), 25, 64, 181, 182, 187; (of Prophets), 254–255, (of Muhammad), 269
Mi'rāj, 38, 410
mithālī (imaginal), 209, 269, 293, 312n; (*mithālī* forms), 40–42, 186, 199; (of divine laws), 272, 273. *See* World of Images.
mīthāq (primordial covenant), 82, 199
mufahhamūn (those who are instructed), 154, 250 (defined), 251–252, 254, 334, 336
mufarridūn (those who retire on their own), 335
Muhammad, 3, 4, 7,162, 184, 253, 330, 351, 354, 366, 367; (Prophet), 211,

212, 310, 335, 411, 458, 478
mujtahid/s, 6, 27, 187, 235, 335, 351, 440, 447, 450, 458, 460, 461, 463; (*muṭlaq*/absolute), 463
Mu'tazilites, 22n, 23n, 25n, 26n, 67, 191n, 192n, 193n, 395, 472

nafl, 14, 402
nafs (lower soul), 12, 61, 80, 154, 155, 158, 165, 207, 215, 216, 218, 220, 222, 228, 233, 239, 244, 290, 348; (soul in general), xviii, 28, 232, 328, 329, 331, 334, 336, 376
al-nafs al-ḥayawāniyya (animating soul), 58
al-nafs al-nāṭiqa (rational soul), 54, 79, 80, 81, 98, 99, 109, 153, 156, 157, 207, 208
nasama (pneuma), 64, 98, 99, 103, 109, 157, 207, 208, 250n
Necessary Being, 26, 166, 175, 177, 180, 181, 182, 184
numbers and amounts, 292–298, 381

oaths, 188, 210, 312, 368, 375, 409, 433

Paradise, 38, 43, 258, 332
People of the Hadith, 23, 40, 192n, 193, 256, 382, 395, 437–450, 452, 460
People of the Sunna, 23, 24, 25, 26
Persians and Byzantines, 306, 308, 318, 342, 343, 352, 355, 368
philosophers (criticism of), 204, 271, 394
piety (*birr*), xx, 28, 172–174, 175, 190
pillars, (of acts of worship), 258, 271, 280 (defined), 281, 320, 322, 324, 369, 378, 379, 384, 401, 414
pleasure and anger of God, 196, 206, 275, 278, 333, 377, 378, 380, 400
polytheists (*mushrikūn*), 176, 182, 184, 187
prayer. *See ṣalāt*.
Pre-Islamic era (*Jāhiliyya*), 18, 28n, 96, 236, 305, 306, 308, 309, 338, 339, 341, 343, 352, 356, 361–369
predestination (*al-qadar*) 197–202, 244, 258, 364; (compelling pre-destination-*al-qadar al-mulzim*), 28, 198
Preserved Tablet (*lauḥ*) 82, 83, 106, 198n. *See dhikr*.
prime matter, 25, 81
prophecy, 247, 261, 268, 303, 308, 340–5, 377; (first type of mission), 252, 255, 359; (second type), 252, 355, 359
prophetic inspiration, 63, 65, 67, 71, 195, 373, (*waḥī jalī*), 410
prophets, 330, 335, 336
prostration, 179, 182, 183, 185, 220; (while reciting Qur'ān), 400
providence, Divine (ināya), 88, 115, 206, 274
purity (ṭahāra), 44, 156, 164, 214, 215, 217 purity (*ṭahāra*) as a virtue, 157, 164, 175, 329, 330, 336

qaḍā' (divine decree), 45
qaḍā', (making up for missed religious obligations), 299–302, 369
qāḍī/s (judges), 135–6, 313, 427, 454, 465, 471; (*Adab al-Qāḍī*-book), 465
qarīna, pl. *qarā'in* (contextual evidence), 413, 417
qibla, 24, 183, 212, 258, 282, 301, 312, 325, 326, 357, 423
qiṣāṣ (retaliation) 12, 258, 403
qiyās (analogical reasoning), 5, 22, 312, 350, 379, 382, 386, 398, 399, 412, 436, 449, 458, 460, 464, 473, 476n, 477; (*iqtirānī*–conjunctional), 398, 398n (defined), 449; (*sharṭī*–conditional) 399, 449 (defined)
qudra (ability), 85
questioning, dislike of excessive, 16, 263, 269–270, 379, 381
Quraish, 363, 367
Qur'ān, 4, 13, 20, 23, 24, 25, 26, 27, 37, 176, 178, 201, 206, 211, 229, 230, 237, 288, 291, 306, 309, 336, 350, 351, 355, 357, 358, 360, 366, 383, 394, 400, 410n, 413, 419, 425
al-quwwa al-'amaliyya (practical faculty), 64, 65

Rafidites, 395
Ramaḍān, 16, 22, 262n, 283, 288, 291, 314, 320, 325, 331, 397, 400 check
al-rashikūna fī-l-'ilm (those firmly grounded in knowledge), 336

rationalists (criticism of), 256
ra'y (personal opinion), 194, 434, 436 (defined), 437–450, 468, 475, 477
al-ra'y al-kullī (the comprehensive outlook), 74, 76, 104, 115, 121, 137, 143, 244, 266, 304, 305; *arā*

juz'iyya (partial/limited outlooks), 137, 143, 250
religion (*dīn*), on one basis, 257, 258, 264; innate, 266, 267; religious change, *see* change, continuity
resurrection, 25, 56, 196, 258, 336; (Day of), 12, 37, 236, 335
ribā (usury), 305, 309, 368, 399
ritual impurity (*ḥadath*), 214, 215
al-rūḥ (the Spirit), 53-56 (defined), 195, 207, 225, 269, 271
al-rūḥ al-aʿẓam (the highest spirit), 106, 108
al-rūḥ al-hawā'ī (the airy spirit), 54, 64
al-rūḥ al-ilāhī (the divine spirit), 54, 55 56
al-rūḥ al-malkūtī (the angelic spirit), 54, 331n
al-rūḥ al-qudsī (the holy spirit), 47, 55, 251, 272
al-rūḥ al-ṭabīʿī (the natural spirit), 58
rukhṣa (dispensation), 279, 299–302, 369, 379, 380, 406, 409, 423

sabab (cause-legal), 277
al-sābiqūn (ones who excel), 152, 154, 334, 337, 338
sacrifice, 227, 475; (Pre-Islamic), 361
ṣadāqa (free-will offering), 222, 337; (virtuous deed), 327, 328
sakīna (tranquility), 250, 323, 324
ṣalāt (ritual prayer), 11, 13, 29, 34, 211, 212, 213, 258, 259, 271, 281, 289, 291, 301, 302, 321, 327, 330, 333, 344, 401, 409, 468, 473; (Inner dimensions), 11, 13, 15, 218-220; *tahajjud*, 283, 289, 290, 321, 405; (for rain), 218; (*basmala*), 468, 469; (*qunūt* prayers), 469, 470; (*āmīn*), 468; (congregational prayer), 314, 322, 344; *'Id*, 322; (*iqāma*), 468; (positions in), 320, 473; *witr* prayer, 398, 400, 405
ṣaum (fasting), 12, 16, 29, 37, 78, 168, 258, 263, 281, 320, 321, 330, 348–9, 397, 466; (*'Ashūra'*), 291, 367; (inner dimensions), 224–226
science of disputation, 66
sensus communis (*ḥiss mushtarik*), 55, 101, 157
shaʿā'ir (emblems of God), 210–213, 228, 257, 262n, 263, 282, 353, 355
sharīʿa, Islamic, xvi, 5, 7, 17, 20, 21, 23, 45, 187, 188, 262, 278, 318, 330, 344, 347, 353, 355, 361, 364, 376–382, 383, 387, 410, 452, 462; (general), xxi–xxiii, 166, 259; (divine laws) xxiii, 11, 21, 28, 142, 156, 159, 181, 198, 205, 206, 209, 234, 236, 257, 263, 264, 267, 271, 330, 340, 342, 353, 355; (of species) 107, 115
sharṭ (legal condition), 259, 271, 277, 379
shirk, 179, 184–9, 339, 361, 362, 363, 354, 366, 367, 369
ṣiddīqūn (truthful ones), 335,
ṣifat. *See* attributes.
sin, 87, 172–, 224, 232–244; (major), 236–237
soul. *See nafs*
species, properties of, 29, 33, 265, 312
specific form of humans (*al-ṣūra al-nauʿiyya*), 61, 69, 84, 105, 106, 107, 147, 149, 208, 265
spiritual exercises (*riāḍāt*), 94, 104, 149, 154, 168, 170, 335, 336,
star-worship, 176, 181, 187, 212, 349
strictness, avoidance of, 282, 283, 300, 322–326, 348–349, 381
Successors (of the Prophet), 23, 24, 380, 384, 386, 388, 392, 394, 400, 413, 414–421, 427, 429, 430, 432, 434, 439, 440, 452, 469, 478
Sufism, Sufis, xvi, xviii, xxiv, xxv, xxviii, xxxv, 21n, 29, 35n, 42n, 64, 153, 208n, 229n, 307, 394, 335
sulūk (spiritual itinerary), 21, 77
sunna, 24, 25, 26, 237, 257, 268, 351, 361, 382, 394, 400, 403, 410, 425, 441, 445, 456, 462, 464; (*al-rāsida*), 248, 250, 361; of Prophet, 16, 27, 361, 382, 384, 386, 430, 438, 442, 443, 446, 457, 458; (mu'akkad), 275, 382
sunna (of God), 49–52, 78, 173, 237, 310
symbolic, metaphorical interpretation, 24, 28, 40–42; (of numbers), 292–298, 332, 379–380; (*ta'wīl*), 24; symbolism in laws, 272, 288, 301, 332, 328, 329

tadallī (descent-as a facet of God's creative activity), 33n, 252n
tadbīr. *See* Divine Management.
taisīr (facilitation), 322–326

tajaddud (renewal), 70
tajādhub (mutual contention), 58, 74, 75, 154, 208, 335, 337
al-tajallī al-a'ẓam (the Greatest Theophany), 71, 196, 204, 208n; (of Allāh), 23; (*tajalliyāt*), 218, 230; (of religious revelation), 268
takaffuf ḥal (supplication of state), 63
takbīr, 280, 281; (variation in *'Id* prayers), 467–468, 470
takhrīj (juristic derivation) 449, 451, 458, 462, 477
taklīf (imposition of religious duties), 9, 197, 199, 238
talbiyya, 184
taqlīd (imitation of previous rulings or law school), xxiii, xxviii–xxxiii, 235, 351, 440 (defined), 451, 453, 454, 455, 456, 458, 461, 464, 466 (*muqallid*) 235, 447
taqrīr (Prophet's sanctioning of an act), 4, 411
tarāwīḥ prayers, 262
tarjīḥ (preference of juristic opinions), 449
tashahhud, 405, 468
tashbīh (anthropomorphism), 166, 169, 181, 192, 194, 232
tauhīd (unity of God), 66, 175–178, 185, 220
taxes, 117, 131, 134, 135, 136, 138, 143; (pre-Islamic), 306; (*kharāj*), 134, 138, 139, 260, 306, 355; (*'ushr*), 134, 260, 306, 475; (*jizya*), 138, 139, 344, 355, 406
Throne (God's), 26, 43, 176, 198, 326
Torah, 93, 262, 349, 351, 352, 355, 367

ulema, 26, 347, 379, 427, 434, 460, 463, 464, 466, 468, 477
Unseen, knowledge of, 47, 64, 179, 225, 252, 265; (Unseen world), 65, 66, 67, 175, 195, 250, 303, 304, 365; (Unseen of the Unseen-*ghaib al-ghaib*), 67–68
uṣul al-fiqh, xxxv, 5n, 255, 325, 379n, 433

variation. *See* change.
veils, xx, 160, 165–171, 229, 231, 273, 377; (of physical nature), 165, 231, 239, 338; (of convention), 165–6, 168, 217, 220, 225, 231, 339; (of misinterpretation), 166, 169, 170, 217, 220, 229, 339
virtues (*khiṣāl*), 148, 151, 156, 284, 334s. (Acquisition of through intellectual regulation), 161-163, 175, 335, (practical regulation), 163, 335
Vision of God, 11–12, 23, 24, 40, 111, 196, 220, 383

Wali Allāh, Shāh, (biography) xxiii–xxvii; 3; (visions of) xxvi, 7, 200, *see also* name index.
waṣf (quality), 473, 475
wijdān (intuitive awareness), 65, 247, 248
witr. *See ṣalāt*.
World of Images (*'ālam al-mithāl*), xx, xxiii, 23, 37–42, 56, 67, 82, 103, 104, 106, 196, 199, 200, 222, 225, 233, 254, 269, 287, 331, 365
wuḍū' (ablutions), 164, 142-217, 315, 408, 470, 474; (things which necessitate), 469, 470, 476

yād dāsht (keeping in mind), 101

Zahirī (literalist), 477
zakāt, 12, 29, 37, 258, 277, 297–298, 312, 325, 378, 379, 400, 407, 475, 476. Inner dimensions, 221–223
Zamzam, 22, 330
ẓann (conjectural opinion), 459, 463
ẓarāfa (aesthetic sensibility), 116n
zuhhād (ascetics), 336

INDEX OF PROPER NAMES

'Abd Allāh ibn Salām 406, 407
'Abd al-'Azīz, Shāh (d. 1823) xxvii, xxxiii
'Abd al-Razzāq ibn Humām (743–826) 390, 393, 431, 439, 454, 472, *Jāmi'* 431
'Abd al-Raḥīm, Shāh (d. 1719) xxiv–xxv, 3
'Abd al-Raḥmān ibn 'Awf 416, 417
'Abd al-Raḥmān ibn Mahdī (d. 813) 390, 439
'Abd ibn Ḥumayd (d. 863/4) 393, 445, *Musnad* 393
'Abdel 'Aal, Khalīl xviii, 351
Abraham (Prophet) 3, 21, 34, 36, 162, 257, 262, 263, 267, 324; his people 212, 347, 354, 361, 368, 369
Abū Bakr al-Ṣiddīq (d. 634) 109, 343, 367, 416, 441, 442, 443, 444
Abū Dāwūd 392, 439, 445, 446, 447; *Sunan*, 417, 439, 446, 447
Abū Dharr 367
Abū Ḥanīfa xxxiv, 425, 431, 448, 463, 464, 465, 466, 471, 477
Abū Huraira 406, 420, 426, 429, 435
Abū Jahl 360
Abū Lahab 198
Abū Mūsā al-Ash'arī (d. 663) 310, 322, 417
Abū Nu'aym al-Isfahānī (948–1038) **394**
Abū Shāma 457–8
Abū Ṭālib al Makkī (d. 996) 445n, 451; *Qūt al-Qulūb* 451
Abū Ya'lā Muḥammad ibn al-Ḥusain (990–1066) 445
Abū Yūsuf, Muḥammad (d. 798) 390, 431, 432, 464, 466, 468, 469, 470, 471; *al-Amālī* 390, 471; *al-Athār* 431, 468
Adam (Prophet) 33, 108, 179, 199, 268, 288
'Ā'isha 19, 25, 321, 324, 374n, 385, 403, 412, 419, 420, 422, 424, 429, 431
'Ali al-Madīnī (d. 849) 439
'Alī ibn Abī Ṭālib xxiv, 19, 45n, 330, 375, 384, 429, 437
'Alqama (d. 680/81) 425, 428, 429
Al-A'mash, Abū Muḥammad Sulaimān (c. 680–765) **53**, 443
'Amr ibn Luḥayy **361**, 366, 368
Aristotle 147n, 190n
'Āshiq, Muḥammad (d. 1773) xxvii, xxxiv, 8
'Āṣ ibn Wā'il 367
'Aṭā' ibn Abī Rabāḥ (d. 732/33) **385**, 424
'Aṭā' ibn Yasār 429
Aurangzeb (d. 1707) xxiv, xxviii
Al-Auzā'ī, 'Abd al-Raḥmān ibn 'Amr (d. 774) 443, 464

al-Baiḍawī 57
al-Baihaqqī, Aḥmad ibn al-Ḥusain (944–1066) **393**, 445, 452, 464
Banū Ubayraq 316
Baqā', Makhar xxix, xxx, xxxi, xxxii, xxxiv
al-Bazzāz (d. 1034), Abū 'Alī **393**, *Fatāwā al-Bazzāziyya* 470
Brunschvig, R. 328n, 399n, 473n, 474n
al-Bukhārī, Abū 'Abd Allāh (d. 870) xxv, 389, 391, 392, 445, **446**, 447, 460; *Ṣaḥīḥ*, xvii, 389, 391, 392, 393, 417, 419, 429, 439, 446

Corbin, Henry 37n

al-Dailamī 394, 445, 452
Daniel 352
al-Dāraquṭnī 445
Al-Dārimī, 'Abd al-Raḥmān (797–869) **415**, 437, 442n, 438, 443, 444n, 445, 448
David (Prophet) 108, 253, 398
Diḥya ibn Kalbī 331

Eve 189

al-Faḍl ibn Dakain (748) 439
Faṭima bint al-Aswad 316
Fāṭima bint Qais 419
Fāṭima, daughter of Muḥammad 25, 316

Gabriel (angel) 38, 39, 44, 195, 291, 331 (as Diḥya)
al-Ghazzālī (1111) xvi, xx, xxxv, 15, 21, 40–42, 46n, 47n, 57, 82, 85n,

147n, 447, 452–453, 473n, 476n; *Iḥyā' 'Ulūm al-Dīn* vii, 41–42, 83, 85n, 255n, 303n, 453n

Al-Ḥākim al-Nisābūrī (933–1014) **391**, 392, 445, 464; *al-Mustadrak*, 391
Hallaq, Wael 397n, 402
Hannād ibn al-Sarīd (d. 857) 439
Harūn al-Rashīd 390, 430, 432, 469, 470
Ḥasan al-Baṣrī (d. 728) 350, 364, 385, 424, 437, 458
Hermansen, M. K. 208n
Hishām *Kitāb al-Nawādir* 465
al-Hormuzān 342

Ibn 'Abbās 15, 19, 364, 384, 402, 405, 415, 420, 421, 424, 429, 437, 443, 447, 468
Ibn 'Abd al-Barr (978–1070) 445
Ibn Abī–Shayba, Abū Bakr (d. 849) **391**, 393, **431**, 439; *Muṣannaf* 393, 431
Ibn 'Adī, Abū Aḥmad 'Abd Allāh (d. 975–6) **394**, *al-Kāmil*, 394
Ibn 'Arabī xxxv, 106n
Ibn 'Asākir (1118–1245) 394
Ibn al-Athīr 392, *Jāmi' al-Uṣūl* 39n, 392
Ibn 'Aun (933) 448
Ibn al-Daghīna 367
Ibn Ḥajar al-'Asqalānī (d. 1449) 192
Ibn Ḥājib, 'Uthmān ibn 'Umar (1175–1249) **429**, 436, 476n; *Mukhtaṣar al-Uṣūl* 429, 436
Ibn Ḥanbal, Aḥmad (d. 855) 193n, 393; *Musnad* 393, 439, 440, 445, 453, 462, 464, 469, 477, 478
Ibn Ḥazm, Dāwūd (994–1064) xxx, xxxi, 455–6, 458, 460, 477
Ibn Ḥibbān (d. 965) *Kitāb al-Ḍu'afā'*, 394
Ibn Jurayj (d. 767) 430
Ibn Khaldūn xix, xxxv
Ibn Mājah 445
Ibn Mājishūn, 'Abd al-Mālik (d. 827) 40
Ibn Ma'sūd, 'Abd Allāh **40**, **384**, 386, 405, 418, 424, 425, 429, 437, 442, 44, 447, 468
Ibn al-Mubārak (d. 797) 191
Ibn al-Munkadir 438
Ibn al-Najjār (d. 1118–1245) 394
Ibn Nujaim 465n, 472
Ibn al-Qāsim (d. 806) 390
Ibn al-Rawândī 22n
Ibn Rustam *Kitāb al-Nawādir* 465
Ibn Sīnā (1037) 65, 175–176n
Ibn Sīrīn, Muḥammad (d. 728) 79, 85n, 349, 443
Ibn 'Umar, 'Abdullāh (d. 692/3) 15, 349, **384**, 415, 420, 422, 423, 424, 425, 429, 431, 435, 437, 448
Ibn 'Uyayna, Sufyān (814) 191, **389**, 430, 431
Ibn Wahb 'Abd Allāh ibn Muslim (743–813) 390
'Isā ibn Ibān 477
Isḥāq ibn Rāhawayh (852/3) 439, **445**, 477, 478
Ishmael, Ismā'īl (Prophet) 28, 262, 266, 355, 361, 361, 366, 368
Ismā'īl Shahīd, Shāh xxvii, xxiv, 183n
Ismā'īlīs 22n
'Iyāḍ ibn Mūsā (1088–1149) *Mashāriq al-Anwār* 392
'Izz al-Dīn Ibn 'Abd al-Sallām (d. 1262) 15, 457

Jābir ibn Zaid (c. 721) 437
Jacob (Israel) 262, 266
Ja'far ibn Abī Tālib 45, 103, 110
Jalbani, G. N. 11n, 163n
al-Jauzī, Ibn Qayyim (d. 1201) *al-Mauḍū'āt* **395**
Jesus 257, 348, 351, Messiah 177, not son of God 177–8
Job 34, 36
Joseph 179
al-Jūzāqanī (al-Juzajānī), Ibrahīm ibn Ya'qūb (d. 870) **394**

al-Karkhī 471, **477**
al-Kaṣṣāf *Adab al-Qāḍī* 465
al-Khaṭṭabī (d. c. 998) 15
al-Khaṭīb al-Baghdādī (d. 1070–71) **394**, 445
al-Khāwarizmī, Abū Ja'far Muḥammad ibn Maḥmūd (d. 1257) *Musnad* 391, 394
Khadīja 367
Khiḍr 36n, 152
al-Khujandī 470, 471
Khurāfa 374
Kufan school 384, 385, 424, 425, 426, 429, 432, 467n
al-Kurdī, Abū Ṭahir, xxvi, xxx

Lot 243

Madinan school 384, 385, 424, 425, 426, 429, 468n, 470
Maimūn ibn Mihrān (d. 734) 441
Makḥūl (730–736) 424
Mālik ibn Anas xvi, xxx, xxxvi, 40n, 191, 389, 390, 428, 430, 431, 432, 437, 447, 453, 464, 470; *Muwaṭṭā*, xxvi, xxx, 389, 390, 392, 430
al-Manṣūr (Caliph) 430
Ma'qil ibn Yasār 417, 418
al-Marghīnānī xxv; *Hidāya* 472
Mary 39, 348
Moses 13, 36n, 46, 108, 152n, 162, 253, 257, 291, law of 258, 331, 354
Mu'adh ibn Jabal 203, 322, 350, 437
Mu'ammar ibn Rāshid **390**
Al-Mughīra ibn Shu'ba (669/70) 416
Muḥammad ibn 'Abd al-Raḥmān Ibn Abī Dhi'b (775)
Muḥammad ibn Salama 416
Mujāhid ibn Jabar al-Makkī (d. 722/3) 257
Mullā Ṣadrā 198n
Musaddad (ibn Musarhad (d. 843) 439
al-Musayyab, Sa'īd (d. 712/13) 109, 385, 424, 429, 435, 437, 470
Muslim al-Nīsāpūrī (875) 389, 391, 392, 445, 446, 447; *Ṣaḥīḥ*, 389, 392, 393, 417, 419, 446
al-Muzanī 458, 464; *Mukhtaṣar* 458

al-Nakha'ī, Ibrahīm (d. 715) **384**, 385, 424, 425, 426, 427-8, 429, 432, 437, 443, 447, 448, 464; *Mujtabā* 392
al-Nasā'ī, Aḥmad ibn Shu'aib (831–915) 392, **418**, 452
Noah 257, 263, 267, 291, 347, 354

Al-Pazdawī (d. 1089) 472, 473
Peters, Rudolph xxxi, xxxii
Plato 129n, 147n

Qāsim 429
Quss ibn Sā'ida 366, 367

Rabī' ibn al-Ṣabīḥ (777) 430
Rabī'a ibn al-Ḥārith 308
Rabī'a ibn 'Abd al-Raḥmān (753) 424, 429
Rahbar, Daud xxx, xl, 451
Rahman, Fazlur xxvii, xxxv, 65n, 352n
Razīn ibn Mu'āwiya (d. 1140) *Tajrīd al-Ṣiḥaḥ* **392**

Sālim ibn 'Abd Allāh ibn 'Umar (725) 429, 448
al-Samarqandī, Abū al-Laith *al-Baḥr al-Ra'iq* 465–466
Samuel 137n
al-Sarakhsī, Shams al-A'imma (d. 1090) 450n, 472; *al-Mabṣūt* 450, 468, 472
Schacht, J. 444n
al-Sha'bī (d. 728) 350, 385, 429, 438, 443–4, 447
Al-Shāfi'ī, xxix, 390, 432, 436, 439, 458, 462, 464, 477; *al-Risāla* xxix, 23n, 416, 417, 423, 428, 433, 434, 435, 468, 474, 476; *Kitāb al-Umm* 433
al-Shaibānī, Muḥammad ibn al-Ḥasan (d. 805) 390, 432, 433, 434, 465, 470; *Kitāb al-Athār* 390
al-Shar'ānī, 'Abd al-Wahhāb (d. 1566) *al-Yawāqīt wa'l-Jawāhir* 463–465
Shuraiḥ **385**, 429, 442
Sindhī, 'Ubayd Allāh xxxviii, xxxix, 4
Solomon 253
al-Suhrawardī, Shihāb al-Dīn (d. 1191) 42n
al-Suyūṭī, Abū Faḍl (d. 1505) xxix, 40, 235n, 270n, 430

al-Ṭabarānī, Sulaymān ibn Aḥmad (d. 970/1) **393**
al-Ṭaḥāwī, Abū Ja'far Aḥmad ibn Muḥammad (d. 933) 391, **393**, 471
Ṭarafah ibn al-'Abd d. c. 564 C. E.) 155n
Ṭāwūs ibn Kaisān (c. 720) 424
al-Ṭayālisī, (d. 818) **393**
al-Thaurī, Sufyān (777/8) 191, 192, **390**, 430, 447, 453
al-Tirmidhī, Abū 'Isā (892) 191, 192, 392, 444, 445, 447, 458; *Jāmi'* 392
Ṭu'aima ibn 'Addiyya 316

'Ubayd Allāh ibn 'Abd Allāh 429
'Umar ibn 'Abd al-'Azīz 443
'Umar ibn al-Khaṭṭāb xxiv, 14, 19, 109, 201, 342, 350, 352, 367, 375, 384, 386, 402, 405, 415, 416, 419, 420, 424, 426, 429, 431, 437, 442, 443, 444, 448

Umm Zara' 374
Ummayya ibn Abī al-Ṣalt 365, 366
'Urwa 429
'Uthmān (Caliph) 109, 429
'Uthmān ibn Maẓūn 349

Walī Allāh, Shāh, cross references in footnotes to other works *Alṭāf al-Quds*, xxxvi, 54, 77, 208; *al-Budūr al-Bāzigha*, xxxvi, 116; *al-Fauz al-Kabīr*, xxxvi, 163; *Fuyūḍ al-Ḥaramain*, xxvi, xxvii, xxx, xxxi, xxxii, xxxvi, 7; *Hama'āt*, xxxvi, 75, 76, 208, 250; *Ḥujjat Allāh al-Bāligha II*, 29, 380, 420, 460, 461, 468, 469, 472, 475, 476; *al-Inṣāf fī Bayān Sabab al-Ikhtilāf*, xvii–xviii, xxxi, xxxiv, xxxix, 437, 451, 452, 453, 454, 455, 463, 472, 478; *'Iqd al-Jīd*, xxx, xxxiv, xxxvi, xxxix, xl, 451, 455, 458, 463, *al-Khair al-Kathīr*, xxxi, xxxvi, 4; *Lamaḥāt* xxxvi, 46, 63, 67, 71, 312; *Muṣaffā*, xxvi, xxx, xxxvi, 23, 416, 454, 461, 466, 468, 469, 471, 476; *Saṭa'āt* 71; *Sharḥ Tarājim Abwāb Ṣaḥīḥ al-Bukhārī* 446, *al-Tafhīmāt al-Ilāhiyya*, xxxiv, xxxvi, 7, 71, 196, 250, 351; *Ta'wīl al-Aḥādīth* 108.

Yazīd ibn Harūn (d. 736) 439
Yaḥyā ibn Sa'īd al-Qaṭṭān (d. 760/1) 390, 424, 439, 444

Zaid ibn 'Amr 366
Zaid ibn Aslam 429
Zaid ibn Harūn 445
Zaid ibn Thābit 15, 19, 374, 385, 403, 429
Zufar (d. 775) 390, 432, 433, 434, 465, 470
Zuhair 340
al-Zuhrī, Muḥammad ibn Muslim (d. 742) 420, 429, 431, 437

TITLES OF BOOKS

Ghayā al-Muntahā 445
al-Mujarrad 465
Tabyīn al-Ḥaqā'iq 472

www.ingramcontent.com/pod-product-compliance
Lightning Source LLC
Chambersburg PA
CBHW031318230926
51848CB00001B/7
9789004444775